The Man Who Invented
the Chromosome

The Man Who Invented the Chromosome

A LIFE OF CYRIL DARLINGTON

OREN SOLOMON HARMAN

HARVARD UNIVERSITY PRESS
Cambridge, Massachusetts
London, England 2004

Library of Congress Cataloging-in-Publication Data
Harman, Oren Solomon.
    The man who invented the chromosome: the life of Cyril Darlington / Oren Solomon Harman.
        p. cm.
    Includes bibliographical references and index.
    ISBN 0-674-01333-6 (hardcover : alk. paper)
        1. Darlington, C. D. (Cyril Dean), 1903- 2. Geneticists--England--Biography. I. Title.
QH429.2.D37H37 2004
576.5'092--dc22
[B]
                                                                2004040603

For Abba and Imma with love

Contents

Acknowledgments ix

Introduction 1

I FROM CHORLEY TO TABRIZ 5

1 An Improbable Birth 7

2 A Rising Tide 21

3 Auspicious Beginnings 32

4 In Search of Tulips and Truth 49

II SCIENCE 67

5 From Cytology to Evolution 69

6 Roots of a Scientific Controversy 89

7 Method, Discipline, and Character 105

Interlude 126

III POLITICS 137

8 The Lysenko Affair 139

9 Marxism and the Slaying of a Mentor 153

10 Science in a Changing World 169

IV MAN   *189*

11 The Conflict of Science and Society   *191*

12 On the Determination of Uncertainty   *206*

13 The Breakdown of Classical Genetics   *222*

14 On the Uncertainty of Determination   *235*

15 One Final Hurrah   *253*

     Conclusion: Paradoxes   *262*

     Notes   *275*

     Index   *319*

Acknowledgments

THIS BOOK HAS ITS ORIGIN IN the years I spent at Oxford between 1999–2001 trying to get a little work done on my doctorate amidst all the fun and games. Paul Weindling was a wonderful supervisor, and has since become a friend, as have Tim Horder, Robert Fox, and Avner Offer, all of whom taught me a great deal. I thank them kindly. In Oxford, far away from my native Jerusalem, the Yudkin family took me in to their cozy home on sleepy Norham Road, adopting me as a son and making me feel welcome: thank you Pat and Michael. As for you, Dr. Benny Yudkin, you are not only my smartest reader, but also my closest friend.

It was the kindness and generosity of Everett Mendelsohn that allowed me to come across the Atlantic to complete my work at Harvard. Here, too, a family away from home invited me into their rich lives and wonderful house, becoming dear friends in the process. Ann and Marty Peretz, I cannot thank you enough, and hope one day to be able to repay your kindness. Thank you Catherine and Eva for all your help and smiles. And thank you Moshe, Michal, and Yasmin Safdie for being wonderful and caring friends. To my buddy, Marty West: it was a pleasure walking the road together.

Many people were kind enough to answer my questions and to offer their assistance in letter and in person. I would like to thank them all: Garland Allen, Jeannine Alton, Geoffrey Beale F.R.S., Yemima Ben Menahem, Sir Allan Bullock, Lionel Clowes, Brian Cox, Rafi Falk, Daniel Hartle, Henry Harris F.R.S., Barry Juniper, Daniel Kevles,

Nancy Kleckner, Dan Lewis F.R.S., Harland Lewis, John Maynard Smith F.R.S., Ernst Mayr, Everett Mendelsohn, Mathew Meselson, the Honorable Dame Miriam Rothschild F.R.S. (and Noga Arikha who introduced us), Sir David Smith F.R.S., Canio Vosa, and Stan Woodell. Special thanks go to Daniel Kevles for his generosity and encouragement, and to Nathaniel Comfort for pushing me to make this a better book. My gratitude to David Roberts, for allowing me to use his beautiful room facing the deer park at Magdalen College, and to Colin Harris, for his kind assistance at the Bodleian. I owe a special and personal thank you to Darlington's living relatives: Paul and Yvonne Harvey, Oliver Darlington, Rachel Richie, and, most of all, Clare Passingham, for sharing their most intimate memories and feelings, and for being enormously generous and kind.

I would like to extend my thanks to the people at Harvard University Press: Michael Fisher who lent his ear to my green suggestion, Ann Downer-Hazell, who stuck it through with me, and was of enormous help, Sara Davis and Sheila Barrett, many thanks. Many thanks, too, to Jennifer Roehrig and GEX Publishing Services, for helping bring the book to its final stages of completion with talent and patience.

Sam Silverstein and the late Donald Cohen have both been wonderful mentors and loyal friends over the years, whose encouragement and teaching have left a lasting mark on me. I love them both dearly, and only wish Donald was still with us today. To my close friends Marilyn and Hugh Nissenson I owe the warmth of their friendship and the creativity of their minds.

I have benefited from the wise words and advice, but especially from the laughter, of many friends as I worked on conjuring up Darlington's tale from the past. Thanks to Ghil'ad Zuckermann, Maya Topf, Eyal Doron, Simon Palferman, Barnaby Yates, the Vickster, Roni Zirinski, Craig Martin, Ben Reis, Kata, the lovely Samantha Power, and, of course, the transcendental Organism: Faust, Koksner, Naitz, Jeno, Ace, Chailik, and Ben Gnaiv. To you, green eyes, I send a knowing smile. Finally, I level my gaze at my family, Savta, Abba, Imma, Danzi, and the Mish: you are my very heart and happiness.

The Man Who Invented
the Chromosome

Introduction

IT WAS A RATHER CHOPPY DAY IN late October 1999 when I got on the train at King's Cross, on my way to Peterborough. Seven miles down from the station I was to find Ashton Wold, the home of the Honorable Dame Miriam Rothschild, now ninety-five years old. On the phone she had not easily revealed the fact that she had written over 350 scientific papers, and had been for years a venerable Fellow of the Royal Society. Her studies of the plague-carrying flea, *Xenopsylla cheopis*, were famous. Getting out of the black taxicab, I saw a large, ivy-covered façade, reminiscent of an old Elizabethan country house, and made of local stone. As the door was open, I walked in and found myself in a spacious library. Large volumes of scientific books, leather-bound and somewhat faded in color, rested on oak shelves. The air smelled antediluvian, regal somehow, with traces of pipe tobacco, tea leaves, cooking sherry, and the earthen aroma of stone architecture. In the corner of the room, a stuffed owl—shot in Egypt by Miriam's father, Charles Rothschild, while collecting exotic specimens that would end up in the British Museum many years ago—was perched on a wooden stand. Beyond the pane of a middle-sized window, the dark waters of a small pond glittered. I approached to have a look. The grounds were sprawling. A herd of deer could be spotted on a grassy hill in the distance. Directly beneath the window, bulbs of every sort and color grew beside the occasional upturned earth marking the small burrows of voles.

Suddenly, a strange, mechanical hum wakened me from my dreamy thoughts. It grew louder and louder until finally Miriam Rothschild appeared, sitting in a small electric wheelchair. She looked upon me with a mischievous grin. "Have you come to speak to me about Darlington?" she pronounced. I nodded. A large smile appeared on her handsome face. Gazing directly at me with her wise, piercing eyes and raspy, English voice, she uttered: "The chromosome? Why, Darlington practically *invented* the chromosome!"[1]

～ CYRIL DEAN DARLINGTON (1903–1981), together with J. B. S. Haldane, Julian Huxley, and R. A. Fisher, belongs to the great school of British geneticists, evolutionists, and biological statisticians produced by a country basking in the afterglow of the eminent Victorians, Charles Darwin and his first cousin Francis Galton. It had been said of Darlington that from the moment he published his first cytological work, *Recent Advances in Cytology*, in 1932 his impact was like that "of Dante with the Divine Comedy. Just as Dante was so far ahead of his time that he stifled the growth of poetry in Italy for many years, so Darlington, in drawing together and almost creating the science of cytology dominated the scene until the biochemists introduced another revolution after the war."[2] His influence was of such enormity that he would ultimately be referred to in obituaries as the "Copernicus" or "Newton of cytology," and his 1939 book, *The Evolution of Genetic Systems*, would be hailed by some as the greatest single contribution to evolutionary thought since Darwin's *On the Origin of Species*.[3] Later in his life, Darlington would turn to the subject of history, becoming one of the most fascinating and controversial exponents of the sociobiological approach to human culture.

But Darlington was forgotten. His life and science, I quickly discovered, had been relegated to the memories of those few still alive who knew him, and to a single article in the *Biographical Memoirs of Fellows of the Royal Society*.[4] Having spoken to Miriam Rothschild, to Darlington's living relatives, and to others who once knew him, and having come across the vast collection of Darlington's personal papers at the Bodleian Library and at Magdalen College at Oxford, I decided his was a story I wanted to tell. In doing so, I seek first and foremost to redress the imbalance brought about by the uncanny, untrustworthy quantity we call historical memory. Darlington's rich collection of papers, his uncensored thoughts in his diaries ("My notebook is the one friend to whom I can communicate without reserve"),[5] and copious correspondence help to provide a highly personalized picture of the development and ideas of one of the most penetrating and controversial biological thinkers of the century. They also help to explain why he was forgotten.

Chronology, so goes the saying, is the last refuge of the feebleminded, and the only resort for the historian. In that spirit, the book advances from Darlington's birth to death, recording four central spheres of activity and themes in his career and life. The first is a portrait of the young scientist, struggling against the backdrop of a new and exciting science, and a venerable, reactionary founding father. Early childhood, education, and scientific beginnings are thrown into relief against the history of genetics in Britain and abroad in the first three decades of the century. Next, Darlington's scientific contributions in the field of cytology, genetics, and evolutionary theory during the 1930s and through the molecular revolution of the 1950s are examined. These would ultimately prove central to our understanding of life, but were not universally appreciated at the time. The strong opposition to and controversy surrounding Darlington's ideas tells an illuminating tale of how new theories, boldly expressed, often need time to be swallowed and digested, and about the particular determinants that guide their progression down this rocky road of initial resistance.

The notorious Lysenko affair in the Soviet Union, a central Cold War episode in which science and politics became hopelessly enmeshed, is the focus of the third part of this book. Darlington's vociferous, divergent, reaction to the affair serves as a key to a much wider political debate in the West about the place of science in modern society. Finally, I consider Darlington's controversial work on man and the evolution of culture, social structure, and race. Darlington's sociobiology, at the height of postwar sensitivities through the radical 1960s and into the reactive 1970s and 1980s, illustrates the importance of political determinants in the formulation and reception of scientific ideas. A fascinating case of the extrapolation of world-view from the translation of genetic principles of plants and animals to man, it serves to close the circle opened with his first glimpse of the chromosomes as a young, unsettled youth.

Idiosyncratic, self-taught, and always controversial, Darlington's life story serves as a new and unconsidered prism through which to view biology and its social relations in the twentieth century. Genetics, evolution, man's place in nature, and the place of science in government and society are all themes upon which Darlington's story sheds light. Probing science's often subtle, always complex relation to politics in its disciplinary, methodological, institutional, national, and ideological twentieth-century setting, our story raises central normative questions pertaining to the production of scientific knowledge and the sociological activity of the pursuit of science. It is in the terrain where the frailties of men are juxtaposed with the transcending laws of nature that our story builds its home.

As part of the celebrations of the centennial of Charles Darwin's *On the Origin of Species* in 1959, Darlington wrote and published an iconoclastic little book—*Darwin's Place in History.*[6] Here he used the opportunity to challenge Darwin's originality, and the manner in which he saw fit to present it. Darwin's strategy, Darlington claimed, was to leave out man, to be loose enough in his definitions and arguments to allow for a line of retreat, and to appear quite new and unique, acknowledging no predecessors. Like all that came from Darlington's pen or his mouth, these claims immediately provoked an outcry. Darlington fashioned his career in direct opposition to that which he perceived spurred Darwin's to world fame. In the process he lived one of the more controversial—but also one of the most reflective and paradoxical—scientific lives of its time.

~ I

From Chorley to Tabriz

～ 1

An Improbable Birth

Wᴵˡˡᴵᴬᴹ Hᴇɴʀʏ Rᴏʙᴇʀᴛsᴏɴ Dᴀʀʟɪɴɢᴛᴏɴ ᴡᴀs a schoolmaster, son of a schoolmaster. He was the grandson of an uneducated Welsh nursemaid and her unfaithful employer, a wealthy Liverpool merchant and cousin of William Gladstone, the Liberal prime minister of Great Britain. One day in 1893 William was introduced to Ellen "Nellie" Frankland. As the lean, jaunty figure strode along the railway platform in Leigh, a small Lancashire cotton and coal town in the English wool country, someone whispered in Nellie's ear: "That's mister D, the master from the grammar school. He is the cleverest man in Leigh. But he is a Socialist!" Nellie had come from nearby Atherton, the product of a family of whitesmiths and solicitors who claimed descent from Edgar the Atheling, twelfth-century king of Scotland. Nellie was the granddaughter of Atherton's Unitarian minister, and the daughter of a madman committed to an asylum. She had finished her schooling at thirteen and soon became captivated by this wiry, eloquent, and learned stranger. For his part, William was almost immediately infatuated.[1] In December he wrote declaring his love:

My dear Miss Frankland,

I had almost said my dear Nellie yet though it were the truth I dared not write it for you would have been annoyed at me, wouldn't you? At any rate, if I have spared you the impertinence this time it is not because in my secret thoughts I have not been guilty of it.[2]

Three years later, in April 1896, William and Nellie married, and in the following year, a son, Alfred, was born. Finances were strained, and the struggling couple resolved not to have any more children. Despite this intention a second son was born on Saturday, December 19, 1903, in the small Lancashire cotton town of Chorley. Cyril Dean Darlington later recalled: "My birth gave rise to a crisis in the family. My mother claimed that she didn't know how it happened. My father on the contrary was quite certain. He applied a rule without formally declaring it that never again would he cohabit with his wife. It was a rule that he never broke in his remaining forty years. Whether he regarded me as an accessory to this mishap or misdemeanor I am still not sure."[3]

For the Darlingtons the economic consequences of Cyril's birth were painful. The salary from running a private school could no longer support a mother, a wife, and two children. In a bind, the desperate young father turned to an old grammar school friend for advice. Reggie Ormandy had met some bright chemists as a pupil of Bunsen at Tübingen in Germany. One of them, Karl Markel, was chief chemist and managing director of the old family soap firm of Joseph Crossfield of Warrington. With Alfred and Cyril now ten and three, Will gave up his career teaching, moved the family to Warrington, and became company handyman and assistant to Markel. Crossfield's was a profitable business, and prospects were good. "My birth," Darlington reminisced, "turned out to be a blessing in disguise and I was perhaps gradually forgiven although never quite on a footing with my brother, seven years older and somewhat more legitimate than I could hope to become."[4]

At five and a half, young Cyril was sent to Heathside Elementary School in Bewsey, Warrington. His early memories were of rubbing his hands with lard; tar invariably decorated them before he reached home from school. One day the school inspector asked, "What do bees live in?" and the cautious youth volunteered, "A hive?" receiving a book as a prize. It was a glorious moment, but fleeting, and one of Cyril's few happy memories. The walk to school led down Tanner Lane, with its "own professional stench, under Pearson and Knowles' smoking chimneys, down Haycock street, where fat women, too big to fit into their own houses, filled every doorway while they peeled their onions and potatoes and their children played and pissed in the gutter."[5] At the age of six, Cyril was deeply conscious of the filth, the ugliness, and the misery around him. Living in the street adjacent to Lovely Lane, he was aware of the profound irony, and oppressed by the dullness of life. Sounds of breaking bottles and the grunts and cries of street fights among drunken men on Saturday nights became familiar, unpleasant denizens of his remembrance. Most salient were the recollections of his father's severity, often finding an outlet in thrashings for raised voices on Sundays. Dinners were a somber affair; the boys were forced to listen to their father talk about Dr. Markel saying this, and Dr. Markel doing that, and were expected to say nothing in return. Alfred and Cyril never really got along. Mother and father slept in separate rooms. The Darlington home was filled with silence. A row in the middle of the night once ended in Cyril's father coming into his son's bed. Visitors had been staying over, and Nellie had unsuccessfully tried to share Will's bed. It was an austere and deeply repressed existence for a young boy. No wonder Cyril looked forward to the annual two-week vacations with family friends, the Cockrams: "No smoke, no tar, no father."

⌒ Dr. Markel was a son of a Lutheran minister from Stuttgart, and after having studied under Bunsen (with Ormandy), was recommended to Crossfield's, where he quickly became successful. Some of his ideas, like publicizing "the factory with smokeless chimneys" with a photograph taken on Sunday, were brilliant, if somewhat less than honest. His administration was also most efficient. "Always sack a man in any new department you are put in charge of," he was known to say. "After that you'll have no trouble." His German appointees, however, had lots of trouble. Markel wanted his foremen to click their heels and salute him. But he went too fast. German chemists were installed in every department, and soon the Crossfield brothers, who had trifling talents and only such science as they could pick up at Harrow in the 1890s, became jealous. They took the opportunity of sending Markel off to a holiday in Japan, dismissing him and his followers in his absence, though they did retain his numerous innovations. Ever-loyal Will decided to follow Markel, and in the summer of 1911 the Darlington family left Lancashire, moving first to London, and then permanently settling in Ealing, on the city's outskirts. Cyril's next five months were spent in blissful freedom. Nellie and Will had yet to find him a school, and he spent most days exploring London in the company of his mother's sister May. Looking back, Darlington considered the decision to follow Markel a disaster for his father, but salvation for himself and Alfred: "A new life had begun on that hot July day."[6]

In January Cyril was finally enrolled at the Mercer's School, a small, crowded tenement in the city with 270 pupils. When the family moved to Ealing in October, Cyril had to stand both ways on the train doing his Latin sentences in pencil, and not daring to reach for a strap—he was too small. In the evenings he would help his father with gardening—holding or fetching tools. Father and son worked silently until long after dark in the winter by the light of the street lamp.

As war spread across Europe, Markel adopted the role of godfather of the German POWs in England, and after the peace became an agent of the German embassy, distributing propaganda and supporting pacifists like Ramsay Macdonald and E. D. Morel with election expenses. This was a remarkable somersault for Markel, a man who had put his whole weight into the National Service League when alliance with Germany still seemed possible. These activities were deeply hateful to patriotic Darlington senior, who spent the rest of his professional life in miserable frustration, torn between loyalty to his longtime benefactor and his own political allegiances.

The war left Cyril the lone son at home. Alfred had joined the Royal Artillery in July 1915, and was sent to France the following February, where he fought at the Battle of the Somme. The atmosphere at home deteriorated. Will remained celibate, frustrated, and resentful of his pacifist boss. His eldest son's fate across the channel was unknown. Emaciated, dyspeptic, dictatorial, and argumentative, he seemed to loathe children. He had reasons for everything he did. "Nellie, it's a principle," he would say. If she noted his inconsistencies, he would escape: "I am like that!" Violent tantrums would often end with the threat to leave his wife. Continual ranting about economic, moral, and aesthetic principles at Nellie and at Cyril, combined with the ever-present danger that Markel might be interned, seemed to dull and sharpen the senses all at once. Life felt sterile, and very precarious.[7]

In December 1916 Cyril sat for an exam for St. Paul's School, but failed to win a scholarship. "I rather think his failure will in the long run have been the best thing that could have happened," William Darlington wrote to high master A. E. Hillard. "I find him much too confident for his abilities."[8] With a favored older brother courageously fighting for Britain abroad accentuating Cyril's own uselessness, an ever-more silent and depressed mother, a strict and stringent father, and an uncomfortably strained home environment, Will's son had begun to escape behind an increasingly cocksure and infallible exterior. He was cultivating a countenance of self-assurance that would leave a lasting mark. On Easter of the following year, Cyril sat for the exam a second time, this time successfully. At last he had done something to win praise from his father (and a raise in pocket money to sixpence).

Cyril and Alfred (top right) with cousins, 1910. The two brothers never really got along. Photo courtesy of Clare Passingham.

~ AT ST. PAUL'S EVERYTHING WAS DIFFERENT. Black clothes were the rule. Decorum was enforced. Sound chaps, self-reliant, lacking emotion, and not too long on curiosity, were what the school sought to cultivate. Excellence at team sports was prized. Thomas Arnold's Rugby was the model; austerity, pride, and a tight upper lip were the goal. Cyril hated it. St. Paul's enjoyed the reputation of one of the most prestigious public schools in London, and yet

teaching was poor and impersonal yet no one worried about it.

The high master Hillard was a tall thin man with an expression of vinegar and a voice of wormwood. He no doubt owed his Episcopal doctorate to his *Life of Jesus* which was compulsory reading on the classical side, for most of the prescribed reading was written by the masters who prescribed them. I went at once to the science side. The tone of the school was naturally set by the "High Man." It was a sterile blend of piety and snobbery. 'Remember,' he said, addressing the assembled school, 'that you are born rulers of men.' Certainly they were born with great advantages— but the future was not to be so easy as the past. Everyone agreed that the public schools had won the last war for us but as Mr. Penny said, back from the war, 'You boys must realize that a pass degree won't be enough for you in the future. The social stratum a little below our own is entering into competition with us. You won't have everything your own way in the future. You will have to work hard!' Some did, but I was not one of them.[9]

Cyril's Latin deteriorated so far that one enlightened Latin master mercifully allowed him to read Shakespeare on his own instead of Virgil, an unheard-of kindness. At St. Paul's German was the alternative to science, and Cyril had to take it in his spare time. Everything taught seemed so utterly disconnected! He enjoyed neither sports nor his studies, and exerted the better part of his energies perfecting methods to avoid both. Life at St. Paul's was dreary and drab. "All classrooms faced north and the whole spirit of the school was arctic." Darlington recalled his chemistry teacher, Master Lillie, asking him whether chemistry would be of any use to him in life, and expressing the satisfaction that it would not. He was always bottom of the class in sciences, and "the science masters expressed the same poor opinion of me as I did of them."[10] The three subjects in which he was least successful—science, drawing, and Latin—were those in which he would excel in later years, a sure sign of complete disaffection.

As he withdrew at school, Cyril found an inspiring teacher at home, in his increasingly lonely father. William was equally at home with chemistry or mechanics, Latin or Anglo-Saxon, German or French, finance or law. He taught his son principles of business and investment, introduced him to the problems of dialect and place names, to languages, to mathematics, and most important of all, to the Latin poet Lucretius, whose classic *De Rerum Natura* made a deep impression on Cyril. The lessons would often take place in the garden, where father and son worked together, digging, planting, and cultivating vegetables and herbs. Cyril would later attribute his interest in plants to the time he and his father spent together gardening the small plot at Rectory House on dreary Warwick Road.

⌇ CYRIL'S CHIEF PLEASURE now became the many hours he spent browsing bookshops along bustling Charing Cross Road. As his knowledge and understanding grew, William began to find in him the intellectual companion he lacked in his wife. Still remaining very much afraid of his father, Cyril was developing an immense respect and admiration for him that he would harbor throughout his life. Father and son shared broad-ranging interests uncommon to the usual products of an ordinary public school and university education. Cyril would later take pride in the fact that he had come from two lines of non-Oxford academic folk. It was due to this heritage, he claimed, that all three generations of Darlingtons shared a singularly unconstrained, self-directed breadth of intellectual interests that included both the arts and sciences.[11]

But quiet moments at home were rare. Will was stern and severe; there was much in him to admire but little to adore. With no outlet at home, Cyril's angers and frustrations were displaced onto all forms of organization and authority at school. His hatred of the pillars of the public school ethos, team sports and academic competition, was creating a singularly self-reliant constitution in the young man— iconoclastic, unorthodox, and nonconformist in nature. "The public schools," he would write in his diary some years later, "give a man a code for his 'equals,' a protection against his 'inferiors,' and the conviction that he has no superiors. The code is called 'fair-play,' the protection is called 'esprit de corps,' and the conviction is just bloody ignorance."[12] As for companions of his own age, they were duly noted in Cyril's 1917 journal, *My Book*, under "friends and their use:"

C. E. Scott—cycling
Greg and Ruck—talking
Kirkpatrick—both.[13]

⌇ ALFRED WAS MENTIONED IN dispatches of 1917, and

wounded in the heel in March of the following year. "The sufferings, hardships, and travails of you poor brave men have been reflected in the minds of hundreds of millions of people in proportion to their thinking powers," fifteen-year-old Cyril wrote to his brother, adding quietly, "I am passing away my unimportant existence here as happily as may be expected under the circumstances, cogitating over the war most of time." From this outpour of sympathy Cyril quickly turned to a dispassionate analysis of Germany's success in the great spring offensive of 1918. How could the Germans have made such a supreme effort after more than three years of grueling fighting, in a manner clearly impossible for any other nation? "Germany must have some power on her side," he thought. Systematically tabulating, considering, and then ticking off the possible factors, Cyril finally arrived at brains and education, a classic "nature and nurture" assessment. And yet his final conclusion—"If the Russians had been educated this offensive would not have happened"—was one which, after his own education had been completed, he would come to reject vehemently.[14]

Meanwhile, at home Cyril was growing increasingly lonely. His 1920 journal sadly reported: "No friends." The very next entry, however, "Theory of Relativity appears. Taking conquest," pointed to a growing interest in science. Having scraped through the London Matriculation Examination in January 1920, Cyril took up biology with a view toward going into agriculture, and eventually emigrating to Australia to work as a potato farmer—an idea that appealed to his disillusioned father. He could not stand St. Paul's any longer, and promptly left at sixteen, as his father had done before him. Unlike his father, however, Cyril could afford to continue his schooling. In the spring he applied to the South Eastern Agricultural College at Wye, founded in 1894 by Sir Daniel Hall.

The college, later to become Wye College of the University of London, huddled in the halcyon setting of the Kent village beneath the South Downs, was a welcome contrast to London's bustle and Ealing's gray suburbanism. Cyril was a weedy, town-bred, home-keeping youth, however, and his new surroundings provided a violent shock. The students were "the scum of the public schools, the dregs of the army, combined with some agricultural oddities and a few bright lads from the elementary schools."[15] It was a far cry from Cambridge, where Cyril's much more conservative brother, back from the war with a Military Cross, was recuperating and studying medicine. Gambling, poaching, and wenching were all unthinkable, repulsive mysteries to the repressed and sheltered youth. And those detested games, again!

Quickly, Cyril's solitary habits once more fashioned him an outsider. In the bullying environment of a rugged, all-boys agricultural boarding school, he increasingly became the butt of jokes. His journal entries began to grow in length and frequency. They chronicle a first dissection of a frog (October 20, 1920), a first shave (November 2), the rescue of a fish from a frozen pond (December 18)—equally salient alongside observations of the national coal strikes, flooding of the mines, and Sinn Fein activities in the Irish national struggle. A note on a boil on the nose (third boil) directly precedes an entry announcing Ernest Rutherford's claim to have split the atom.[16]

CYRIL WAS BEGINNING TO show some academic promise. Pressed to settle on a degree course, he chose zoology and geology, but at the end of the first year failed his intermediate science examination on botany, and only passed a year later after having committed all of D. H. Scott's *Structural Botany* to memory. Painfully aware of his social immaturity, and still uncomfortably scrutinized by his demanding father, Cyril increasingly adopted an audacious confidence in his intellectual abilities. "I pride myself so much on my intelligence that I don't regard the *Daily Mail* as quite up to my mental standards," a selfsure eighteen-year-old wrote to his father in one of his many letters home.[17] To his mother he complained: "Many people here have no better wit than filthy language and stories, I fear that if I told a good tale it would not be appreciated by them for the point would be too fine for them to see."[18] Inwardly and outwardly, an air of superiority that would characterize his dealings with people throughout his life was beginning to take form. Added to this were an extreme antiauthoritarianism and a jealous defense of independence of thought. "I am not disposed to reveal my beliefs," he wrote, "but nothing galls me so much as to have other people's beliefs forced down my throat."[19]

While academic success remained elusive, his troubles with his fellows continued. The remedy—boxing—was suggested by his father. Cyril reluctantly took up the gloves, and fought his first fight at the annual Wye festival in March 1922. He was now a strapping nineteen-year-old lad, six foot and three inches in height. Although beaten (by a short Welshman named Downey), Cyril's was by far the best fight of the evening. It quickly brought about a complete inversion in his social standing among the boys. He was soon taking pleasure in bathing nude in the river Stour, swearing with the best of them, and in October becoming the school's liberal candidate in the annual mock election.

But a promising agriculturist he was not. Darlington remembered being regarded as a highly unlikely candidate in the bachelor of science exam in October 1923. He had twice won the Paton-Figgis Scholarship, and once a silver medal for a flower collection that he had stayed up preparing many nights "until 3 or 4 o'clock." Still, it would nevertheless have been ludicrous in 1923 to entertain the thought that he would one day succeed the founder of his college as director of Britain's premier horticultural institute, and far surpass him in scientific achievement.

BEFORE THE RESULTS OF THE EXAM WERE IN, Cyril put in an application to the Empire Cotton Growing Corporation for a scholarship to go to Trinidad as a farmer, but was rejected. It was then that a master from Wye, Professor E. S. Salmon, who was also a member of the governing body of the John Innes Horticultural Institution, recommended that he apply for a scholarship. The John Innes's venerable director, William Bateson, had recently announced scholarships of £250 for work in genetics. At Wye Cyril had learned about seed testing and horse doctoring, methods of analyzing milk and soil, construction of farm implements; about rents, and wages, and agricultural law. All these, added to the botany, chemistry, and biology he had taken, had made little impression on him. One thing, however, did capture his attention. These were the lectures on Mendelian theory given by Mr. Brade-Birks. The new ideas sparked a curiosity that sent Cyril to the college library, where he came across Thomas Hunt Morgan's *The Physical Basis of Heredity*.[20] He was immediately struck by the force of Mendelism, and by the stark contrast between Morgan's methods and arguments and those he had heard discussed in connection with the breeding of crops and livestock. He wondered whether Morgan's work on flies did not apply equally to every plant and animal, even to man. Did not all organisms possess these small chromosomes of which Morgan spoke?[21] Not yet twenty, and still hoping to travel to the colonies when he was a bit older (he put down his rejection to his aversion to sports but also to young age), Cyril found his teacher's suggestion appealing. The John Innes was one of the only places in Britain carrying out research on heredity and genetics, and he could temporarily investigate problems of interest before reapplying to the cotton farming scholarship overseas. Resolute and hopeful, Cyril promptly sent a letter to Bateson asking to be considered for the studentship in genetics.[22]

⌐ ALMOST CERTAINLY UNKNOWINGLY, Darlington had applied to join Britain's most vibrant biological community. The John Innes had been founded at Merton in Surrey in 1904 by a childless London merchant who bequeathed his name and fortune—over £300,000—to establish a school for the improvement of horticulture. In the succeeding five years a trial of strength was waged among various interested parties. Innes's shunned heirs claimed the bequest was proof of an unsound mind, and argued that the money should be divided among them; the Ministry of Education and Ministry of Agriculture had different ideas about what the "improvement of horticulture" meant; the charity commissioners and trustees found it difficult to agree on anything at all. After a magistrate rejected the family's claims in 1906, a compromise was finally struck in which the controlling interest was ceded to the Ministry of Agriculture. The institution was to be administered by a council consisting of three trustees and nine members, appointed by the Ministry of Agriculture and Fisheries, the Royal Horticultural Society, the Fruiterers' Company, the National Fruit Growers' Federation, the Universities of Oxford, Cambridge, and London, and the Imperial College of Science and Technology. As it turned out, the delay between 1904 and 1910, when the institution was finally ready to appoint its first director, was of great significance, for in that period William Bateson had seized the imagination of the scientific world by his success in demonstrating the general validity of the recently rediscovered laws of heredity set forth by Gregor Mendel. Hope for the future of this new science was high, and it was decided that besides being a training school for gardeners, the institution would focus on genetics, a term Bateson had coined just a few years earlier.[23]

Bateson was by this time a famous man.[24] His early embryological work on the worm Balanoglossus, showing it to share a common ancestor with vertebrates, had become a textbook classic, and won him a Fellowship at St. John's, Cambridge, in 1885. After spending a few years in the steppes of Western Central Asia in an attempt to prove how environmental changes could give rise to variation in species, Bateson published his landmark *Materials for the Study of Variation Treated with Special Regard to the Discontinuity in the Origin of Species* in 1894, and was immediately elected a Fellow of the Royal Society. As the subtitle of his magnum opus suggested, Bateson's research had led him to a conclusion opposite to that which he had initially set out to prove. He was thereafter a firm advocate of discontinuous, saltationist variation in nature, a view contrary to the prevalent belief that "nature did not take jumps," but that changes occurred in a continuous and gradual fashion. In *Materials*, Bateson had amassed a wealth of evidence for sudden jumps. These mutations, as they would later be called, were preserved in subsequent breeding, he found. They were a result of changes in the inherited "factors" corresponding to physical traits. They were internal and unaffected by the environment. In Bateson's hands they became a challenge to the Darwinian notion of evolution through continuous variations produced by natural selection. These views brought about bitter debates with the Biometrical School led by Karl Pearson and Bateson's old Cambridge friend W. F. R. Weldon, who rejected the recourse to a notion of heredity based on unseen, theoretical "factors," defining inheritance instead in terms of Francis Galton's statistical Law of Ancestral Heredity, which is based on perceivable, continuous, and measurable physical variations.[25]

As his contributions to the Evolution Committee of the Royal Society, published in 1902, clearly show, Bateson was now close to independently establishing the principles of Mendelism. Traveling on the Great Eastern Railway from Cambridge to London in May of that year to deliver an address at the Royal Horticultural Society entitled "Problems of Heredity as a Subject for Horticultural Investigation," Bateson had come across a translation of Mendel's paper, rediscovered independently in 1900 by Hugo de Vries in Amsterdam, Carl Correns in Tübingen, and Eric von Tschermak in Vienna. By the time he stepped off his train at Liverpool Street Station, the forty-year old Bateson had become a convert. The usually cautious lecturer immediately incorporated Mendel's laws into his talk, and had the Bohemian monk's papers translated into English in the Royal Society's journal. Weldon's attack on Mendelism in the first volume of *Biometrika* in 1902 provoked Bateson's *Mendel's Principles of Heredity, a Defence*,[26] which became the first textbook presentation of the principles of genetics in any language. Never again did Bateson mention how close he had been to reaching them himself. Years later J. B. S. Haldane would write that "if the Proceedings of the Brunn Natural History Society had been a little rarer Bateson would be lying in Westminster Abbey."[27]

⁀ BETWEEN 1902 AND 1910 Bateson and his collaborators Edith Rebecca Saunders, Reginald Crundall Punnett, Leonard Doncaster, Charles C. Hurst, and Florence M. Durham presented the results of breeding experiments confirming Mendel's laws and extending them further to a number of different plant species (Mendel's successful work was exclusively on *Pisum*), as well as to animals.[28] Bateson was now firmly placed at the epicenter of biological thought. A special chair of biology was created for him at Cambridge in 1909, anonymously funded by Alfred Balfour, leader of the Conservative Party, in memory of his deceased younger brother, a fellow scientist and former friend of Bateson's. But Bateson quickly found that there was not enough money for research. Disappointed with his superiors' refusals of his frequent requests to establish a center for genetic research, he was disillusioned with his home institution and cast about for a way to make an exit from Cambridge. When the John Innes offered a large salary, a house, and six acres of land at Merton, together with a £5,000 annual endowment, Bateson was therefore more than pleased to accept. He insisted on bringing his poultry, canaries, and bees with him to the newly established plant institute, the council gave in, and the deal was sealed. Ironically, in 1912 the Cambridge chair was re-endowed and renamed—Chair of Genetics. By this time Bateson was settled in at the John Innes, and politely declined Balfour's second offer, recommending instead his friend Reginald Punnett, who thus became the first professor of genetics in Britain.[29]

⌒ CYRIL'S STUDENT application reached Bateson at the John Innes almost twelve years later. When the director received it he wrote back asking the pretentious young man to specify exactly what he wanted to do at the John Innes and how he proposed to go about doing it. Cyril's answer was so unsatisfactory (he didn't know) that no further reply came.[30] Persisting, he called Bateson and inquired whether he could meet him. "When will you come?" asked the old man. "Tomorrow," replied Cyril.

The next day Cyril ran up a long muddy lane near Wimbledon, and found that he was lost. At last, he reached a gate with no name on it and entered. Before him stood a small building with two laboratories and four greenhouses. He was late. Ushered into a large square room that appeared to be a library lit by a skylight, he found a big white-haired man with a prominent moustache wearing carpet slippers. "The rewards from genetics are slight," Bateson said in a rich, mellow voice as he held out his enormous hand to the breathless youth. Only those who attained unheard of heights of achievement could ever hope to make a living out of it. It was all far from encouraging, but Cyril insisted, and managed to persuade the unimpressed Bateson to give him a three-month probationary period as an unpaid volunteer worker.[31] A jubilant young man made his way back from Merton to Rectory House. Though he hadn't succeeded

Darlington ran up a long muddy lane near Wimbledon, and found that he was lost. A John Innes greenhouse, 1929. Photo from John Innes Foundation Historical Collections courtesy of the John Innes trustees.

in getting the scholarship, he was resigned to work hard in order to win it. Besides, he would be at the foot of Britain's most prominent and revered biologist—in Cyril's mind, the founder of genetics himself. Finally, he could return home with news worthy of his father's ear.

Starting November 30, 1923, Cyril journeyed to Merton every day. For a week he counted pea seeds, recording their shapes and colors. When there were no more pea seeds, he counted *Campanula* seeds. And when there were no more *Campanula*, he counted *Antirrhinums*. They were tiny, and Cyril was glad when he was finally asked a fortnight later to go out into the open to count plum trees instead, and make drawings of their winter buds. There were more than a thousand trees, and the task lasted him another week. This was December, and besides, no one had yet explained to the young worker what all this tedium was supposed to accomplish. Ignorance made the winds seem even colder. In reality, there was much more that Cyril was oblivious to, for his hero Bateson was in fact by now a broken man.[32]

~ 2

A Rising Tide

I F BATESON WAS A PIONEER and world leader in genetics in 1910, by 1915 his arguments were made in the face of overwhelming evidence to the contrary, and by the time Darlington had arrived at the John Innes in 1923 they had been soundly refuted and surpassed.[1] At the center of the debate were the tiny thread-like substances in the nucleus of every living cell. After Oscar Hertwig's discovery in Germany in 1875 that fertilization, the basis of sexual reproduction, involves the cell nucleus, these tiny threads, observed inside the nucleus, naturally became a focus of interest In 1884 the Swiss botanist Carl Wilhelm von Naegeli published the first systematic attempt to discuss heredity as inherent in a definite physical basis contained within the cell. But he made no attempt to locate or identify the particular hereditary material with any morphological constituent. Shortly afterward, Rudolph Albert von Koelliker, Eduard Strasburger, August Weismann, and Hertwig himself all arrived at the same conclusion: that the hereditary substance could be identified with the chromatin in the nucleus.

A few years later, in 1888, the newborn "nuclear theory of heredity" finally named its progeny when Wilhelm Waldeyer in Germany christened the tiny nuclear habitants the "chromosomes," a name reflecting the fact that the thread-like structures absorbed colored stain readily. Waldeyer thought that if the name "is practically applicable, it will become familiar," but added: "otherwise, it will soon sink into oblivion."[2] Although Waldeyer's name stuck, and despite a wealth of variable theories concerning the particular basis of heredity that didn't,[3] no hard evidence was available to confirm that the chromosomes were indeed the physical bearers or inducers of hereditary traits. In fact, until the rediscovery of Mendel's laws in 1900 there was no dependable conception of hereditary transmission against which the cytological facts could be compared.[4] Furthermore, at the turn of the century the problem of just how the latent adult characters lie in the germ cells and how they are activated as development proceeds remained a complete mystery.[5]

In 1902, however, a big step toward the union of the particulate conception of heredity and the cytological evidence was made. Walter Stanborough Sutton, a young graduate student working in the lab of celebrated cytologist Edmund B. Wilson at Columbia University and studying the morphology of chromosomes of the lubber grasshopper, concluded that the phenomena of germ-cell division, as observed through the microscope, and heredity, as evidenced by breeding experiments, have the same essential features—that is, purity of units (chromosomes, characters) and their independent transmission. The significance of this insight was immediately understood: if cytology and genetics were describing the same phenomena, then the chromosomes could in fact be the physical material of heredity. They could contain the secret of life. Like the rediscovery of Mendel, this too was an instance of coincidental theorizing, for the same conclusion was reached simultaneously by Theodor Boveri, Correns, and de Vries, and was finally named the Sutton-Boveri hypothesis by Wilson.[6] Cytogenetics may be said to have begun with this first consistent application of observations on chromosomes to the results of experimental breeding.

In the following years a number of researchers, Wilson prominent among them, presented strong evidence that sex determination was connected to the presence of a whole chromosome and followed Mendelian patterns of heredity. Thus a cytological school, led by Wilson in America, argued the central position of the chromosomes in heredity and development, based on observations at the cell level. Across the Atlantic in England, a second, genetical school led by Bateson was arguing the central position of Mendelism in heredity, based on the confirming results of breeding experiments.[7] Before long, the first of these would give rise to the greatest enterprise in biology of the century.

⌒ IN 1905 BATESON AND PUNNETT MADE an important discovery—a clear exception to one of Mendel's two laws. While independent segregation was being confirmed and upheld in breeding experiments, this discovery rendered free assortment provisional: certain "elements" carried in all somatic cells and passed to the germ cells appeared more frequently than expected in tandem, and would be passed on to the next generation together.[8] Bateson and his coworkers called this phenomenon "gametic coupling" and explained it with a theory they called "reduplication," whereby the characters appeared together because gametes containing them underwent additional post-meiotic divisions; the explanation excluded any recourse either to the chromosomes or to the term "linkage" (which was to eventually be applied to the phenomenon), based on the assumptions of the chromosomal theory of heredity.[9] Meanwhile, in Wilson's department at Columbia, Thomas Hunt Morgan began work on breeding experiments in fruit flies, which he got from Frank Lutz at Cold Spring Harbor. Like Bateson, Morgan's background was embryological. Like him, he was skeptical of Wilson's chromosomes and challenged Darwin's notion of natural selection, deeming it teleological. But unlike Bateson, Morgan was a staunch materialist and therefore very much opposed to unseen and, in his view, unsound "factors" postulated as the base of heredity. Although he paid respect to de Vries's mutation theory (to be discussed later),[10] he was by no means certain of Mendelian heredity, and, when in 1907 his pupil Fernandus Payne asked him for a problem, he offered the Lamarckian task of breeding Castle's *Drosophila* in the dark for sixty generations.[11] While Payne failed to diminish the flies' eyes by keeping them in the dark for two years, Morgan, with a hand lens, was able to find a mutation, a single male fly with white eyes. The responsible factor proved to be in the unpaired sex chromosome, and from this one fly (admittedly part of an extensive set of experiments) Morgan could therefore demonstrate Mendelian segregation, sex determination by chromosomes, and sex linkage. A note to *Science* quickly corrected all of his prior misapprehensions and announced his conversion to Mendelism.

Then, on July 7, 1910, Morgan sent a letter to Bateson from Woods Hole offering to share with him his stocks and the theory they might carry with them. Bateson, however, had come to distrust Morgan's judgment, and declined the offer and the theory.[12] Having failed to enlist Mendelian support in Britain, Morgan turned to the chromosome side, to Wilson, in the room just next to his. Wilson pointed his colleague to the work of the Belgian cytologist and Jesuit priest Frans

Alfons Janssens, and recommended three pupils whom he had taught chromosome theory and had learned about Mendelism from the books of Bateson disciples Punnett and Robert Heath Lock. They were Calvin Bridges, Alfred Sturtevant, and Hermann J. Muller, and together they would breed and study Morgan's flies.

In the hands of Morgan and his students, Sutton's chromosome theory was transformed into a theory of the gene.[13] Reading Janssens's observation of homologous chromosomes and their wrapping around one another, an occurrence producing tangled, crosslike structures called *chiasma* (meaning "cross") during cell division, Morgan immediately understood that here was the cytological corollary to a genetic theory of crossing over, and the true explanation of linkage.

THE CHIASMATYPE HYPOTHESIS

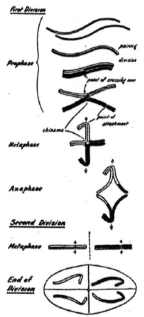

Diagram showing the history of two homologous chromosomes at meiosis according to the chiasmatype hypothesis. The four daughter nuclei all contain chromosomes of different origin owing to crossing over. From Darlington, Recent Advances in Cytology (London, 1932), p. 259.

If homologous chromosomes wrapped around one another at reduction division, this could be the tentative yet sought-after evidence that they recombined genetic material between them, resulting in the phenotypic changes he had observed in his flies![14] Whereas Bateson rejected a material, particulate basis for heredity, Morgan took linkage (the "gametic coupling" phenomenon Bateson himself had discovered) and crossing over (an inference based on Janssens's cytological observations and his own breeding experiments) to be the most powerful evidence for a model in which genes were lined up like beads on the chromosomes.[15] Meanwhile, Danish biologist Wilhelm Ludvig Johanssen's 1909 distinction between phenotype and genotype was proving instrumental in helping to clear up a lot of conceptual confusion.[16] These developments spurred Morgan and his team, representing a novel form of research organization, to one of the most fruitful collaborations in the history of biology, one that in the following decade produced compelling evidence in favor of the notion that chromosomes were the physical bearers of heredity, that genes were distributed linearly along them, and that crossing over and recombination between such genetic elements took place during meiosis. In 1913 Sturtevant published the first genetic map, based on the idea that the frequency of genetic crossing over, a quantity determinable from breeding experiments, depended on the distance on the chromosome separating the concerned factors. Two years later the group published *The Mechanism of Mendelian Heredity*, a book that firmly established the chromosomal theory of heredity and quickly became the great monument of classical Mendelian genetics.[17]

If BATESON AND HIS SCHOOL'S EFFORTS and renown had kept Britain in the forefront of Mendelian genetics in the first decade of the century, the war was by now solidifying a dramatic shifting of the tide across the Atlantic, where Morgan and his team, Wilson, Edward Murray East, Rollins Adams Emerson, and the budding cytological-genetical schools were growing institutionally and in number, and producing a wealth of solid, original research. The appearance in 1915 of *The Mechanism* was a testament to the great advances American genetics had made over Britain. Bateson had given the young new science its name, but the Americans were snatching it away.

Bateson's leading position in genetics had done little to alter the existing departmental structure in the British universities. Between 1900 and 1910 there was no serious research done in zoology departments either in experimental breeding or in chromosome genetics.[18] This trend continued well into the 1920s: the young lepidopterist E. B. ("Henry") Ford remembered that when he sought to do research in this field in the middle of the decade, "more than one person said 'surely you are not now going on with *Genetics* when we are seeing what can be done in experimental embryology!'"[19] Indeed, those who had made a great name in zoology at the time, like Ford's teacher Edwin Steven Goodrich, were almost wholly nonexperimental and chiefly concerned with comparative anatomy. Institutional organization both determined and reflected such predilections, and was slow to change; there was little experimental breeding and chromosome cytology of which to speak.[20] Across the Atlantic, the rapid ascent of the American school of genetics was being made possible by institutional developments such as the growth of graduate training and the natural interest in the discipline by farmers and breeders who supported its pursuit at the newly established agricultural experiment stations attached to the land-grant colleges and universities.[21]

Bateson's move to the John Innes was a sign of his strong belief in the new science and a testament to his resolve to see genetics succeed despite the impediments presented by the universities. Quickly, he was able to assemble a talented group of young genetical workers shortly after accepting directorship. A fruitful collaboration with Reginald Gregory on the genetics of *Primula sinensis*, and with William Ormston Backhouse on inbreeding and outbreeding in a wide range of horticultural plants, seemed to point the way to a promising future. A genetics institute of the highest caliber was growing under Bateson's tutelage.[22]

But Bateson's reply to Morgan's letter from Woods Hole attests to a major caveat. For the same man who perhaps more than anyone else had consolidated and firmly established the principles of Mendelian genetics, and who had discovered the crucial phenomenon of linkage, now adamantly refused to grant victory to the chromosomal theory on genetic evidence alone.[23] Reviewing *Mechanism* Bateson wrote: "It is inconceivable that particles of chromatin or any other substance, however complex, can possess those powers which must be assigned to our factors. . . . The supposition that particles of chromatin, indistinguishable from each other and indeed almost homogenous under any known test, can by their material nature confer all the properties of life surpasses the range of even the most convinced materialism."[24] Whereas Morgan the embryologist had recognized that solving the problem of genetic transmission would have to entail its temporary separation from that of the problem of differentiation, Bateson, of similar scientific background but of an entirely different cast of mind, was unwilling to make such a concession. For him, as was the case for nineteenth-century post-Darwinian biologists of the European school, heredity and development were inseparable components of one grand operation.[25] It was essential that somatic differentiation find its parallel, because there was its cause, in germinal differentiation. If the complexity of the chromosomes were to be explained as a function of their architecture, different chromosomes would have to be molecularly, and thus structurally, distinct. This implied that structural differences must entail visible chromosomal expression, but somatic chromosomes were not visually distinguishable from zygotic ones, or from each other! How then could the chromosomes be responsible for heredity or development?

While Morgan was willing to extrapolate from the cytological and genetical associations a material gene as a solid working hypothesis explaining their connection, Bateson was adamantly not.[26] He insisted that correlation and causation be separated, and saw the chromosomes as a secondary manifestation whose association with inheritance merely indicated that heredity and the chromosomes were both cellular phenomena. Bateson accepted the reality of what others called "linkage"— after all, he had discovered it. That genetic crossing over implied material exchange between synaptic chromosomes he understood. But Bateson called attention to exceptions, which he famously preached to treasure,[27] and pointed out that there was no crossing over in male *Drosophila*. This exception seemed to him especially alarming in that it manifested itself in *the* experimental organism used by the proponents of the chromosomal theory! Moreover, work at the John Innes and elsewhere on chimeras and bud sports seemed to indicate the cytoplasmic (hence nonchromosomal) segregation of apparently Mendelian characters in plants. Thus Bateson rejected *in toto* the acceptance of the chromosomal theory of heredity, claiming that it had been generalized from inadequate evidence, that crossing over and chiasmata were derivatives, and that inheritance must explain development as well as heredity and variation. After all, while chromosomes persisted unmodified over cell generations, the soma was being totally transformed. And so while in America the chromosomal theory of heredity was generating highly fecund experimental results, in Britain the leader of the new genetics was channeling his immense energies into trying to disprove its very reality. Bateson was the dean of British genetics; his refusal to accept the chromosomes played a large part in stunting the study of heredity in that country.[28]

Still, at this stage the new genetics was just being born, and it is possible that Bateson could have had a change of heart. After all, a number of other leading British biologists had already accepted the chromosomes. Gregory, a brilliant young geneticist, despite being initially influenced by his teacher's rejection of the chromosomes,[29] had amassed copious evidence to the contrary and was beginning to shake-Bateson's resolve.[30] Added to this were the endorsement of the chromosomal theory by two of Bateson's students, Robert Lock and Charles Hurst, as well as Wilfred Eade Agar (John Innes), Geoffrey Watkins Smith (Oxford), and Leonard Doncaster's (Cambridge) cytological work addressing the connection between Mendelism and meiosis.[31]

But what hard work, commitment, and brilliance had gained, war quickly squandered. As young men flocked to fight the war in Europe, the John Innes emptied and was soon left almost entirely devoid of a male presence. Owing to this disruption (and perhaps to some degree to Bateson's tightfistedness in matters of money), by war's end almost all the director's brightest men had left: Arthur Bailey went to Egypt, Backhouse to Argentina, James Lesley to Cambridge and then to America, Carrington Williams to Trinidad; others never returned.[32] Then, just fourteen days before the armistice, Bateson suffered the loss of his son Jonathan. Just a few months later, both the promising geneticist Gregory and Bateson's beloved garden superintendent, Edgar John Allard, succumbed to the influenza epidemic that struck London immediately after the war. Crushed, despondent, and increasingly isolated in the rarefied atmosphere of his institution, Bateson lapsed into an indefensible rejection of the chromosomes.

There was some fight left in the old man, though, and, in an attempt to rally, Bateson established the Genetics Society in 1919. This, however, did little to salvage the reeling science. Like the John Innes, the universities had been struck hard by the war, and had lost many fine young researchers. Agar went to Melbourne in 1919, Doncaster (who had played an important role in the discovery of sex-linked inheritance in the early years of the century and who had tried, unsuccessfully, to convince Bateson of the validity of the chromosomal theory of heredity after the publication of *Mechanism*) died in 1920, Lancelot Hogben left London for Edinburgh and turned to other fields in 1921, and both Lock and Smith were dead by 1918. Genetics in Britain, now more than ever, and to a greater degree than in the United States, was practiced and dominated by a high proportion of breeders and horticulturists, rather than professional scientists.[33] Under Bateson's directorship, work at the John Innes continued to concentrate on finding exceptions to Mendelian inheritance with the explicit purpose of disproving the chromosomal theory of heredity. The lead Britain had enjoyed in the first decade of the century, in great part due to Bateson's own efforts, had now long since been squandered.

⌒ THAT A MORPHOLOGIST SHOULD obstinately reject a new and notable generation of organic elements, may indeed seem surprising. Of aristocratic stock, schooled at Rugby and St. John's in Cambridge, and priding himself no less on his connoisseurship of Japanese art and continental philosophy than on his science, Bateson shared Samuel Butler's rage against machine-age materialism, crass professionalism, and unabashed utilitarianism. Science for Bateson was not a profession, but rather a higher calling, part and parcel of intellectual and aesthetic cultivation. It derived its worth not from its utility, as Karl Pearson, the socialist lawyer, statistician, and eugenicist, and Bateson's scientific adversary, would have his followers believe, but rather from its refining qualities. To use genetics as a tool of social reform was as much anathema to Bateson's sensibility as the call to abolish Greek and Latin entrance examinations to Oxbridge.[34] Such dispositions were on the agenda of a new breed of men Bateson just could not stomach. His old friend and Grantchester neighbor, the philosopher Alfred North Whitehead, expressed both men's opinion best, referring to the period from 1880 to 1900 as dull, efficient, and commonplace, as but an echo of dreary eighteenth-century materialism: "It celebrated the triumph of the professional man."[35] If anything professional, Bateson would regard himself as a *Bildungsmensch*, a cultivator of culture. To him, matter smacked of the new utilitarianism taking hold of old Victorian Britain. It offended his austere intellectual disposition and delicate aesthetic sensibilities. It reflected in men such as Pearson, and in the reformers of the newly established Eugenics Society who sought to replace the old social order with a new one. It showed its face overseas in a new breed of specialized geneticists, who, like Morgan, betrayed the organism as a whole with their reductionist, crudely materialist programs. For Bateson, deeply conservative and far removed from modern-day mores, no phenomenon could be meaningfully or usefully reduced to matter alone. Locating heredity in the chromosomes seemed to him philosophically naïve at best, and more to the point an instance of scientific crudeness of the very worst kind. As an alternative, Bateson adopted a vibratory model borrowed from the physical sciences, defining heredity, variation, and ontogeny not in terms of a particulate substance, but rather in those of mechanical separation. Force replaced matter as the ultimate cause of all biological phenomena.[36]

But Bateson was fighting a losing battle. Bridges's demonstration of nondisjunction of sex chromosomes, in which the rare white-eye phenotype in female flies was shown to correspond to a chromosomal mechanical failure whereby two X chromosomes did not separate in cell division (thus producing aberrant XXY female progeny in some cases), was a stark confirmation of the predictions of both sex determination and character determination of the chromosomal theory. Bateson was by now well aware that the thrust of genetic research had surpassed him and traveled across the Atlantic. He therefore journeyed in 1921 to New York to see for himself what the Morgan school had accomplished. The visit was a tragic one for him. Having spent the better part of two decades trying to disprove the notion that the chromosomes were the physical bearers of heredity, Bateson was now compelled to acknowledge the strength of the evidence brought before him. To his wife, Beatrice, back in England he wrote: "I see no escape from capitulation on the main point. The chromosomes must be in some way connected with our transferable characters."[37] Less than a week later, at a meeting of the American Association for the Advancement of Science at the York Club in Toronto, Bateson announced his conversion on this main point. "All the time I can't help seeing that people here *have* got ahead of us," he wrote Beatrice on New Year's Day, 1922. "I was drifting into an untenable position which would soon have become ridiculous. . . . Cytology is a real thing—far more important and interesting than I had supposed." In a saddened, somewhat desperate tone he added: "we must try to get a cytologist."[38]

Everywhere he went, Bateson was treated with the utmost respect and veneration. He met old friends George Harrison Shull, the eminent geneticist from Princeton, Director of the Carnegie Institution of Washington's Department of Genetics at Cold Spring Harbor Charles Benedict Davenport, botanical geneticist Albert F. Blakeslee, Harvard's Bussey Institute geneticist Edward Murray East, as well as the young Sewall Wright, a future important evolutionary synthesist. He was also introduced to the cytologist John Belling, who had been working on *Datura* with Blakeslee and who had found a whole set of forms distinguishable by eye, each caused by an extra chromosome, describing him as "the sort you might see behind a boulder in a Rackham drawing." While the visit was no doubt intellectually stimulating it was also profoundly gloomy and depressing, for Bateson now realized that the novel mathematical, quantitative methods of genetics based on the chromosomes were both pointing the way to the future and painfully beyond his comprehension. On the train ride from Buffalo to Toronto, Bridges and others "talked genetics of the utmost technicality," while he "detached to play bridge to the great surprise of the whole party."[39] Bateson left America a humbled man.

Upon his return to Britain, Bateson immediately hired cytologist Frank Newton, a young and promising student of Lancelot Hogben's from London. Before Newton was officially appointed as a member of staff, however, Bateson suffered a second family tragedy, with the suicide of his son Martin in April. Shortly afterward, Morgan and Bridges visited England, demonstrating nondisjunction in a dramatic Croonian Lecture. Bateson had been resolutely refuted on his home field. A few days later in Piccadilly, E. B. Ford met Leonard Darwin, Eugenics Society president—and son of Charles—who had just seen Bateson going into the Athenaeum. Ford wondered what Bateson's reaction had been to the demonstration, and the two went into the club to ask him. "The chromosome theory is true," Bateson said, "and all my life's work has gone for nothing."[40] Having suffered the death of two of his three sons, witnessed the usurpation of the science he himself had coined, and been humbled by a modernity anathema to his very being, Bateson was by now a crushed man. All this was completely unknown to the young Darlington as he made his winter bud drawings in the John Innes gardens.

～3

Auspicious Beginnings

ONCE A REVOLUTIONARY AND A MAVERICK, Bateson was now the grand old man of an isolated horticultural institute, run like a family affair, and very much remote from the new genetics.[1] He was, in fact, alarmed by the possibility of his privacy being invaded by a new type of professional scientist who was bound, in his view, to be a careerist. Bateson wanted his institution to be a personal and dedicated establishment. He therefore fixed his scale of salaries at a level that led many of his male pupils, as soon as they had completed their training, to disperse chiefly to the remote colonies, and surrounded himself with less mobile, often well-to-do ladies, who worked for very little pay and created a local society more and more congenial as criticism became less and less so.[2] The women were led by the formidable Miss Caroline Pellew, acting as his "lieutenant, secretary, mentor and foil." Alongside her, Dorothy Cayley, Alice Gairdner, Irma Anderson, Dorothea de Winton, and Aslaug Sverdrup staffed the "ladies' labs." Most of the work there concentrated on crossing *Primula* and *Pisum* in an attempt to further generalize Mendelian principles. Across the way in the "men's labs" were Newton, Reginald John Chittenden, Ernest Jacob Collins, and Morley Benjamin Crane, working on cytology, genetics, botany, and pomology, respectively. On December 19, 1923 Darlington turned twenty and, when not working on tasks delegated to him by the women's labs, spent much time in the institution's library. By January he had read all of Bateson, Morgan, Doncaster, East, and Jones.[3]

As Bateson was locking up one evening amid the "humid scents of glasshouses, chirping of grasshoppers, and the breaking stillness of the night,"[4] Cyril approached him with a Morgan book in his hands and asked hesitantly, "I suppose it is sound?" "Yes," Bateson replied. "I am afraid we have to admit it."[5]

~ IN THE LABORATORY, Darlington had been sitting next to a man with long greasy hair falling over his eyes, appointed to the John Innes only a few months before him. It soon became evident to him that Newton, the cytologist, held less than orthodox views. He did not believe in Darwin, did not altogether believe in Morgan, and by no means believed in Bateson, for despite his conversion speech at Toronto, and admission to Darlington, the director continued to pursue investigations aimed at disproving the reality of the chromosomes. When affiliated plant geneticist Reginald Ruggles Gates asked him to appoint a Miss Latter to the John Innes as a second cytologist in February 1925, Bateson replied: "I see no prospect of our appointing another cytologist for a considerable time to come."[6] If Newton was certain of one thing, it was of these tiny nuclear bodies.

In March Bateson took Cyril with him to his lecture on "somatic segregation" at Imperial College, explaining to him on the way about chimeras and how they were proof of hereditary mechanisms that had little to do with the nucleus and its stringy contents. Bateson could not concede entirely to Morgan, even if he admitted that his work on flies was sound. He held on to the exceptions he found in plants, hoping they would ultimately discredit the chromosome interpretation. Reading the American literature on the one hand, being tutored by his venerable director on the other, and soaking up Newton's skepticism on the third created a bewildering confusion for Darlington. Evidentiary certainty was very scarce in this new world of science into which he was being inducted. Doubt lay on every side.

~ ON OCCASION NEWTON WOULD SHOW the chromosomes to Bateson, who would peer down the microscope with his left eye, covering his right one, and, as Darlington suspected, seeing equally little with either.[7] One day Newton took Cyril to the microscope and had him peer down the lens. "Sure enough, there they were as clear as daylight," he remembered. They were 24 Hyacinth chromosomes, and

In the laboratory Darlington had been sitting next to
a man with long greasy hair falling over his eyes. A
drawing of Frank Newton from the Darlington Papers,
Bodleian Library. Reprinted with permission.

Newton told the young initiate: "They are of different shapes and sizes.
They are constricted like strings of sausages, sometimes like two or
three in a string and sometimes like single sausages. But there are three
of each kind." Darlington asked how that could be since every plant and
animal had one chromosome of each kind from its mother and one from
its father, and got in return his first lesson in *polyploidy*—the existence of
more than two homologous chromosomes in the chromosome complement. From that day on he began to look at the chromosomes, whose
mysterious movements, constrictions, and structures he found infinitely more fascinating than anything anybody else was working on.

At the end of his probationary period, the John Innes Council
decided to hire Darlington as a Minor Student and offered to pay him
£150 per year, Bateson having overlooked that he possessed a degree.
Cyril protested and was finally promised £200 as a compromise,

although he did not get his first salary for seven months. "Starvation was Bateson's method of trying to discover whether you were really interested in your work," he remembered half jokingly. "If I'd have been quite penniless, I (or my parents) would have failed under the test. Bateson was really choosing the people who were somewhat well to do."[8] Whether this was true or not, the old chief did agree to buy Darlington his own secondhand brass microscope. He could now make and examine his own preparations beside his new friend Newton. The fruits of these labors were beginning to ripen: in June Darlington recorded his first discovery—the tetraploidy, or possession of four identical copies of each of the different individual chromosomes of the sour cherry. Nevertheless, he kept up his plant breeding, mainly to please Bateson, but also to have a second string to his bow in case the chromosome studies proved unsuccessful.[9]

BATESON WOULD OFTEN INVITE Darlington to the lavish dinner parties held in his home, where the young man met many of his chief's famous international friends. He remembered on one occasion being introduced to the embryologist Hans Driesch, a close friend of Morgan, though a vitalist, and on another, listening in on a lively discussion about "presence and absence" with the great Swede Johannsen. The "impressive Prussian" Erwin Baur, Germany's first professor of genetics and a close friend of Bateson's, was another visitor. These occasions were feats of hospitality. Retiring to the library, guests would enjoy port and cigars as Beatrice Bateson played the cello, while her husband displayed his collection of Japanese prints. Bateson the aesthete despised politicians. When Lloyd George, interested in forming an agriculture policy, inquired whether he might come see the work at the John Innes, Bateson brushed him off. "Nothing practical here," he reported, hoping, despite the interests on the council, that indeed there was not. Balfour, the "philosopher," was another matter, and after a tea party in the garden one evening, as the noble lord departed, Darlington remembered Bateson calling him "a very credible politician."[10]

Bateson also took Cyril to meetings of the Genetical Society, and to conversaziones at the Royal Society. On one occasion, he was given the opportunity to show his chromosomes to its president, Sir Charles Sherrington. On another, Bateson and Darlington walked into George

Bernard Shaw on a crowded London street. "You are the fellow who knows about Mendelism, aren't you?" the long-bearded Shaw grunted at Bateson. "Well I've bred rock pigeons and can tell you there's nothing in it! You may as well save yourself the time."[11]

～ BUT DARLINGTON WAS STRUGGLING. The science into which he was being inducted did not have as long a history as breeding, or development, or even physiology, but it was vast, complicated enough nevertheless, and a daunting task for a novice. The central book initiating Darlington into the field was American cytologist Edmund Wilson's first edition of *The Cell in Development and Heredity*,[12] an excellent gauge of where cell theory and general biological thought had arrived at the turn of the century. Dedicated to his friend, the biologist Theodor Boveri, Wilson's work of synthesis brought together all current knowledge and became a monument to the great advances cytology had made in the last three decades of the nineteenth century.[13] Darlington spent many hours struggling through it, trying to comprehend.

On the eve of the Mendelian rediscovery, Darlington now learned, cytology had much of which to be proud. Naegeli, Hertwig, von Koelliker, Strasburger, and Weismann had all contributed to an understanding that put the nucleus at the center of heredity, and its constituent chromosomes in the limelight. Researchers before them, and Friedrich Anton Schneider prominent among them, had noticed a complex pattern of movement of tiny bodies within the nucleus of cells during cell division. These regular, mysterious nuclear events were immediately recognized as important phenomena to be studied. The problem was that the tiny bodies, not yet named chromosomes, could just as easily be mere artifacts—after all, they had never been observed in living cells, just in fixed, stained, *dead* ones. If real importance were to be attached to the nuclear dance, it would have to be observed in living cells. This was accomplished by cytologist Walther Flemming sometime around 1882. Flemming had discovered that the chromosomes of epidermal cells of salamander embryos were relatively large and could be observed doing their nuclear ballet in the living. Even though he was not the discoverer of the process of cell division called *mitosis*—Schneider was—Flemming was the first to describe it with rigor. What he offered in 1882 remains until this day, with minor alterations, the basic picture of our understanding of mitosis.[14]

Flemming found that the drama in the nucleus, though continuous in reality, could be described by a number of discrete stages. During the *resting stage* cells did not undergo mitosis, and chromosomes, and for that matter any kind of internal structure within the nucleus, were not observed. Changes and movement within the nucleus were the sign that mitosis was under way. Suddenly, at a stage Flemming called *prophase*, tiny nuclear threads become apparent. When they begin to condense into shorter, denser bodies during *metaphase*, the nuclear membrane disappears entirely, as the chromosomes migrate and assemble at the center of the cell. This they do within an elongated fibrous structure called the spindle, at the ends of which are tiny granule-like bodies called *centrioles*. At this stage, called *anaphase*, the chromosomes separate into two groups and begin to move to opposite poles of the spindle. When they arrive, at *telophase*, they become less and less distinct, and finally disappear, as the nuclear membrane forms around each polar group once again. By this time, the cell has returned to its resting stage, though there are now two cells in place of the original one.

Observing salamander cells, Flemming was able to show how a daughter cell is produced. But if this daughter cell was to be identical to the original cell from which it came, it was clear that the chromosomes must first double in number before they could be divided. To Flemming's great delight, this is exactly what they did. Sometime between their disappearance in the previous telophase and their reappearance at prophase, each one of the chromosomes doubled, producing an entirely double chromosome complement. This was wonderful for explaining how cells *divide*, but it immediately became apparent that it would run into a major difficulty when explaining how cells *unite*. This was no small matter; after all, biological reproduction depends upon fusion of sex cells from the female and the male. If cell division produced cells identical in their chromosome number to the mother cell, the union of sperm and egg would result in a cell with twice the amount of chromosomes. This cell would then produce cells with an identical number of chromosomes that would yet again fuse with cells of the opposite sex to produce quadruple chromosome progeny, and the process would continue ad infinitum. Before long, organisms would be comprised of cells with millions upon millions of chromosomes, but observations with a microscope had shown that this was never the case. The number of chromosomes in any given species remained constant from one generation to the next. Clearly, some kind of special division, different from mitosis, was needed to produce sex cells. And yet not even the most meticulous observation had succeeded in detecting it. Flemming was stumped.

Where the sharpest of eyes had failed, the sharpest of minds would succeed. August Weismann had begun his scientific career looking at cells through a microscope, but due to his deteriorating eyesight had gradually turned from observer to theoretician, becoming one of the most important biological thinkers of his day. Taking up where Flemming and company had left off, Weismann surmised that there had to be a corresponding process of reduction sometime in the life cycle of the cell by which special sex cells—*gametes*—with half the number of chromosomes of a regular cell, would be produced. These *haploid* gametes, in turn, would combine in the act of sexual union to produce a *diploid* daughter cell—or *zygote*—with just the right number of chromosomes, half from its mother and half from its father. This was a grand, almost irresponsible speculation on the part of the nearly blind Weismann, but he nevertheless sent it in a letter to *Nature* in 1887.[15] Although it would prove to be one of the most important pieces of theorizing in the history of biology, the exact manner in which Weismann supposed this reduction division took place was to be a source of great confusion for the researchers who came after him.

Shortly after Weismann's letter to *Nature* was published, cell biologists began searching for the elusive reduction division. Before long, they had found it—a clear instance whereby a hypothesis proved necessary for observers' ability to *see*. By 1890 meiosis was *known and seen* to consist in the occurrence of two successive divisions of a nucleus in which the chromosomes divide once instead of twice as they would in two ordinary mitoses. (This was actually slightly more complicated than what Weismann had predicted, which was a single nuclear division without a chromosome division.) In such a manner, where the distribution of chromosomes was regular, the four daughter nuclei each have half the number of chromosomes of the parent nucleus, just as Weismann had predicted. During the first division, the chromosomes come together in pairs, and each one seems to split down the middle, producing two *chromatids*. These chromatids are attached at a point later to be called the *centromere*, where one chromosome existed beforehand. This was true for mitosis as well, with the only difference being that the chromosomes in mitosis metaphase did not line up in pairs, but merely singly, one next to the other. Thus, while two chromatids are associated in pairs in mitosis, four are associated at the metaphase of the first division of meiosis. No one understood exactly how or when the chromosomes split into chromatids, or how they acted once this had

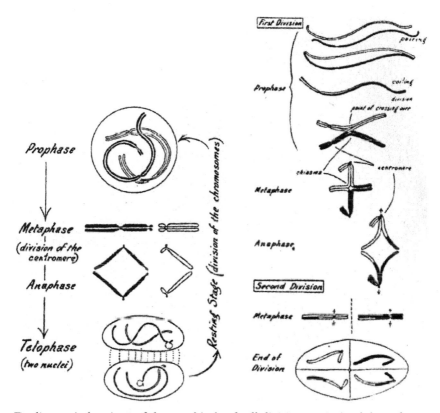

Darlington's drawings of the two kinds of cell division, meiosis, right and mitosis, left. From Dan Lewis, "Cyril Dean Darlington, 1903–1981," *Biographical Memoirs of Fellows of the Royal Society*, 29, 1983, p. 130.

happened, but the validity of Weismann's hypothesis had been generally confirmed. Haploid gametes were produced by a special reduction division similar to mitosis, but also different in important ways.

When Wilson sat down to write his cytological tome in 1896, it was believed by many that heredity was somehow accomplished through the union of the chromosomes of cells produced by a special reduction division. Many, but not all. It remained unproven that the small thread-like contents of the nucleus had anything to do with heredity. Observing their behavior was one thing; articulating a coherent mechanism of heredity and linking it to the chromosomes was altogether another. With no dependable mechanism of heredity yet articulated to guide him, Wilson set out to explain to the best of his knowledge what the chromosomes did and why. His three most salient conclusions were the following:

- The constancy of chromosome numbers throughout the individual, even throughout a species, was proof that these bodies had existed in the resting nucleus when they were invisible. The chromosomes, therefore, had a characteristic continuity necessary for anything supposed to be the receptacle of heredity.

- The equality of division of the chromosomes at mitosis by which a linear thread splits into two exactly equal daughter threads is proof that each of those threads is linearly differentiated (i.e., it consists of a series of different particles, each one of which is necessary to the daughter chromosomes). Once again, this was an indication that the chromosomes have a structure that would be expected for the organs responsible for the business of heredity and development.

- The sperm and the egg are equal in their chromosome content, whereas in any other respect they are different.

All three conclusions were designed to show the importance of the nucleus in the workings of the organism—of heredity, development, and evolution—and all three were patently, even scandalously speculative. Most certainly, they could all very well have been true. With no hard evidence to prove this, however, they could be nothing more than a figment of the theoretical imagination.

Darlington began working on chromosomes exactly thirty years after Wilson's book was published. Immediately he was drawn to Weismann. With the Sutton-Boveri hypothesis (described in Chapter 2) articulated, the first decade of the century saw the idea that the chromosomes were somehow responsible for heredity quickly gain currency. Weismann's speculation of the biological necessity of a reduction division—first called *meiosis* by John Farmer and John Moore in 1905—made sense of it all.[16] If the chromosomes were in fact the bearers of heredity, they would have to divide to create haploid gametes in order to allow for the union of the hereditary complements of the two parents. When Morgan's fly team, by observing chiasmata and breeding results and correlating the two, understood that this meant the chromosomes would have to carry genes ordered linearly along them, modern classical genetics was born. What remained unknown, however, were the precise mechanics, and ultimate theory, behind the chromosome movements during cell division. How did the homologous chromosomes, all tangled up in the crowded tenement that is the nucleus, actually find each other in order to pair up during metaphase of cell division? How, wrapped around each other in chiasmatic formations, did they exchange genetic material? Why did it make sense for them to do so? And how had the cell developed two forms of division for two separate functions—somatic replication and sex-cell formation—both necessary for life? These were the questions that needed answering.

Joining Newton and a small international community of cytologists, Darlington immediately inherited two fundamental problems yet to be solved. These were the problem of chromosome pairing and the problem of crossing over and chiasma formation, both features of meiosis. The first of these was derived from observations of the chromosomes, all tangled up in a knot, in the prophase nucleus.

Two rival schools came to understand these confusing configurations in opposite ways. The first, advocating the parasynaptic theory of chromosome pairing, claimed that what the chromosomes were doing during metaphase of meiosis was pairing side by side in preparation for segregation and reduction division. The second, advocating telosynapsis, claimed they were pairing end to end in preparation for being cut crosswise, which would result in segregation and reduction. This argument was of crucial importance to the second problem—the manner in which crossing over between homologous chromosomes took place—because the entire notion of crossing over during meiosis upon which Morgan and his school's success had rested depended on side-by-side pairing: if the chromosomes were not ordered side by side, how could they exchange and recombine genetic material?

Careful cytological preparations of later stages of meiosis had shown the chromosomes attached to one another at their ends.[17] The outstanding example was Hugo de Vries's evening primrose *Oenothera*, in which end-to-end chromosomes created stark, ringlike formations. Lancelot Hogben, who at the suggestion of Leonard Doncaster had turned his attention to chromosome cytology, had shown cockroach chromosomes to pair side by side in a series of papers in 1921.[18] Some years earlier, Farmer and Moore had asserted in a celebrated paper that no parasynapsis occurs in *Periplaneta*. When Farmer, a Fellow of the Royal Society and considered Britain's premier cytologist, heard of young Hogben's results, he went to Imperial College, examined the first available preparation, and immediately conceded—though only for the

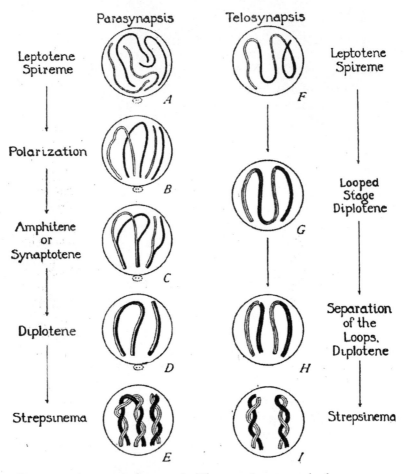

Parasynapsis versus telosynapsis. The question as to whether
chromosomes lined up side by side or end to end had immediate bearing
on whether genes are exchanged through crossing over during cell
division in all organisms. Darlington helped settle the debate in favor
of the former. From E. B. Wilson, *The Cell in Development and Heredity*,
3rd ed. (New York, 1925), p. 557.

cockroach. The stakes were so high that Morgan and Muller crossed
the Atlantic to look at Hogben's slides, happy to garner cytological
support for an enterprise based largely on genetical inference.[19]

The debate continued to rage as telosynapsis found allies in Reginald
Ruggles Gates, who had been working for years on *Oenothera*, and L.
Digby, who together with Farmer had arrived at this view through work

on *Primula kewensis*.[20] Wilson, who sided with the parasynaptists, now exhibited caution uncharacteristic of his 1896 book in its 1925 third edition: "On the whole . . . the theory of parasynapsis has gained ground since 1900, when the possibility of such a process was first considered. Parasynapsis has, however, been attacked even by recent writers, and we are by no means in a position as yet to assert its universal occurrence."[21] As late as 1930 a venerable Finnish zoologist produced a very large book on the subject of the controversy, enumerating all those who had held one view or the other, adding them up, and concluding that the excess of votes for the one (although minuscule) must be an indication of the fiction of the other.[22] And so it stood, one group arguing side by side, the other arguing end to end.

The second and related problem Darlington faced upon entering the field concerned the question of crossing over. Janssens's 1909 suggestion that the points of contact, or chiasmata, he observed between homologous chromosomes at meiosis might represent the sites of crossing over—of the exchange of genetic material—between those chromosomes, had been taken up by the geneticist Morgan and was at the foundation of his chromosome-mapping enterprise. Morgan believed that such crossing over at each reduction division was responsible for the production of four slightly differing gametes, which, when fertilized, produced the phenotypic variability he and his team observed. But like Wilson before him, Janssens's chiasmatype theory was based on speculation, for the cytological evidence of the exchanges was lacking—what he had actually seen were merely chromosomes twisting around one another.[23] Neither Morgan nor Wilson nor Janssens himself knew for sure whether the chiasmatype theory was true. Most cytologists, in fact, took it to be false, adopting instead a "classical theory" that did not require breakage and rejoining of the chromatids, but rather simply the exchange of partners between the four chromatids. Whether chiasmata implied breakage and crossing over, what their exact structure was, and whether they were the cause or perhaps the consequence of genetic crossing over remained complete mysteries.

This was the state of affairs into which Darlington entered when Newton first began to show him chromosomes under the microscope at the John Innes. Newton, a pupil of Hogben (who had since moved to Edinburgh to work on comparative endocrinology), was anxious to establish the universality of the theory of side-by-side pairing on which Morgan's coordination of Janssens's crossing-over theory and

Drosophila breeding had been based.[24] Farmer, Hogben's former chief and a telosynapsis man, had argued that the double structure observed at prophase of meiosis was not of two homologues pairing side-to-side but rather a single chromosome that had divided. At about this time, Newton had discovered a triploid variety of the tulip (with three copies of each chromosome instead of just two) and realized that if a third chromosome paired alongside the other two, Farmer's explanation would be put out of court. Darlington, for his part, did his best to keep pace.

⌇ IN THE SUMMER BATESON WOULD often invite everyone to play croquet on the lawn, an invitation Darlington invariably declined due to his ingrained dislike of all sports. Bateson had taken up chess after learning that it was Mendel's favorite pastime, and would invite Newton to a game after tea. A better player than his boss, Newton was careful not to go ahead in the tally of which the old man scrupulously kept score. Meanwhile, the cytologist and his protégé were growing closer, and it was through talks with him that Darlington began to espouse a religiously materialistic view of the chromosomes' role in heredity. If there was one certainty in all this confusion, it was to be found in the observation of the behavior of these tiny bodies. If there was one escape from the demands of the "ladies' lab," the ringing insults and impossible expectations of his father, and the somber moods of his mother, it was to be gained by peering into the silent world of these bearers of life. No disturbances were possible there. In January of the new year [1925], Cyril left his parents' home, settling into digs at 2 Manor Road, close to Merton.

⌇ IT SOON BECAME CLEAR that if belief in the chromosomes divided the two young cytologists from Bateson, a shared contempt and disrespect for the work of three prominent British biologists created a deep bond of sympathy between them. These were the plant cytologists John Farmer (UCL) and Reginald Ruggles Gates (King's), and the embryologist E.W. MacBride (Imperial): the former two for their incompetent cytological observations and deductions, and the latter for his anti-genetical, Lamarckian views. The particular feud with MacBride reached a climax surrounding the affair of *Alytes obstetricians*, the midwife toad. Paul Kammerer, a Viennese zoologist, claimed to have proven that acquired traits are inherited. The crucial experiment consisted in breeding a rare variety of toad that mates on dry land (most do so in water) by forcing it to mate in water for several generations. When on land, the male clasps the body of the female with his forelimbs without difficulty. Yet when amphibian skins are made slippery by water, the male needs something with which to hang on. After several generations of coital discomfort he will, Kammerer exclaimed, acquire it. Little dark protuberances, or "nuptial pads," will emerge on the forelimbs and will henceforth be heritable.[25]

Kammerer's claims were challenged by Bateson, and an acrimonious debate between the two men soon followed.[26] MacBride defended the Austrian vigorously, unleashing a vituperative assault on Bateson and his genetics.[27] That same year an indignant, animated Kammerer crossed the Channel to Cambridge to exhibit his toad and gain the respect for his theory and person he believed were due. Though Bateson missed the meeting, MacBride did not, taking advantage of the opportunity to launch yet another public and derisory attack on Bateson and the science for which he made such great claims. Ultimately, Gladwyn Kingsley Noble of the American Museum of Natural History and Hans Przibram, director of the institute where Kammerer worked, examined the famous specimen in Vienna in 1926. They reported their sensational findings in adjacent papers in *Nature*.[28] The acquired characters, which Kammerer claimed to have made hereditary, turned out to be India ink. Just days later, after admitting the fraud in a letter to the presidium of the Communist Academy of Moscow (though claiming to be personally innocent of deception and ignorant of the identity of the perpetrator), Kammerer put a bullet through his head on an Austrian mountain path.[29]

Before all this happened Darlington, Newton, and Bateson had many laughs at MacBride's expense: Bateson would often joke about MacBride copulating, like the toads, with his thumbs out, and was especially gratified one day when he saw a cartoon Darlington had drawn of MacBride on the blackboard of the lab. "That boy's got talent!" he remarked with a great roar of laughter, admitting to Newton jokingly that he might be forced to give him a raise.[30] Despite his ambivalence over the chromosomes, Bateson remained a strong believer in the genetic "factors" and their constancy, and such sentiment became imbibed by Darlington early on. Scientific grounds united with personal, conditioning factors to consolidate opinion. Thus did the young Darlington become exposed to the old doctrine of the inheritance of acquired traits, and thus did he become firmly resolved against it.

❡ Darlington was quickly coming into his own. A journal entry from February 16, 1925—"Develop a theory of inverse truth of deistic ethics. Reading the history of the caliphate"—testifies to a twin fascination with religion and history, and a religious atheism that was to leave a lasting mark. Besides his microscope work, he was now turning to read the wide speculations of Jacques Loeb in *The Organism as a Whole*[31] and the more practical suggestions of Edward Murray East and Donald Jones.[32] In June Darlington completed his first scientific paper, "Chromosome Studies in *Scilleae*," and gave it to Bateson to review. Bateson complained of the punctuation, but even young Cyril was adept enough to notice that he was "rather at sea in the subject."[33] As Darlington went about verifying the hybrid origin of sweet cherries and observing the chromosome behavior of hyacinths, Bateson traveled to Leningrad to visit his old friend and student Nikolai Vavilov, who had spent a year before the war at the John Innes and was now president of the Lenin Academy of Agriculture.

Bateson was the granddaddy of genetics in England. Nikolai Vavilov had been a student of his at the John Innes during the war, and had since become a Soviet hero and world-renowned scientist. Vavilov, left, old-man Bateson, center, and an unidentified Russian, circa 1924. Photo from the Darlington Papers, Bodleian Library. Reprinted with permission.

In January David Prain of the John Innes Council wrote Bateson asking him to stay on as director. "When the centre of chief interest on Genetics shifted from work of my own type to that of the American group," Bateson replied revealingly, "I was already too old and too much fixed in my ideas to become master of so very new and intricate a development. It had taken me years even to assimilate the new things, and I recognize that the Institution has a right to a younger man in my place." Bateson, however, had not succeeded in producing such a man and promised to make this his first preoccupation. He accepted the invitation to stay on a bit longer "with pride and gratitude."[34]

Less than a month later, on February 2, 1926, Bateson attended his weekly dinner at the Athenaeum and collapsed at Earl's Court Station on his way home. He died six days later. Darlington's first publication appeared that same day in the *Journal of Genetics*, alongside Bateson's last. While the young initiate presented a study of chromosomes in the *Scilleae*, Bateson pointed to yet another exception, Strangeway's evidence that chromosomes in cell divisions of tissue cultures disappear altogether after telophase and pass into solution, as evidence that they could not be the bearers of heredity. "If the genes are, or are attached to, particles of material grouped in a more or less permanent arrangement in the chromosomes," he wrote, "and if these particles return to their serial places after dissolution of the chromosomes, in preparation for the next mitosis, we may have to do with a behavior of matter hitherto, I believe, unrecognized by physics."[35] The changing of the guard was under way. Bateson's last words to his young recruit were in keeping with his social and professional scientific outlook. They were in reference to a very good post for the breeding of cattle in South Africa.[36]

⁓ BEHIND HIM BATESON HAD LEFT the name of genetics, its terminology, his journal, the Genetics Society, and a mixed bunch of followers of the creed: Julian Huxley at King's, John Burdon Sanderson Haldane and Punnett at Cambridge, Francis Albert Eley Crew in Edinburgh, Gates and MacBride, and the breeder Charles Hurst. In addition, there were three students, Roland Henry Biffen in England, East in the United States, and Vavilov in Russia—all primarily plant breeders. Besides this were his writings, already overtaken by chromosome theory. At Merton he left twenty acres, six houses, six labs, ten greenhouses, an excellent training school for gardeners, a rich genetics library, and a governing body largely ignorant of the work being done and divided in interest. On it were MacBride, the staunch Lamarckist, John Heslop-Harrison, who was going that way with his experiments on melanism in moths,[37] and Farmer, Britain's premier cytologist, espousing highly misleading theories. At the center was a staff of ten in two groups: five relatively anonymous women researchers—de Winton, Cayley, Gairdner, Anderson, and Sverdrup—led by Caroline Pellew, and four similarly unknown male investigators: Crane, Newton, Chittenden, and Darlington. There was a botanist, Collins, who lectured to the gardeners but never really understood Mendelism, a congenitally deaf botanical illustrator, Herbert Charles Ostertock, who made drawings of the plants, and two keen research assistants who were later to make a name—William C. Lawrence and Len La Cour. The fate of this legacy was put in jeopardy with Bateson's death. When after Biffen refused to succeed his teacher, Collins became acting director, it seemed to young Cyril and his companions that anything might happen. For six months their fate hung in the balance.[38]

~ 4

In Search of Tulips and Truth

Iɴ Jᴜɴᴇ ᴛʜᴇ Jᴏʜɴ Iɴɴᴇꜱ Cᴏᴜɴᴄɪʟ ᴀᴘᴘᴏɪɴᴛᴇᴅ one of its own, the sixty-two-year-old Sir Daniel Hall, F.R.S., as director. Sir Daniel had been the first principal of Wye, Darlington's old school, and in 1902 became director of the Rothamsted Experimental Station. A chemist and geologist by training, he had for forty years produced innumerable editions of books on soil and food, and saw that science had a great deal to offer agriculture. His practical bent was a stark contrast to Bateson's theoretical disposition, as was his love for administration and committee work. "By the time he reached the John Innes he was showing acute signs of real or feigned deafness," Darlington remembered. "He always spoke loud enough when addressing two or three for an audience of 200 or 300. He was in fact in an advanced stage of committeeosis, the disease of the scientist turned administrator under the English system."[1]

In the manipulation of committees, in making friends and influencing people, Sir Daniel was assisted by his position as Governing Trustee of Long Sutton, a charity school set up for the education of boys and girls, preferably those whose parents were connected with agriculture. As at Long Sutton, he quickly proved to be a visiting director more than anything else. He would attend every morning to his correspondence from 9:30 to 10:00, and often leave things to his competent librarian/secretary/research worker Brenhilda Schafer when away on his extensive travels abroad. Hall was remarkable for his liberality and encouragement. The only persuasion from him was for work on his beloved tulips, which turned out to be excellent material for

chromosome studies.[2] Otherwise, Sir Daniel did very little in the way of
direction. Although an able agricultural administrator, Hall had scant
experience with genetics and sought the advice of Julian Huxley, then
professor of zoology at King's College. Huxley recommended to Sir
Daniel that he secure the services of J. B. S. Haldane, a biochemist from
Cambridge with an interest in Mendelism and the application of mathematics to evolutionary problems. Hall approached Haldane, offered
him direction of the genetic work, and gave him a verbal promise that
he would succeed him in not so many years upon retirement. Haldane
was engaged in March of that year to visit the John Innes every two
weeks for a day and a night during the Cambridge terms, one month
during Easter, and another during summer vacation, and to be free to
work over Christmas.[3]

〜 IN AUGUST DARLINGTON and his friend Chittenden took a
trip to the Jura in France. Before Bateson had died, Chittenden began
working on variegated plants, and the two had constantly disagreed on
the interpretation of what they were finding. Refusing to give anything
to the nucleus, Bateson allowed nothing to the cytoplasm either and
found himself caught in a trap by the cross between oil and fiber flaxes,
with its conflict between nucleus and cytoplasm. He extricated himself with what turned out to be a verbal fantasy, a piece of abstract
Mendelism he called "anisogeny."[4] In hermaphrodite plants, male and
female gametes sometimes had different effects on heredity. Bateson
claimed this was due to a segregation of factors in the development of
the flower. Chittenden, on the other hand, reinterpreted this case,
reported by his boss and Alice Gairdner in 1921, to be due to "the
cytoplasm or some cytoplasmic constituent other than the plastids." He
found it was completely in keeping with the view that heredity was
carried in the chromosomes in both sides, male and female, but that
the female germ cells also carried this cytoplasmic constituent that
interfered with their action. He concluded that species and races could
differ in cytoplasm as well as in nuclear constituents, and that, like the
chromosomes, the cytoplasm was capable of variation and "may have a
definite effect on the ultimate result of the reaction of nucleus and
cytoplasm."[5] Just a fortnight after Bateson's death, Chittenden had
explained this interpretation to his colleagues at lunch, bringing Miss
Pellew to tears because her beloved chief had been immediately discredited. They finally agreed to publish a joint letter to *Nature* giving
the correct interpretation of what Bateson had misunderstood.[6] On

vacation, the two young scientists discussed the notion of cytoplasmic inheritance and hereditary particles other than the chromosomes playing important roles in variation and development. These were ideas that Darlington would return to, but they would have to wait. Presently, he was called back from France by Hall. Newton had undergone a serious operation, and Cyril was needed to prepare the triploid tulip experiments he had designed. Unknowingly, Darlington had started on a track that was to direct his efforts for the next thirty years. While Newton was trying to solve the problem of how chromosomes paired at meiosis, this work would ultimately lead his young student to an interpretation and understanding of the very nature of meiosis itself.

In NOVEMBER, Darlington went to hear Julian Huxley deliver the Norman Lockyer Lecture at the British Science Guild.[7] "At present," the speaker warned, "20% of the population in Great Britain gives rise to 25% of the next generation; and the average of this 20% is neither physically nor mentally so good as that of the other 80%. We are thus confronted with a process which is retrograde in its effects— dysgenic instead of eugenic." To remedy this dangerous situation, Huxley recommended a rational birth control policy as a prelude to a rational eugenics through mating control and artificial fertilization, when its technique becomes perfected. "We have to stop pretending that comfortable mediocrity is our ideal," proclaimed the ambitious grandson of Darwin's bulldog. "We are told that this infringes the sacred rights of the individual and prejudices the idea of personal liberty. Such utterances are but another example of the unfortunate tendency, apparently inherent in the primitive human mind, of demanding and pretending to find absolute sanctions for ideas which are not in any sense absolute."

Cyril was outraged. At home he recorded his reaction in a notebook under the heading "Concerning Eugenics."

> Where is the fallacy in the eugenic argument? First, the distinction between desirable and undesirable classes is arbitrary, being usually based on material circumstances, and of all distinctions— as policies teach us—most susceptible to personal prejudices. Second, however truly classes were divided the characters which might distinguish them would not be inherited *en bloc* but in a race so heterogeneous as the British a lower group and an upper group would appear in each class in nearly the same proportions as in the

original classes. The eugenist is obsessed by transmission, he neglects segregation. Thirdly, in no experiment in heredity has the effect of any variation of environment comparable to that in human life been tried: for this reason that it cannot be tried. The complexity of the intellectual development of the human organism is of **a different order** than that of any kind of physical qualities whose inheritance has been studied. . . . I wish to protest against the doctrines being accepted as the final word of authority. Pronounced originally *ex cathedra* by much repeating they come to be accepted and their underlying assumptions thought axiomatic. I suggest myself, a humble eugenist, that in the present state of our knowledge we should be extremely diffident in interfering with those secular processes of Nature to which we owe our existence.[8]

Searching for an explanation, Darlington concluded that the eugenicist's position is not surprising in that "he is usually a well-to-do man." He ended on a more personal note, echoing George Bernard Shaw, who, having inserted himself into the debate over eugenics, had made a strikingly similar comment: "My feelings are all the more deeply moved against Professor Huxley's philosophy because I feel that if his method had been enforced about ten generations ago I could never have been born." Albeit for entirely different reasons, Darlington's sentiment was being strengthened by what he had picked up in-house under the tutelage of his now deceased director. "The eugenist and geneticist will, I am convinced," Bateson had written in 1919, "work most effectively without organic connection, and though we have much in common, should not be brigaded together. Genetics are not concerned with the betterment of the human race but with a problem in pure physiology, and I am a little afraid that the distinctness of our aims may be obscured."[9] Six years later his attack on eugenics took a fiercer turn: "To real genetics it is a serious, increasingly serious, nuisance, diverting attention to subordinate and ephemeral issues and giving a doubtful flavor to good materials."[10] Bateson was no friend of eugenics.

∽ IN DECEMBER, WHILE TRAVELING IN the underground with Alice Gairdner, Darlington had his very first scientific insight. The two were discussing the effect of polyploidy on fertility when he spontaneously formulated what later became known as Darlington's Rule: there is a negative correlation between the fertility of the polyploid and that of the diploid from which it arose.[11] This was a simple deduction from the pairing or lack of pairing, and homology or lack of homology, of the parental chromosomes. The insight stands as an early example of the method Darlington was to adopt as a scientist: genetic deductions based on cytological assumptions—where possible observed, where not, inferred. This reflected an interesting marriage between Bateson, the staunch Mendelian, and Newton, the chromosome man. A theoretical disposition meant that for Darlington the genetic premise and hypothesis would come first; its observed corroboration was almost invariably second. *Nature* published his letter on the matter two months later. It was regarded as a mistake by all Darlington's colleagues but ultimately became the basis for the study of fertility in species, hybrids, diploids, and polyploids. Work at the John Innes was at this time divided as though in two adjoining university departments. On the one side there was breeding of plants without looking at their chromosomes; on the other there was looking at chromosomes without breeding the plants.[12] Innocently unaware of the rules, Darlington was now breaking them.

⌒ᐟ IN EARLY 1927, Darlington's spirits were on the rise. On February 20 he wrote to his brother:

> I think the next few years are going to give me great opportunities as they will you. I hear that Hall is retiring in two years and will perhaps be followed by Haldane who is very straightforward as well as enlightened. Newton is coming back at the beginning of March and is apparently quite a new man. If only we knew that he could live!

"I have become so ambitious," he ended, "that I have come to think a Doctor's degree may be almost a hindrance to me so that I shall still be plain Mr. for a few more years to come."[13]

But Darlington's optimism was soon to suffer greatly. "I fear he is on a path he will not retrace," Cyril wrote to Chittenden of his teacher Newton.[14] While their joint tulip experiments proved parasynapsis to both men's satisfaction, Newton, from his bed at the Nelson Hospital in Surrey, continued to ponder the second problem of the relation of crossing over to chiasmata. Growing weaker by the day, he wrote a letter to *Nature* that he asked Darlington to sign, in which parasynapsis was affirmed but any relationship between chiasmata and crossing over, either as a cause or a consequence of the other, was denied.[15] Newton was dying, but he remained as cautious as ever. Confused by his struggle to understand and by his fears of loss, Darlington tended loyally to his waning mentor, unable yet to grasp the science, and even less able to face his impending death.

By December, the cancer that had attacked his body had metastasized uncontrollably, and the soft-spoken Newton, just thirty-three years of age, had finally succumbed. It was a horrible blow for Cyril. Their friendship had lifted him into a new world. Darlington had been brought up to believe in reason but had taken reason on faith as expounded by his father. Newton took away this certainty and left him floating in a sea of doubt.[16] Now he was gone. "We are not predestined to be fatalists," Darlington wrote in his diary, trying to find some comfort in understanding.[17] But was everything else really determined? Were a man's traits, his intelligence, personality, and *fatal diseases* all etched into the tiny threads of his nuclei? Could Huxley be right after all? Was free will nothing but a grotesque illusion, a soothing yet deeply deluding mental placebo?

Darlington's condition was dangerously precarious. Not only had Bateson and Newton died within a year of each other, leaving him with no guidance or direction, but the very scientific ground on which he stood was brittle, challenged as it was, all around him. Ruggles Gates had expressed claims to the directorship of the institution after Bateson's death and was angling for a Royal Society Fellowship. Farmer was a dominant member of the governing body of the John Innes, as was his friend MacBride. The little training Darlington had had was enough to make him see that the cytology of the former and the Lamarckism of the latter were grossly mistaken. But who was there to vouch for an inexperienced, essentially untrained and unheard of youth? Darlington had gained confidence in the chromosomes by using them as morphological tools for taxonomic purposes.[18] But much of the debate concerning crossing over was well beyond him, and, unfamiliar with the extant literature, he was inadequately prepared to make any kind of judgment. If anything, Newton had left his young student with the lesson of doubt in an extreme form, in particular of everything that had been written on his own subject.[19]

Comfort remained elusive. In a letter to Newton before his death Cyril had admitted: "Belling's hypothesis for crossing over . . . is unfortunately beyond my understanding, but you will at least know what he thinks."[20] In his teacher's absence, Darlington had attempted to read the literature on chromosomes and cell division, but had failed miserably. The technical language of "karyokinesis" and "heterotypic" had left him completely baffled. With the death of Bateson, and now

Newton, he effectively became a scientific orphan, deprived of all guidance. Lonely, depressed, and unsure of his way, Darlington was having second thoughts. "Cytology into which . . . circumstances have driven me is now rated very low, and I am rated a Cytologist."[21] Darlington's complicated relationship with Brenhilda Schafer was increasingly fraught, and he was getting along badly with a new Hall recruit, the Canadian Charles Leonard Huskins. Letters to his old teacher Salmon and to directors of agricultural experiment stations in the United States testify to his unhappiness. Darlington was looking for a way out, his growing sense of confidence momentarily shattered.

~ THE OPTIONS ABROAD AND AT Wye quickly fell through. Coincidentally, Hall recommended to the council that his young cytologist be promoted. It was Christmas 1928, and Darlington's old confidence was returning. This was not unconnected to the many hours he was now spending in the presence of Haldane, who had arrived at the John Innes in the fall. Heavy set and commanding, the Cambridge man was "as physically impressive as intellectually formidable: thickish, sandy moustache and a head almost innocent of hair had already begun to give him the air of a playful alert walrus."[22] Haldane's popular essays and books were widely read, and his Periclean speeches at the Royal Society, Cambridge's iconoclastic Heretics Society, and Hyde Park were famous. An eminent scientist, a philosopher, a writer, and a political speaker, Haldane was a Jack of many trades. *"Ce n'est pas un homme, c'est une force de la nature!"* one scientific colleague had said of him.[23]

Haldane's interest in experimental genetics had been acquired early on. A letter dated March 18, 1915 and addressed from 1st Black Watch, France, read:

Dear Professor Bateson,

I have been doing some work on (coupling) reduplication in mice, along with A.D. Sprunt and my sister, who is now carrying on, as we are both at the front. The factors concerned are C colour and E dark eye. Our latest from Ce.cE x ccee was 3 dark eyed, 24 pink eyed (albino and Cee) suggesting 1:3:3:1 repulsion between C and E. If I am killed could you kindly give my sister help if she wants it.

Yours Sincerely,

J. Haldane 2/Lt[24]

Ten years later, and still very much alive, Haldane had been removed from his post as Reader in Biochemistry by a committee known as the *Sex Viri* on grounds of infamous conduct. He had had an affair with a woman not yet divorced and was now fighting the decision of the committee. "I am not a prude," Darlington remembered Bateson saying about the disgraced alumnus of his alma mater one day over lunch in the lab, "but I don't approve of a man running about the streets like a dog." Clearly Haldane had made a bad impression on the older generation of Cambridge men.[25]

In the precarious period following Bateson's death, however, Haldane's arrival was viewed as a godsend by the young workers at the John Innes. He kept the institute going by the sheer force of his Herculean personality. Not everyone believed his claim that he'd gone into the wrong examination at Oxford by mistake and won a scholarship for classics instead of mathematics, but the story suited him, and Haldane lived up to it. His early scientific successes and recognized genius had brought out in him a determined sense of independence and self-sufficiency. "J. B. S. was against—against authority, and against the government, any authority and any government; if possible in the cause of reason, if not as a matter of principle."[26]

Such qualities immediately appealed to Darlington, who found in Haldane a paragon reflecting his own developing sensibilities and a living testament to the viability of iconoclasm and rugged individualism. Sitting side by side in the laboratory, like two parasynaptic chromosomes, the men spent many long hours discussing genetics, eugenics, politics, and philosophy, combining and recombining ideas. Haldane was an intellectual type Darlington had not come across at Wye. The sense of proportion and ability to deal with big issues that was a mark of the "Great Man" made a big impression on the young initiate. Haldane, for his part, took a fancy for the clever, ambitious youth, recognized his scientific acuity, and encouraged him.[27] Although he did not receive any real scientific guidance— Haldane was neither an experimentalist nor cytologist—Darlington gained from Haldane both a confidence in science itself and in his own opinions still lacking in the void the deaths of Bateson and Newton had created. More importantly, Haldane brought home to him, with his work on mathematical population genetics, the incredible power of theory. This was a lesson Darlington would learn well.

What Haldane could not offer in the form of experimental guidance, Darlington found elsewhere. Two months with the visiting world-famous cytologist Karl Belar helped hone his technical skills with the microscope and chromosome slide preparation. A visit to the Fifth International Congress of Genetics in Berlin in 1927, the first of its kind for the budding scientist, helped define for Darlington a world into which he was beginning to enter and gave him the confidence that comes with belonging. The old buoyancy, self-assurance, and cockiness soon returned. In his diary he admitted: "At the age of 18 most of the world seemed stupid and annoyed me; at 24 I know it is stupid and it ceases to worry me."[28]

～ QUICKLY, DARLINGTON WAS GAINING a grasp of the scope and particular problems of his science. Working in an institution run by a man with little knowledge of genetics and remarkable for his liberality and minimal direction on the one hand, and having no real professional teacher on the other, Darlington chose to stick to what he knew. If crossing over was yet beyond him, he would concentrate on pairing, a problem he had already worked on with Newton. To begin with, could he be certain that the chromosomes never paired in threes? From observation Darlington knew that in certain tetraploid plants (such as *Primula sinensis*), while each of the sets of four chromosomes would be expected, were likeness the sole condition of pairing, to associate in fours at meiosis, nearly always some of these groups failed to be formed, their chromosomes appearing merely paired, or in threes. This behavior was not explicable on the affinity theory, whereby chromosomes with similar structural properties (such as length and shape) were those that paired up. The four chromosome homologues should be either *all in fours* or *all in pairs*.[29] Darlington also knew that in triploids the third chromosome often becomes coiled around the other two, so that it is difficult to know for sure that the three are not all paired. This was all very confusing, and Darlington desperately wanted a generalization that made good genetic sense. Pairing two by two made good genetic sense: after all, that would allow for orderly segregation in meiosis; without orderly segregation, organisms would often die or have serious genetic problems, like not being able to cross or breed. He therefore decided—this being his first independent scientific decision—to begin to tackle the problem of *why* chromosomes pair by settling once and for all the problem of *how* they pair, choosing a plant in which chromosome pairing could be observed easily and clearly—tetraploid *Hyacinths*. If the four chromosomes in this clear instance paired two by two, Darlington reasoned, striving for a lawful generalization, then he could discount the possibility that in the triploids they did so in threes, and surmise that sometimes in the primula something had gone wrong. This he did: the four chromosomes always paired two by two. The confusing configurations were just that: confusing. Genetic logic and the clear case of hyacinth tetraploids rendered disparate observation subordinate to theory. Chromosomes had to always pair two by two, even if it sometimes appeared that they didn't. When in fact they didn't, this was an exception to the rule.

Next, Darlington turned to *trisomics* in *Tulipa*. In plants it is not uncommon to find one chromosome out of the complete complement represented, instead of twice, three times. In the various triploids and tetraploids they examined (having respectively three and four complete sets of the entire chromosome complement in their somatic cells), American cytogenetic botanists John Belling and Alfred Blakeslee claimed to have found just the number of trisomics one would expect from the whole number of chromosomes.[30] *Datura stramonium*, the American jimsonweed, was an example: it has twelve pairs of chromosomes and twelve easily distinguishable trisomic forms, in each of which one chromosome is represented three times, and each of which produces a slightly different phenotype. Hyacinth triploids, Darlington found, were another example: the eight-chromosome plant produced eight tri somics of different kinds, each with one of the eight chromosomes represented thrice. This was puzzling. Why, Darlington asked himself, should one get certain chromosomes that failed to pair while others, which were equally like their partners, paired? It seemed obvious to begin looking for a solution by examining the associations between the chromosomes at meiosis. There Darlington quickly observed two different kinds of such associations: first, there were the chiasmata of Janssens; second, there were certain terminal associations. But the relationship between these two types of association remained unclear.

If chiasmata were in fact the structural evidence for genetic crossing over, as Janssens's chiasmatype theory held, it followed that there should be a correspondence between both the *frequencies* of chiasma formation and of crossing over, and the *position* of the chiasma on the chromosome and the corresponding genic factors known from chromosome mapping. Darlington had already helped to establish the first of these relationships—namely that chiasmata and points of crossing over show a constant ratio of occurrence[31]—but had been stumped, like other cytologists, by the inability to prove the latter correspondence. He now concluded that matters would be much simpler if one were to suppose that the terminal associations between pairing chromosomes were in fact all derived from interstitial ones of Janssens's kind, which had run along the chromosome to its ends, like a zipper opening, a process he called *terminalization.*

This would explain those occurrences when the chiasma appeared not to correspond physically to the known genetic factor that had crossed over. It would also explain why there appeared to be a reduction in the

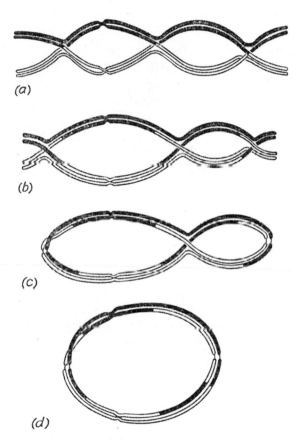

(a)

(b)

(c)

(d)

Terminalization of chiasmata, showing how three interstitial chiasmata (a) become two terminal chiasmata (d) by way of a zipper-like action. From H. L. K. Whitehouse, *Towards an Understanding of the Mechanism of Heredity* (London: Edward Arnold, 1969), p. 129.

number of chiasmata between prophase and the metaphase of the first meiotic division (imagine two knots tying two ropes together—when the knots are pushed downward by the opening of the ropes, they suddenly appear to be only one knot).

A pair of chromosomes that failed to form a chiasma on Darlington's terminalization scheme would simply be left unpaired, possibly leading after segregation to a threefold representation of a single chromosome. This, Darlington argued, is precisely what happens in his and Belling and Blakeslee's trisomic forms. Darlington knew that, at a stage within meiosis prophase called *pachytene*, paired chromosomes begin to repel each other and are pulled in opposite directions in order to complete the division of the cell into two. Therefore, chiasmata had to be the *condition* whereby chromosomes that had been paired should *remain* paired at metaphase. If this were true, parasynapsis, and not telosynapsis, was sure to be true, as Newton and Darlington had already argued. In addition, longer chromosomes with more chiasmata would tend to have a better chance to remain paired than shorter ones with fewer chiasmata between them. The first of these consequences was quickly accepted, and the second prediction was soon borne out by observation—there was a higher frequency of failed pairing among shorter chromosomes.

Striving for a generalization, Darlington proceeded to examine and review the whole literature of chromosome behavior of diploids and polyploids, hybrids, animals, and sex chromosomes. What he found was that this law held true for them all: chiasmata were the condition for all chromosome associations following pachytene. It had, of course, been clear from the beginning to both Janssens and Morgan that crossing over *within* chromosomes and segregation *between* them were the two components of Weismann's and Mendel's recombination. Now, however, Darlington was showing that these two components were not independent. They were regularly, mechanically, and inherently combined: crossing over was tied to pairing, to segregation, to reduction, and to the whole integrated process of sexual reproduction.

What began as a tentative attempt to buttress his deceased teacher Newton's work was now leading Darlington down a path of deduction, to the expression of chromosome behavior in terms of laws induced from direct observation and genetical inference. By the time Darlington had written his joint paper with Newton on parasynapsis in triploids, he had returned to his own hyacinth preparations and arrived at his chiasma theory of pairing. However, Newton's widow,

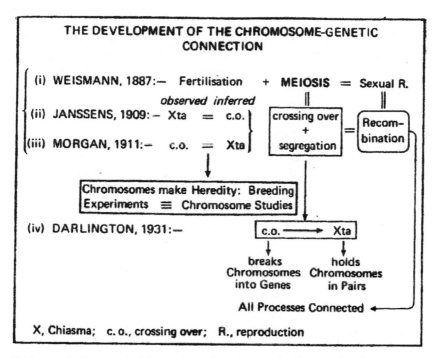

Weismann's, Janssens's, Morgan's, and Darlington's schemes. Darlington's
advance was to comprehend that crossing over was a necessary precondition
for pairing, segregation, reduction, and therefore ultimately for sexual
reproduction. From Darlington, "The Place of the Chromosomes in the Genetic
System," *Chromosomes Today*, 1973, p. 6.

Lily, herself a botanist, did not approve of associating her late husband
with the young Darlington's theoretical views. After all, on his death-
bed, Newton himself had refused to make the very claims his student
was now making. Inhibited, Darlington was forced to divide the paper
into two. They remain today a testament to the coming of age
and assertion of independence of a young scientist. In "Meiosis in
Polyploids I: Triploid and Pentaploid Tulips" (with Newton), parasynapsis is affirmed, but exposition stops short with regard to chiasma formation and pairing. In "Meiosis in Polyploids II: Aneuploid Hyacinths
(chiasma theory of pairing)" (Darlington alone), Darlington began to
bite at his deceased mentor's agnosticism.[32] By now, the orphaned scientist had already defined what seemed to him to be two universal laws:
(1) attraction between chromosomes is always in pairs, and (2) chiasmata are the condition for orderly segregation in cell division.

⁓ IN NOVEMBER 1928, before any of his meiosis papers had been published, Sir Daniel called Darlington to his room and told him he was sending him to Persia to collect seeds of *Prunus* and bulbs of *Tulipa*.[33] From *Prunus* had emanated all the economic forms of plums, cherries, almonds, peaches, and apricots, Hall's application to the Marketing Board had explained. The original wild species, however, remained unknown. It was from these primitive forms that it was hoped to find material for the improvement of the present varieties by hybridization; they might help to solve some of the difficulties with self-sterility and incompatibility caused by polyploidy in the domesticated species.[34] Darlington's guardian was to be John Macqueen Cowan, a forty-year-old retired Indian forest officer, presently of the Royal Botanic Gardens, Kew. Together they were to sail to Port Said, travel on land through Palestine to Babylonia, en route to the mountainous back regions of northern Persia. Cairo, Jerusalem, Damascus, Baghdad, Palmyra, Isfahan, Tabriz—all were on their path. Mostly, however, they would travel off the main roads, staying in small village inns or sleeping in the wild—below the stars and between the ravines. With increasing anticipation and trepidation, Cyril spent the next few months completing experiments, writing up his paper on meiosis in polyploids, and shopping around London for equipment with Cowan. Finally, on February 22, 1929, both men bade farewell to their families at Tilbury and embarked the boat for Port Said. The excitement of seeing Gibraltar, Stromboli, and Messina for the first time allayed Cyril's worst forebodings. "I feel a new man," he wrote to his parents, as the ship pressed for Africa.

All the while he was struggling at all costs to *understand*—to understand man and his motivations, to understand free will and behavior, religion, culture, and history. Above all he sought to understand the role the chromosomes were playing in all these. "Consciousness of future and past is the basis of all man's spiritual ideals," he wrote,

> and the source of his inherent tendency to self deception. It is this which makes him believe in a God, take interest in what has happened before him and what will happen after him, and seek and admit moral values that are not determined by his own necessities. For consciousness of the past makes him ask for a cause; consciousness of the future makes him realize death. If men were not illogical and uncalculating life would be impossible: the affection of parents for offspring, the interest of a dying man in his ideals' future—are absurd and unprofitable delusions. . . . Religion, says

Professor Whitehead, is an attitude of respect to what is outside oneself. Granted. But what inspires it? Faith, many would say, but it is largely lack of faith in any purpose of one's own existence. This is the first offspring of Intelligence which always remains with it. But the religion it inspires dies with credulity and is replaced with philosophy—another form of self-deception. Both of these gone, skepticism is left.

Was there no escape from this somber conclusion? Was there no ultimate cause with which to allay uncertainty, with which to truly achieve *understanding*? Darlington did not yet know, but he had a hunch, a gut feeling. "Perhaps we are predestined to everything," he wrote to himself, "but a belief in predestination."[35]

⌒ ON MARCH 4 THE SHIP SAILED into Port Said. "I worry seriously how I shall feel about my work when I come back," Darlington wrote home. "If my imagination is sufficiently fired by the East I may find it less thrilling than I thought it would be. However, Bateson came back from Turkestan to breed sweet peas so there is nothing impossible in my returning from mountains to microscopes perhaps."[36] Perhaps, but the pungent odors, the souks, back alleyways, and majestic vistas of the Orient immediately captivated the Lancashire lad. They were a far cry from old Tannery Lane, Warwick Road, and the John Innes garden tea parties. For the first time Darlington met a world of uncontrollable, seething chaos; deprivation and disease hopelessly and mysteriously blended with grandeur, beauty, and wealth. Traveling in the Ford Lorry Cowan had arranged to be sent from Hammersmith, Darlington took in all his senses could allow. He recorded in his journal local history, custom, politics, architecture, agriculture, religion, language—nothing escaped his watchful gaze. In Jerusalem he marveled at the panoply of "well dressed Polish Jews; queer looking bearded Russian Jews; ancient and multi-collared Moroccan Jews," all reciting prayers in Hebrew at the Wall of Lamentations. Noticing signs in Assyrian, Hebrew, Arabic, Persian, French, and English, decorating the colorful streets of Baghdad, Darlington observed the close correlations between language, religion, profession, and marriage patterns. On the Iraqi frontier he encountered nomadic Bedouin and was invited for black coffee in their black tents; at Tabriz he met two golf-playing blond Germans with their caddies; and in Urmia he noted with wonderment how even uneducated Persians interlace their conversation

Darlington, standing on left, John Macqueen Cowan, right, and the Ford Lorry, on the Persian expedition, April 1929. Darlington couldn't stand Cowan, who "treated the servants like dirt." Photo from the Darlington Papers, Bodleian Library. Reprinted with permission.

with long quotations from thirteenth- or fourteenth-century poets. Then, in Kurdistan he watched the triumphal march of a Persian force returning from a successful expedition on behalf of their westernizing shah, who was trying to pacify the resisting mountain tribesman. "12,000 soldiers, khaki-clad, and wearing the new regulation 'kepi,' filed down a mountain pass, while their bands played the Persian national anthem."[37]

Where did all this great diversity come from, he wondered? How was it maintained? How, in any event, could one species have produced such immense variety—physical as well as in speech, culture, custom, and religion? What was it that made men so different from one another? If it was all just historical accident, then what was it that determined history? On the backs of mules, and accompanied by two escorting soldiers, Cowan and Darlington pressed on in their hunt for tulips. The country people were courteous and helpful, but obviously considered that only mad Englishmen would take so much trouble to collect weeds. It was by now the middle of June, and Cyril was suffering from bouts of high fever. He was tired and hating Cowan, who "treated his servants like dirt." After an expedition to Asteria, they arranged to sell the car, and Darlington continued alone into Russia, to visit the plant breeding station at Sukhum. Bateson's old student Vavilov had helped arrange the necessary documents for the visit. In his homeland, the Russian was nearly as widely known as Amundsen or Lindbergh in the West. In 1924 he had crossed the inaccessible Hindu Kush without maps or guides. He was the first European to study the mountain outskirts of Afghanistan, the Kafiristan. In 1926 he had crossed Ethiopia in a small caravan. Presently, he was returning from China and Japan, and was well on his way to eventually bringing back to his beloved homeland 50,000 samples of varieties of wild plant and 31,000 wheat specimens.[38]

Vavilov was a Soviet hero, well loved and well respected at home and abroad. It would have been unthinkable then to imagine the fate that would ultimately befall him.

Having himself now experienced a long, rugged journey, Darlington's admiration for Bateson's old student grew even more. His, however, was to be an intellectual life after all. "I am made for talking, thinking, reading, writing—not for action," he wrote. "This life has incapacitated my initiative more than ever."[39] He now longed to go home. His brother Alfred had met a woman and the two were planning to get married in the autumn. "I have missed you almost as much—perhaps I should say indeed more than you can possibly have missed me," his father wrote to him. "Mine is a very lonely life, intellectually. . . . I have no one to exchange thoughts with. Mother yearns for you to be back safe and sound. Glory counts for little with her besides that—and little more with me."[40]

On July 6, Darlington sailed from Batum to Constantinople, arriving on the 10th where "the armies of Genghis Khan and Tamerlane had passed with their conquering hoards, and the Turks had advanced to Islam and Roum." Most importantly, news had arrived from Brenhilda Schafer that his solo paper, "Meiosis in Polyploids II" had appeared in the *Journal of Genetics*. Bridges, Morgan, Sturtevant, Belling, and Otto Renner had all been sending him reprints to the John Innes. The expedition itself had not been a great success, but this mattered little to the homesick youth. He embarked on a boat to Vienna, caught the train to Danzig, and finally boarded the forty-three-hour express to London, Victoria, where he was met by his anxious and much-relieved parents. The next few weeks were spent in a hospital, recuperating from a nasty case of malaria. As he sat up in bed writing a paper on *Prunus*, Darlington could stop to think for a moment of the great changes that had affected his life in the short six years since he had arrived at the John Innes. From doctoring horses and testing seeds in an agricultural college, he had fallen into the fray of the latest controversies at Britain's premier genetics institute. Through Bateson, its venerable director, he had been inducted into the world of science. From a younger teacher, Newton, he had learned the lesson of skepticism, been introduced to the chromosomes, and become a materialist. Before both men died, he had made up his mind about Lamarck's faulty theory of heredity on the one hand, and had been exposed to the possibility of cytoplasmic inheritance on the other. A scientific orphan, he took his first steps working alongside the giant Haldane, from whom he gained a sense of proportion in dealing with large issues, and an important lesson on the power of theory. Through talks with him and other great men like Huxley and Vavilov, Darlington began by example to think of politics, philosophy, and evolution, and about what science had to offer the world. Finally, traveling to a faraway land, he became captivated with man himself, in all his splendid diversity, and wondered how he could come to understand his history, behavior, and social arrangements. A life that had commenced in something less than legitimate circumstances was now beginning to take a momentous turn. Darlington had left England in February an unknown and unconsidered youth. In July he returned home, suffering, to be sure, from amoebic dysentery and fever, but known to experts all over the world as a leading worker in his field.

~ II

Science

~ 5

From Cytology to Evolution

$\mathbf{B}$ACK FROM PERSIA AND INTO THE FRAY, Darlington knew if he was to build a name for himself, the time had come. If only the science was that straightforward! Never mind, boldness and self-assurance would dispel crippling doubt—or so he determined. Chromosomes would have to become his best friends, perhaps his only ones; the microscope, a loyal mirror into their actions and motivations; stains and fixatives—trustworthy informants. Recovering from his Persian malady, Darlington's mood rapidly improved. If anything, the mysteries of the East had only strengthened his resolve to discover and to connect; to pacify his thirst for understanding with the libation of knowledge. He was eager and focused. The prospect of long hours at the bench peering at the mischievous nuclear dwellers did not scare him. In fact, it seemed rather inviting. It was as if the chromosomes themselves, at the other end of the ocular lens, could feel it: Darlington was hungry.

Quickly he set to work, picking up where he had left off. A powerful urgency soon took over his mind. If chiasmata were seen to be the condition of orderly chromosome segregation, as he himself had suggested, a considerable simplification would be possible in stating the relationship of mitosis and meiosis. This was important because, after all, the purpose of trying to understand the chromosome gymnastics during meiosis was to comprehend how—as distinct from the regular division of mitosis—the reduction division of meiosis is accomplished. Darlington reasoned as follows: throughout the prophase of mitosis, until pachytene, the chromosome threads are held together by an attraction in pairs. The same rule applies to meiosis since the evidences of failure of pairing of fragments, of odd chromosomes in triploids, and of the four chromosomes of a type in tetraploids all pointed to the chromosomes having no present attraction at metaphase. They are merely held together by the chiasmata—that is, by the attraction between the pairs of half chromosomes and the exchanges of partners among them; and this attraction exists equally at mitosis. On the principle of Occam's Razor, then, meiosis and mitosis were on the same mechanical plane of attraction in pairs at all stages. This being so, the essential difference between the two types of nuclear division would be found at the prophase stage. Since the same mechanical principles were at work in both types of division, the difference evidently lay in the *timing* at which the chromosomes split into two chromatids. At mitosis, Darlington presumed, it is probable that this has already happened before the chromosomes appear at prophase. At meiosis, it does not happen until pachytene, well into prophase. If there is a universal attraction of threads in pairs at the prophase of any nuclear division, as is at mitosis, it follows that this condition is fulfilled in meiosis by the pairing of homologous chromosome threads when they are still single, and their separation after pachytene—at a stage called *diplotene*—when they have at last come to divide. After all, Darlington already knew that after pairing as single threads, the chromosomes in meiosis draw apart as double threads: they do not attract; they repel one another after they become double at pachytene. Only the chiasmata, already formed and in place, hold them together.

The decisive difference between mitosis and meiosis would therefore appear to be in the singleness of the early prophase threads in the latter. This singleness, Darlington thought, may be attributed to one or both of two causes: (1) a delayed division of the chromosomes, (2) a precocious onset of prophase. The second of these seemed a more likely explanation, on account of the short duration of the pre-meiotic prophase in some organisms. In addition, if meiosis and mitosis are essentially identical, divided only by a slight temporal precocity in the former at prophase, it made perfect sense to suppose that the one had evolved from the other, most likely in a single mutational step. In a burst of observation given shape by theoretical considerations, Darlington's precocity theory of meiosis was born.[1]

⌐ FINALLY HE WAS READY TO TACKLE the second question he had initially chosen to sidestep: the problem of crossing over. "It was absurd," Darlington wrote some years later, "when I found as I believed that the chiasmata were the vital condition of chromosome pairing at meiosis and therefore of all segregation, of all reduction, of all genetical behavior, of all fertility, of all sexual reproduction and evolution: it was absurd to leave open the question as to what the chiasmata really were."[2] Were the chiasmata the result of crossing over or were they the cause? Newton had died claiming there was no direct relationship between crossing over and chiasma formation. Returning to his triploid hyacinth preparations, Darlington was now able to show that a chromosome may show two chiasmata (A and B) with one homologue, and a third chiasma (C) with another homologue at a point between A and B. To account for such a configuration, it was necessary to assume that at the earlier pachytene stage there were changes of pairing partner. On this assumption, chiasmata A and B must involve the same four chromatids, and therefore chiasma C must involve breakage and crossing over. Critical evidence was not available for chiasmata A and B, but Darlington deduced that there is no reason to suppose they differ from C, which he observed in his triploid hyacinths. Since the pachetyne associations in all polyploids appeared to be strictly confined to pairs, it followed logically that all such intermediate chiasmata with a third homologue (C) must be the consequence of previous crossing over. It was inconceivable that having formed the pachytene configuration the chromatids could reshuffle to form chiasmata: they had to break! This critical configuration convinced Darlington that crossing over was primary, and that chiasma formation was its inevitable consequence.

Immediately a wealth of conclusions was thrust upon him. If in all organisms all chromosomes remained paired by chiasmata, and if all chiasmata were the result of crossing over, then all chromosomes in all organisms must undergo crossing over. Upon reflection this made sense: there could not, otherwise, be any genic structure of the hereditary material unless there was crossing over. If there were no crossing over, the whole of each chromosome would behave as a permanent block and no genes could be revealed either to the geneticist by his experiments or to nature by her selection. The only means of recombination of chromosomes would be by breaking them as Hermann J. Muller had broken them with X-rays, and recombining them in that way.[3] The whole of evolution, then, would be completely stopped at the beginning by the absence of this system. For the system of crossing over correlated with the formation of chiasmata as the basis of pairing, making sense of the entire business of sexual recombination, which is another way of saying the whole business of the structure of chromo-

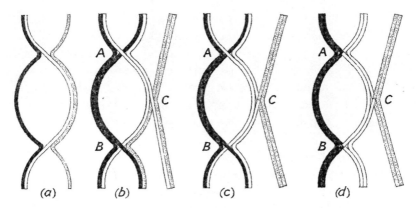

Trivalent chromosomal associations illustrating Darlington's demonstration that crossing over must precede chiasma formation. (a) pachetene stage, (b–d) alternative explanations of diplotene configuration with three chiasmata; (b) with all chiasmata due to changes of pairing partner without breakage (classical hypothesis), (c) with chiasmata A and B classical, but C due to breakage and crossing over (chiasmatype hypothesis), (d) with all chiasmata due to crossing over (chiasmatype hypothesis). From H. L. K. Whitehouse, *Towards an Understanding of the Mechanism of Heredity* (London: Edward Arnold, 1969), p. 121.

somes in terms of genes. *Pari passu* the evolutionary perspective was beginning to force itself into Darlington's argument. Like in a miniature locket carrying ancient, untold truths, the secrets of evolution and heredity were beginning to emerge from the chromosomes.

⌒ IT WAS NO ACCIDENT that evolution began creeping into Darlington's mind, first as a subplot and increasingly as the main design in the script of his prime actors, the chromosomes. Ever since the Fifth International Congress of Genetics in Berlin in 1927, when he first encountered its magnificent chromosome rings, *Oenothera lamarckiana* had piqued his curiosity.[4] The evening primrose had a long and illustrious history, Darlington quickly discovered. In 1886 the Dutch biologist Hugo de Vries had found *Oenothera* growing abundantly on what had formerly been cultivated land. Among the almost pure strains, de Vries noticed a number of aberrant forms, which he promptly collected and propagated in his experimental garden in Amsterdam. Because he was certain *O. lamarckiana* was a pure species, and not a hybrid, de Vries postulated that the newer and aberrant forms represented spontaneous, discontinuous jumps of nature, jumps he called mutations. Such mutations, de Vries claimed, accounted for the origin of variation, a long-standing mystery at the heart of evolution. After a great deal of crossing experiments, de Vries published his Mutation Theory.[5] A 1904 address to the University of Chicago captured the excitement of the discovery:

> Fortune has been propitious to me. It has brought into my garden a series of mutations of the same kind as those which are known to occur in horticulture, and moreover, it has afforded me an instance of mutability such as would be supposed to occur in nature. . . . I have brought it into my own garden, and here, under my very eyes, the production of new species has been going on, rather increasing in rate than diminishing. At once it rendered superfluous all considerations and all more or less fantastical explanations, replacing them by simple fact. . . . The fact is that it has become possible to see species originate, and that this origin is sudden and obeys distinct laws.[6]

De Vries's theory that species originate through sudden, spontaneous mutations was entirely based on *Oenothera*, and on the assumption that the mutants were not hybrids, but rather pure species. If it were true, it would put to rest both Darwin's disputed theory of speciation by natural selection acting on small gradual changes, and Lamarck's even older idea of the inheritance of acquired traits. In the long-standing evolutionary battle between internal factors and the environment, or nature and nurture, the former could finally be declared the winner. Dramatic, sudden, internal mutations—not small, slow, environmentally induced variations—were the motor of evolution. Solving the problems of swamping, the target of natural selection, directed evolution, evolutionary time scale, and that of species definition, all in one broad stroke, de Vries's theory was able to serve as a banner around which both disenchanted Darwinians and anti-Darwinians could rally.[7]

But it was not to be. Already between 1910 and 1912 Bradley Davis showed that the *Oenothera* "mutations" actually followed Mendelian laws of inheritance. Some years later, Otto Renner showed that *Oenothera* was not a pure breed at all, but rather a complex heterozygote—a hybrid. Renner demonstrated that the evening primrose produced two different chromosome complexes, which he called "gaudens" and "velans." When self-fertilized the heterozygote (with a gaudens/velans complement) survived while the homozygote (with either a double gaudens, or double velans complement) did not; this was a balanced lethal system. Consequently, the progeny of self-fertilization always resembled the parents, appearing to breed true, but when outcrossed showed normal hybrids.[8] De Vries, therefore, had not been witnessing mutations at all, but rather producing recombinations of existing genetic materials. In his hands the innocent evening primrose had generated both a false theory of mutation and a false theory of evolution.

ALTHOUGH HE HAD BEEN TOLD BY NEWTON that a good deal that had been claimed for *Oenothera* was probably bogus, when Darlington saw Ralph Cleland's slides in Berlin in 1927, he realized they were perfectly genuine.[9] There were the extraordinary rings that had buttressed Ruggles Gates's telosynaptic theory of chromosome pairing. Darlington was now well aware that if his generalizations of chromosome behavior were of any worth, they would have to explain the mysterious *Oenothera* rings. If they couldn't, they would be no rules at all; they would not be universal. He quickly returned to Belling's work on trisomics.

In his work with Blakeslee, Belling had found a trisomic in *Datura* in which some cells showed a trivalent (or three-chromosome association) in the form of an open chain of three, the chromosome at one end being smaller, and evidently not a full homologue of the other two, though homologous with the others at one end. To explain the situation, Belling developed the concept of "segmental interchange:" an interchange of segments between two nonhomologous chromosomes had taken place in the ancestry of one of the parents of the hybrid, so that the hybrid had received two interchanged chromosomes from one parent and two unmodified ones from the other. The trivalent appearing in the progeny of a hybrid plant received two unmodified (normal) chromosomes through both gametes plus an interchange chromosome, one of whose ends was homologous with one end of each of the two unmodified chromosomes. In this way, if the original hybrid had received chromosomes 1*2 and 3*4 from one parent, and 1*3 and 2*4 from the other (where two digits connected by an asterisk stand for a chromosome, the digits representing the terminal segments of the chromosome), one of its progeny derived by selfing might have received 1*2 3*4 through both sperm and egg, *plus* 1*3. In the sporo-cytes of this trisomic, a trivalent would often be formed with the

1*3 occupying one end of the chain. Since the exchanged chromosome was smaller than the others, its position in the chain was easy to observe.[10]

The idea of reciprocal translocations, or segmental interchanges, was new. It seemed improbable, almost incredible, that a phenomenon yet unknown in nature was responsible for the mysterious rings in *Oenothera*. Darlington, however, accepted Belling's hypothesis with enthusiasm, for it allowed him to do away with the telosynaptic hypothesis and bring *Oenothera* into line with his parasynaptic interpretation of meiosis. The false theories of mutation and evolution to which the evening primrose had given rise had both already been corrected. Now Darlington strove to correct the false theory of chromosome behavior it had been supporting.[11] By developing the idea that a given chromosome in a circular chain is homologous at one end with one chromosome, and at the other end with a different chromosome, he was able to show how a circle of fourteen chromosomes could be built out of at least six interchanges. Indeed, by a succession of interchanges one could get a ring gradually built up from four to six, from six to eight, from eight to ten, from ten to twelve, and finally fourteen, as was

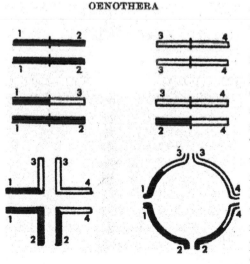

OENOTHERA

Diagram showing Belling's hypothesis of segmental interchange. Nonhomologous chromosomes 1*2 and 3*4 exchange segments to produce 1*3 and 2*4. In an individual that receives 1*2 and 3*4 from one parent and 1*3 and 2*4 from the other, pairing of corresponding segments in synapsis will produce a circle of four chromosomes. The conservative Belling found this phenomenon in *Datura*, but refused to apply it to *Oenothera*. Darlington, however, did, helping to solve the long-standing mystery of de Vries's garden primrose. From Ralph E. Cleland, *Oenothera Cytogenetics and Evolution* (London: Academic Press, 1972), p. 58.

observed in *Oenothera*. Thinking from an evolutionary point of view, Darlington saw that this would be the case only on the assumption that each of the interchanges would itself have a selective advantage over its predecessor. He could now therefore reach two important conclusions. First, that pairing in *Oenothera* is side by side, and that the mechanism that holds the chromosomes together in a ring amounts to terminalized chiasmata resulting from crossing over. Second, that the generalizations of chromosome behavior he had made could now be put on a universal footing. In this manner, evidence that was used in the past to support telosynapsis was turned on its head.[12] Darlington could now begin to think of how a system of breeding and a system of chromosome organization could relate to natural selection and evolution.

⁓ Darlington's plan of unification seemed to be an incredible advance in theory. From a chaotic array of disconnected observations, the laws of chromosome behavior were beginning to take form. And yet, like a worm in the heart of an apple, a major exception threatened to spoil the entire plan: in male *Drosophila* crossing over was entirely absent. If, as Darlington claimed, chiasmata were the condition for proper segregation, and the necessary result of crossing over, how could male *Drosophila* be explained? Either the chiasmata were not responsible for segregation, or they were not the result of crossing over. Darlington could not have them both.

As it turned out, however, Darlington's work on the snakeshead lily, *Fritillaria meleagris*, had familiarized him with the occurrence of two chiasmata so close together that they compensated for one another and that no crossing over would be detected.[13] He now sensed the possible connection. Boldly extrapolating this phenomenon to the male fly— after all, he hadn't actually *seen* it in the fly—Darlington could explain away the apparent exception. Localized, *reciprocal* chiasmata, though constituting an exchange, would produce no visible crossing over. Once again, a quick stroke of theory had come to the rescue.

⁓ Darlington's assumptions now underlay several divergent fields of biological thought. First, there was the *reductionist* level of unifying all chromosome behavior on a mechanical basis. Second, there was the *genetic* level of unifying all breeding behavior reconciling *Drosophila*, *Oenothera*, and other organisms with a single mechanism of crossing over and segregation. Increasingly a third, *evolutionary* level was emerging. Darlington had virtually narrowed his world into that cylindrical space between the eye, resting gently on the ocular aperture, and the slide, just millimeters below the lens. Few distractions—even papers of fellow colleagues in the field—had been allowed to invade that sacred, silent space. But Darlington knew it was time to make his voice heard.

In January 1931, Darlington began writing what would become his masterpiece, *Recent Advances in Cytology*. The expressed intent was nothing short of producing a general analytical account of the chromosomes.[14] Darlington now read much of the general literature in his field for the first time and found that "the whole of cytology (on top of which genetics had to be built) was a jumble of sound and unsound theory and observation, inconsistent with one another and with genetics."[15] He could begin to ask himself a number of questions that seemed to him had never been asked before. How, he wondered, could the chromosomes, which were responsible for organizing heredity and development in every detail, be pushed about and changed from a mitosis to a meiosis and controlled in this way, if they were actually responsible for organizing things themselves? How could nature have produced a hybrid such as *Oenothera* in which chromosomal rearrangements were selected for in such a manner as to change the breeding habit of the species?

With Haldane working at the desk beside him, Darlington pushed ahead, staying late at Merton and many nights writing until dawn. And he wrote quickly. In the first fifteen chapters of the book, completed by the fall, Darlington summarized the state of cytogenetics, boldly interpreting the data in terms of the universal rules of chromosome behavior he had defined in the course of his work between 1927 and 1931. From his experiments on crossing over and cell division, he had come to understand the importance of these mechanisms in generating diversity. They were at base recombinational genetic devices. Consequently, they exposed the dynamic properties of chromosome behavior and their role both in the genetics and evolution of organisms. Darlington nailed his flag to the mast in the preface:

> Previous descriptions of the properties of the chromosomes have consisted in a detailed morphological account combined with demonstrations of the explanatory value of the chromosome theory of heredity. The importance of the chromosome as determining the hereditary functions has placed them outside the ordinary field of evolutionary enquiry. They have been considered as

the very fount and origin of adaptive change and therefore not themselves capable of adaptation. Now they can be shown to be subject to the genetical variation which they themselves, by changes of their parts, determine. In this way it becomes possible to consider the chromosomes to a great extent like other structures in their relation to the organism as a whole, and to attempt a causal explanation of the whole range of chromosome phenomena in genetical terms. We have a coherent system of inquiry into the hereditary mechanism which is less dependent on the techniques and secondary assumptions of breeding work, simply because it is directly built, as all biological studies must be, on the primary genetic postulates.[16]

In the final chapter, "The Evolution of Genetic Systems," Darlington outlined these ideas in full force.[17] Not only did heredity *lead* to evolution, he argued; heredity itself was *subject* to evolution.

Darlington set out the foundations of this theory in the "premises" in the beginning of the chapter as five postulates: the theory of the permanence of the chromosome, the chromosome theory of heredity, the theory of genotypic control of chromosome behavior, the theory of mutation and differentiation, and the theory of structural change and crossing over. From here Darlington turned to a historical explanation of the origin of *genetic systems*, which he defined as the genetic control and adaptive connections between (1) hereditary structures (genes and chromosomes), (2) the reproductive process (meiosis, fertilization, gestation, etc.), and (3) the breeding habit (sexuality, mating behavior, incompatibility, and length of life cycle). Beginning with an explanation of the first self-replicating molecule, or "naked gene," Darlington described his theory of the evolution of genetic systems, beginning with an original method of reproduction that consisted in the accretion of several similar particles simultaneously, but which was survived by one in which one particle is formed at a time—the old particle having divided into two. The essential point here was that reproduction in pairs was probably correlated with the universal property of genes attracting one another in pairs.

In order to evolve from here to the next stage of higher organisms, Darlington postulated that an aggregation of the genes had occurred. In this manner, genes were sorted into groups, and any mutation would lead to differentiation that could in turn be selected for. Darlington suggested that the mechanism of mitosis had evolved in order to coordinate this aggregation and differentiation. As the reproductive material grew in quantity, gene-packaging bodies evolved to ease distribution: these were the chromosomes. A spindle apparatus then came in to being, allowing for orderly division of the chromosomes. In turn, the chromosomes themselves underwent structural changes, including fragmentation. When two such organisms fused, either by accident or mutation, a diploid organism was created, and meiosis evolved as a mechanism to divide its genetic material. Darlington concluded:

> Whether meiosis arose as a new mutation in a diploid, or was already conditioned in the haploid, the resulting cycle would be the same: the first division after fusion would take the form of meiosis and, in the haploid produced, a series of effectively normal mitosis would occur. . . . This mechanism, for all its advantage, would obviously be very imperfect, and there are two directions in which it would be developed, for the very irregularities arising from its imperfection would provide the genetic material for adaptation, and evolution would take on a new *tempo*.[18]

Meiosis, moreover, served to reshuffle or recombine the existing genetic material:

> The opportunity of recombination is therefore the essential advantage of meiosis. But it is an advantage that is restricted to hybrids, i.e., to organisms in which the genes of the pairing chromosomes differ in quality or arrangement. It follows that meiosis has no virtue except in hybridity. Hence the countless physiological and mechanical devices which have been developed to promote hybridity wherever meiosis occurs. It may also be remarked parenthetically that hybridity has no virtue except through meiosis.[19]

A variety of chromosome changes could occur during meiosis. Darlington believed that modifications of chromosome behavior were themselves adaptive to the organisms because they introduced recombination. After discussing the relative merits of the haploid-diploid condition, Darlington explained how hybridity (without which meiosis would have no effect) could be stabilized by the existence of certain

advantageous genetic systems such as apomixis (a form of asexual reproduction that precludes zygote formation), ring formation (end-to-end
chromosome-binding such as in *Oenothera*), and polyploidy. Since
hybrids were often sterile because their chromosomes failed to pair at
meiosis, apomixis would allow such a hybrid to propagate itself; ring
formation could maintain hybridity by producing balanced lethals
through complexes, as had happened in *Oenothera*; and polyploidy
would do the same by increasing the variability of hybrids.

As a final jab at the problem of evolution, Darlington attacked
one of its most persistent problems: the origin of variation. Darwin
himself had pondered this problem, but failed to solve it. Now Darlington would try. He numbered three ways in which variation arises:
(1) through gene mutation, (2) through structural and numerical
change resulting from meiotic irregularities, and (3) through proportional changes brought on by secondary structural changes. The nature
of these changes was made clear in a simple diagram (see page 81).
Darlington stressed that the nature of the changes varied in
importance:

> While all variation is ultimately derived from gene change, pro-
> portion changes are evidently more important as immediate
> agents of variation. Gene change is usually disadvantageous, while
> proportion change is often advantageous. Gene change is as ran-
> dom as the chemical constitution of the gene permits, and it is
> biologically untested. Proportion or quantity change makes use of
> tested materials and introduces nothing new, rather transposing
> the key of melody than changing the individual notes. The impor-
> tance of proportion change, however, must depend on the fre-
> quency of hybridisation; for while the success of new variants from
> both sources requires a trial in different combinations, which are
> procured by hybridisation, proportion changes to a great extent
> owe their origin to the recombinations that arise at meiosis in
> hybrids. This, in fact, is the important work of hybridisation in
> evolution.[20]

Understanding genetic systems was important for understanding the
nature of species. Darlington stressed the probability of speciation
through *genetic* isolation, alongside the *geographic* and *behavioral* kinds.
Indeed, isolation could be brought about by distribution, by change of
habit, or by genetic sterility factors; it could also be brought about by

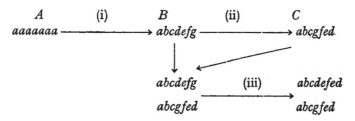

The ways in which variation arises. From C. D. Darlington, *Recent Advances in Cytology* (London: Churchill, 1932), p. 479.

structural and numerical changes in the chromosomes, which at the same time determine genetic changes. Genetic systems could therefore become important taxonomic tools for the systematist:

> The description as species of all these different types has been found convenient by the systematist, and clearly systematic convenience is the sole criterion that can as yet be applied universally. Consistency, whether morphological, genetical, or cytological, if possible in theory, would be impossible in practice. Rules can, however, be applied to the study of the individual species when its systematic character, method of reproduction and chromosome constitution are known.[21]

By 1932 Darlington had come a long way since writing a dissertation on the utility of chromosome morphology for taxonomy five years earlier. Having begun by looking at chromosomes for their own sake, he could now begin to understand evolution itself as a function of their behavior. Most important of all, Darlington now grasped that the combination of the material basis with the evolutionary framework is the only means of making sense of biology as a whole. Genic, structural, numeric, and proportional changes were all material phenomena, but each and every one of them had a history, and each created a new history as it changed. Along this axis, species grew and evolved.

Darlington's bond with his late teacher Bateson had been forged to a great extent through a shared contempt for the Lamarckian ideas of MacBride on the John Innes Council. Now a more mature Darlington could apply his independent years of research to construct a solid argument with which to disprove a notion he could formerly contradict only by instinct.[22] Darlington chose to end *Recent Advances in Cytology* with a frontal attack on Lamarckian inheritance:

> Now the Lamarckian hypothesis assumes that evolution proceeds gradually from the accumulated effect of changes in the body of the organism, adapted to abnormal conditions, in modifying the properties of the germ-cells to give the same changes. This means that the body of the organism has some special reaction to "abnormal" conditions of environment which does not occur in "normal" conditions and which effects the germ cells in a specific way. The principles of change shown by observations of chromosome behaviour conflict with every assumption in this theory. They show that the important changes do not arise for the most part under abnormal conditions, but under normal conditions. And finally they show that the changes are never adaptive to the conditions that produced them, but that adaptation arises by selection, most of the original changes having been disadvantageous ones.[23]

No one had expressed this idea better than Darlington's childhood hero, the Latin Epicurian philosopher Lucretius: *"nil ideo quoniam natum est in corpore ut uti possemus, sed quod natum est id procreat usum."*[24] For the advantage of meiosis is something that only appears in the next generation. Indeed, it appears only in the course of evolution itself, as do the consequences of breeding, intimately bound up with it. Darlington argued that since meiosis, recombination, and breeding habit—the constituents of the genetic system—cannot be adaptations to the immediate environment, but are rather adaptations to the environments of future generations, a Lamarckian interpretation of evolution marshaling use and disuse and the effects of the environment would have to be discarded. Species do not adapt themselves locally, and to order, to changes in their environment. If nature could be said to possess a special property of foresight, it was rather to be found in the very crux of the genetic system: the chromosome. Species are prepared by the very nature of their chromosome-centered genetic systems for unexpected events not yet ushered in by time.

Having firmly planted himself in the neo-Darwinian camp, Darlington could now conclude:

> The Lamarckian assumptions, which are plausible in explaining the evolution of gross structure, break down in face of observa-

tion of genetical mechanisms. On the other hand, the essential Darwinian assumptions, of variation followed by natural selection, hold good. In two directions only they require modification: Variation, as Bateson first maintained, is discontinuous, whether by gene or by structural or numerical change. Variation may be described as random in a Darwinian sense when taking place by structural change, but it is random in a chemical sense, and almost entirely deleterious, when taking place by gene change. Yet from the few gene changes that prove satisfactory all organic variation is derived.[25]

⌇ DARLINGTON COMPLETED *Recent Advances in Cytology* on June 4, 1932, one day before sailing to America, where he was to spend twelve months as a Rockefeller Fellow.[26] In his letter of recommendation on Darlington's behalf, Haldane had written: "I have never come across his equal as regards power and originality of thought and output of work. The only other British scientist under 30 years of age whom I should place in the same class as him is P. A. M. Dirac."[27] There was much of which the twenty-eight-year-old Darlington could be proud. In a short four and a half years since Newton's death, and just three years after his return from Persia, Darlington had become a major force in the world of cytogenetics. Back from the Persian mountains, he studied the tiny nuclei of the many-colored flowers he had collected under his old brass microscope. Observing the chromosomes, Darlington followed their every move, learning their likes and dislikes. With a sharp eye and distrustful mind he had tamed the chromosomes, imposing order where others had seen only disorder; he had joined radical theory and careful observation to arrive at a new view of life. Chromosome mechanics had led all the way to the origin of sex and to an explanation of the breeding habit; mechanisms of heredity both controlled and were manipulated by selection, shaping evolution and, in turn, being shaped by it. With fifteen people studying under him, he had created a school the largest of its kind in the world. Across the ocean the greatest names in his field were awaiting. Morgan, Muller, Bridges, Dobzhansky, Belling, and Babcock were sure to have read the advance copy of his book by the time of his arrival. As the ship made its way across the Atlantic, it must have seemed to Darlington that a new life—a new world—beckoned.[28]

But the reception of Darlington's book would be far less smooth. It would signal the beginning of one of the most controversial scientific careers of the century. Stepping off the plank in New York on a hot summer day in 1932, Darlington walked straight into a cold wall of resistance and outrage. Hampton Carson, a young graduate student, remembered the reaction to *Recent Advances of Cytology* in his biology department at the University of Pennsylvania: "The older members of this strongly cytological department received the Darlington book with stiff attitudes of outrage, anger, and ridicule. The book was considered to be dangerous, in fact poisonous, for the minds of graduate students. It was made clear to us that only after we had become seasoned veterans could we hope to succeed in separating the good (if there was any) from the bad in Darlington. Those of us who had copies kept them in a drawer rather than on the tops of our desks."[29] Traveling upstate for the Sixth International Congress of Genetics at Ithaca, Darlington was given just five minutes to defend his views, and was shouted down by a storm of critics. On the West Coast, an ailing Belling was working zealously on a scathing review of *Recent Advances in Cytology*. And yet, oddly, Darlington was being ushered into the company of the best and the brightest of his field. Spending a month at Woods Hole's Marine Biological Laboratories, Darlington met the aging Wilson, who "treated [him] nicely" and told him jokingly but with a glimmer of truth that Morgan's two greatest discoveries were Bridges and Sturtevant.[30] Across the country, in Berkeley, California, he worked in Belling's room in Ernest Brown Babcock's department, and at Cal Tech collaborated with Theodosius Dobzhansky in Morgan's laboratory, befriending George Beadle and Calvin Bridges. These were the great names in genetics and cytology, and Darlington was one among them. It were as though beyond the exasperation his work had engendered, a genius had been recognized, one that must be reckoned with if not embraced. Darlington's ideas irritated and enticed all at once, leaving no one indifferent. The otherwise friendly Morgan expressed his scientific disdain unabashedly, asking Darlington to pay for the typing of his papers when he was a visitor in his lab. Yet, he was a visitor in Morgan's lab.

"This was seven years too soon or too late. I was known enough for my views to be disliked," Darlington remembered.[31] Though pleased to learn that Haldane, on a lecture tour in the United States, would be joining him at Pasadena for a week of King-Arthur-Court-in-Yankee-land fun, Darlington was feeling the discomfort and yearned to get away. In truth, this had more to do with a woman than with science.

As he had grown older, Darlington's rebellion against his austere, puritan home education had increasingly manifested itself in frivolities of a sexual kind. This was the heyday of an earlier sexual revolution that saw the publication of Marie Stopes's groundbreaking sex manual, *Married Love* (Stopes was married to Gates and a botanist herself), the growing women's movement, and the general sexual promiscuity among the liberal-minded youth of the 1920s. Still, Darlington's sexual rebellion was of a particularly provoking "stick it to them" nature and was consciously, almost maliciously, directed at his parents. Having already engaged in numerous affairs at the John Innes (with a young Indian researcher named Edaveleth Kakkat Janaki-Ammal, with another colleague, Eileen Erlanson, and with the institution's secretary, Brenhilda Schafer), and with another woman by the name of Audrey, Darlington let loose the ultimate shocker in a letter to his parents from Woods Hole in August 1932. Following a lengthy discussion of his meetings with Morgan and Wilson, and banal descriptions of sight-seeing, Darlington added in a short P.S.: "I am afraid when I tell you that I have been secretly married since July 27th you will be very worried. The girl, Kate Pinsdorf, is two years older than I. She teaches history at Vassar college. . . . She's of German extraction, charming, graceful (not American); speaks Spanish, Portuguese, French, and German, has travelled in Brazil, Argentine, California, Italy, and Germany; sings beautifully in all languages, facially resembles Alfred and yourself, is physically robust and healthy, is well informed in literature and affairs, has a delightful accent and musical voice. . . . She has similar views on marriage as myself, and the liaison is not intended to be a permanent one. We can get divorced if necessary any day. . . ." The geneticist in Darlington hastened to add: "She will be a perfect mother because she is well built, intelligent, and (what is more rare), emotionally normal and equable."[32] In fact, the fetching Pinsdorf had approached Darlington with a marriage offer for the purpose of having a child by him just a day or two after they first met on Cape Cod. Finding her attractive, and not thinking much besides, he gladly agreed, and a few days later in a small church ceremony in Barnstable, near Hyannis, the two smiling and sunburned strangers were married.

Nellie and Will were deeply dismayed. Just a day before Darlington had sent his letter, Nellie had written in her usual, edifying tone: "Darling, don't sacrifice your moral side for the physical. If you do what is right, you can look the whole world in the face and you would be much happier—you would radiate sunshine wherever you

Darlington and his American bride Kate Pinsdorf, 1933. The couple were married just several days after having met. Photo courtesy of Oliver Darlington.

went." Having heard the news she now wrote back: "I scarcely feel myself yet. . . . How could you be so impetuous? I used to think you were so courageous and strong-minded, and you are still except when women are concerned. She may be all you say, and I hope she is, but I can't admire a woman who rushes into the most serious thing in life after such a short acquaintance. Father is grievously hurt." At the bottom of the letter, Nellie added: "I have come to the conclusion that the sun has affected you."

In a reply to a barrage of incoming letters ("Have you lost all your moral sense? My dear boy, think a good deal of what you are doing before it is too late! . . . Father has no respect left for you, and I am made to feel it. Your loving and affectionate mother. . . ."), Darlington chose only to further taunt his grieving parents. "On the 31st we bathed in the sun entirely in the nude and got a little sunburned!" To himself he wrote, in a journal: "Promiscuity only flourishes under condemnation. Mankind is too un-enterprising to indulge in it without this encouragement."[33] Finally, when he could stand the reprimands no longer ("I don't like when you say that you won't settle down with her. Do you cast off a wife like a shirt or a collar??"), Darlington wrote back angrily: "I wonder whether you will ever learn not to expect your son to behave the same way as your next door neighbor's sons?"

But the forbidden liaison would be short lived after all. Kate was beginning to show signs of earnestness. The frivolity and spontaneity, it seemed to Cyril, were rapidly, terrifyingly disappearing. He was now frantically seeking a way out, the quicker the better. Applying successfully to the Rockefeller Foundation to continue his fellowship year in Japan and India, Darlington boarded a boat for Tokyo in early May. By mid-month, en route to the East once again, he was already hula dancing with a new girlfriend in Waikiki, and allusions to Kate in letters to his anguished parents grew less and less frequent. Before long, any mention of his legal wife had disappeared altogether. Within a few years, he would sign the divorce papers she would send him, calmly insert them in an envelope, and send them back to America.[34]

In June, Darlington's ship touched land. Almost immediately, the intellectual and aesthetic passions ignited in Persia were reignited. As he moved in to work on rye and oats in the laboratories of Kihara and Kuwada in Kyoto, Darlington marveled at the Buddhist temples and Shinto shrines, at the colorful markets and mysterious Gion nights of the ancient city. The gentle aesthetic of Zen calligraphy amazed him, and he wondered whether and why the Japanese were inherently more artistically sensitive than other peoples. Before long, he was on his way to Trivandum, India, where he would work with his old John Innes lover Janaki-Ammal, but not before he had filled countless small notepads with dense, urgent scribbling on the marriage patterns, customs, arts, history, and language of Japan.

India held another sort of fascination altogether. If Japan was austere, discrete, almost silent, India, in its dilapidated, bustling, unabashed poverty and grandeur was close to being its opposite. Everywhere he turned, Darlington saw countless shades of skin, countless ethnic tribes. The rigid caste system had frozen ancient peoples in time. It had typed them, like a plant breeder types his chosen varieties, generation after the next. Awed by the splendor of diversity, Darlington's imagination was titillated. Could breeding of men really be compared to that of plants? Could similar genetic mechanisms be working in both, shaping culture and tradition as much, and as easily, as color and form?

In both countries Darlington's *Recent Advances* was already widely read and acclaimed, and the Japanese and Indians hosted him like a dignitary.[35] Any American reserve on account of his views was in the East entirely absent, any harm done to his delicate ego assuaged. In August, Darlington sailed to Marseilles and was back home on September 5, 1933, in time to find that he had been included in *Who's*

Who of that year. At twenty-four he had become a scientific orphan. Now, not yet thirty, he had completed a masterful cytogenetics treatise, turning an uncontrollable wealth of observation into a fledgling science. Darlington was taming nature but his own personal and professional future remained painfully unclear. He was championed by some and ridiculed by others. The path ahead curved out of sight, like a nectar-filled pitcher plant beckoning a bee.

～ 6

Roots of a Scientific Controversy

FROM THE MOMENT HE SET FOOT on American soil on that hot summer day in 1932 until his dying days nearly fifty years later, Darlington's name was to be embroiled in controversy. Later forays into politics and social science understandably got him into deep waters, as we shall see, but how was it that a cytogenetics treatise, of all things, should cause such a stir? What sacred toes had Darlington tread upon so that a respected colleague, Charles Leonard Huskins, should use the occasion of the 1935 Botanical Congress in Amsterdam to publicly rail against him in all sessions, whether he had been speaking or not, and to privately offer to come across the Atlantic and punch his head off![1] In the relatively tranquil world of chromosomes and genes, Darlington had done something to make a lot of people very mad.

To understand something of the nature of cytology and cytological claims is to begin to understand the reasons for the strong reaction against Darlington's ideas. To a greater degree than in physiology, biochemistry, or genetics, in cytology processes of scrutiny and inference often require kinds of experience not easily communicated to those who have not participated firsthand in the observations. They demand a special and extensive training of the eye, a product of a personal intimacy between the researcher and his materials. In the case of cytology, the assumption of an intrinsically precise "scientific language," understood by all practitioners even when they disagree, is misleading.[2] In the small confines of the cell, most experiments rely on "tacit knowledge:" personal skills difficult to articulate—like riding a bicycle.[3] Ironically, precisely *because* of the intuitive and ambiguous nature of "seeing," cytologists in the first decades of the twentieth century became known as "show me" men.[4] For reasons that will be made clear, cytologists set a high standard for burden of proof regarding observational claims, creating a paradox whereby in a science notorious for its intrinsic slipperiness—a "strange and difficult kind of visual chemistry with rules only dimly perceived," as one practitioner called it[5]—little could be legitimately claimed that was not immediately obvious.

A letter to Darlington from leading American cytologist Jack Schultz, then at Cal Tech, illustrates the problem well. Reading Darlington's papers on *Fritillaria*, in which reciprocal chiasmata had been shown, Schultz felt resigned to write simply: "I am not sure I understand them."[6] Insofar as these observations had served as the basis for Darlington's explanation for the male *Drosophila* anomaly, a failure to see the process in the flower meant an unwillingness to extrapolate to the fly. This, in turn, meant that the universality of Darlington's laws of chromosome behavior remained challenged.

When Darlington had finally broken the bridle of Newton's agnosticism, claiming that chiasmata result universally from genetic crossing over and are requisites for segregation, he was making a generalization many leading cytologists felt unwarranted. One of these was Karl Sax at Harvard's Arnold Arboretum, who from 1930 through 1936 waged a tireless battle against that very claim. A former student of Edward Murray East, Sax used *Paeonia suffruticosa* to argue that not only was homologous chromosome pairing sometimes accomplished without chiasma formation, but that when chiasmata are found, their structure points to the fact that they do not arise from previous genetic crossing over.[7] Curt Stern in *Drosophila* [8] and Harriet Creighton and Barbara McClintock in *Zea* [9] had already succeeded in showing a convincing correlation between genetic crossing over and an actual physical interchange of chromosome segments. Finally, cytological proof could be firmly attached to Janssens's 1909 chiasmatype theory—until then a speculation. Still, Darlington was making universal claims for chromosome behavior and its relation to genetics, and Sax was not prepared to support his generalizations on *cytological* grounds. At the 1932 Ithaca Genetics Congress, with East on the Executive

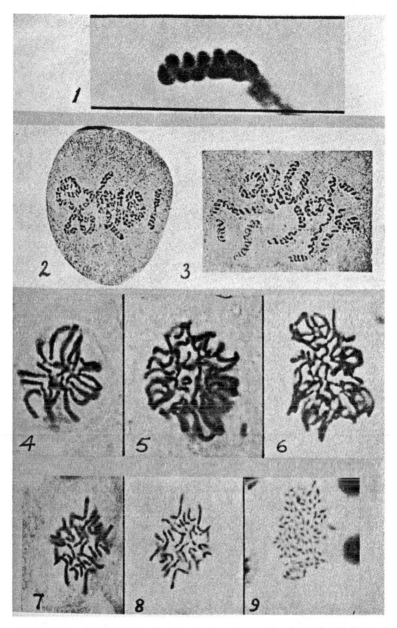

A "strange and difficult kind of visual chemistry with rules only dimly perceived." Photographs of chromosomes. From Darlington, *Recent Advances in Cytology* (London: Churchill, 1932), p. 34.

Council planning the sessions, Sax allowed Darlington little room to respond to his claims.[10]

Sax was also skeptical about Darlington's supporting observations, notably the quantitative agreement between observed frequencies of crossing over and counted numbers of chiasmata. He felt that counting of chiasmata could be carried out in a decisively convincing manner only in a few very favorable objects, and that these unfortunately did not include any forms in which there is a considerable body of evidence on the total frequency of crossing over. The demand for a high standard of proof resulting from the elusive nature of observation in cytology was once again acting to Darlington's detriment.

A third criticism loomed even larger. Chromosome pairing without chiasma formation, or *achiasmatype* meiosis, in male *Drosophila* threatened to disqualify Darlington's rule of crossing over as the requisite to orderly segregation and gamete formation. In the male fly there is no crossing over, and in conformity to the correlation between crossing over and chiasmata, its regular, *autosomal* chromosome pairs show no chiasmata. Nevertheless, the X and Y sex chromosomes enter into a combination that results in a normal, cross-shaped, four-legged configuration, known as a *tetrad*. If Darlington's rules were correct, this means that a chiasma is involved, but why then was there no genetic evidence of crossing over? Darlington's explanation was ingenious: crossing over does indeed occur in the sex chromosomes, but it is always restricted to a single region devoid of active genes. Further, since a single crossing over would interchange an arm of the X with an arm of the Y—which certainly does not occur—Darlington postulated that there is always a reciprocal crossing over, similar to that which he had claimed to see in *Fritillaria*. In this way an apparent exception to the rule was shown to constitute a striking confirmation.

Still, Darlington's scheme was imperfect. In the first place, it failed to explain achiasmatypy in the autosomes, in which no tetrad formation occurred. Recognizing this phenomenon to be an exception to his rule, Darlington was unable to explain it away. In the second instance, the theory of reciprocal chiasmata in the sex chromosomes involved three major assumptions as well as eight steps in the cytological argument that admitted alternative explanations. It failed on Occam—like a conspiracy theory, it was simply too complicated to be true.[11]

Cytologists also criticized the very claims upon which the precocity theory of meiosis was based. Darlington had generalized that it is the split and unsplit conditions of the chromosome into chromatids that determine its behavior with respect to homologous chromosomes.

Although this did not explain what the underlying forces might be (in other words, how the chromosomes actually find their pairs and move in each other's direction), merely to establish the existence of such a numerical rule of behavior was immensely important. Aside from its immediate applicability to genetic reasoning, it would seem to point out the direction that the final attack on this crucial question in cellular mechanics must take. Unfortunately some cytologists were convinced that the factual findings were erroneous. Were the chromosomes really single at the beginning of prophase? Darlington was sure they were, but this was really a guess. Others were not as certain and it was impossible to tell beneath the microscope. If they, and not Darlington, were right, then the law of affinity in pairs could not be true (this, after all, had been a bold speculation based more on genetic logic than on cytological observation), nor could the precocity theory remain valid. If it were not the singleness of the chromosome which underlay its attraction to another single homologue, then, as Darlington's former John Innes colleague and rival Huskins stated: "The general problem now appears to be not to what degree the chromosome is subdivided, but how it comes to behave as a bipartite unit in inheritance."[12]

To a certain extent, initial resistance to Darlington's rules of chromosome behavior was based on these reservations. And yet the cytological objections are not sufficient to explain the rancor and acrimony that ensued. Eventually, Sax's reservations were taken back one by one, and by 1936 he had retracted nearly all of them and accepted Darlington's general scheme. The *Drosophila* explanation was also generally accepted for a time as an elegant solution to a troubling enigma. As for Huskins's reservations, they were articulated most forcefully more than a decade after *Recent Advances in Cytology* had appeared, when evidence began to mount that there is a subdivision into half chromatids earlier in meiosis than Darlington had suggested. Before that time, however, Darlington's scheme had become orthodoxy. The initial shock his book elicited, had ultimately given way to acceptance, and by 1939 Alfred Sturtevant and George Beadle, in their standard text *An Introduction to Genetics*, spoke of it as marking "the unification of chromosome cytology. From then on," they wrote, "it has been possible to see the subject as a whole, united by a series of working hypotheses."[13] Sturtevant later remembered that "Darlington's was the most ambitious attempt at a general scheme, and it was very generally accepted and for a time came to be considered the very backbone of cytogenetics."[14] The scientific critique of Darlington's cytological work, after an initial period of resistance, seemed to slowly fizzle out. Ultimately, it rested less on *what* he was claiming and more on *how* he had arrived at his claims. Even still, there were a number of conceptual difficulties into which Darlington's ideas quickly ran. Knowingly, he had set about to challenge two of the greatest enterprises of the biology of his day: Morgan's classical genetics, and mathematical population genetics.

    MORGAN'S *Drosophila* enterprise was based on the notion of genes strung out like beads on the chromosome, adhering to linkage relationships he and his group could observe in the phenotypes. More than Morgan, it was Hermann J. Muller who extrapolated a real, physical gene as the important and irreducible hereditary element.[15] Morgan, the staunch empiricist, reluctantly followed Muller's lead. Genes were inferred to exist from breeding experiments with "characters," from studying the inheritance of differences of character when different parents were crossed. Yet the whole of heredity, of the genotype, was supposed to be made up of genes added together! Morgan's enterprise was concealing a gap between the analytical or differential gene and the integral genotype. For Morgan, the chromosome did what the fly wanted it to do, or in any case, what it required in terms of *genes*. In his hands, therefore, the chromosome itself could be taken for granted. But Darlington's genetic systems put the chromosomes in the center of the evolutionary picture. Like some concepts in physics, it reversed the common-sense priorities of the observer. Darlington's scheme sketched a dynamic process whereby the mechanism of heredity itself was subject to natural selection and constantly changing in adaptive ways. Such a conception's goal was, in fact, to bridge the gap between the differential gene—Morgan's gene—and the integral genotype, by way of the chromosome. Morgan's insistence that Darlington pay for his own paper while a visitor in his laboratory at Cal Tech reflected a resistance to the upstart's challenge.[16]

Darlington's genetic systems were incompatible with yet another great enterprise of the 1930s. This was mathematical population genetics, widely viewed today as the primary tool for the unification of the formerly estranged ideas of Mendelian genetics and Darwinian selection theory.[17] The synthesis and restoration of faith in the power of selection was primarily due to the work of three men, Ronald Aylmer Fisher and J.B.S. Haldane (working alongside Darlington at the John Innes) in England, and Sewall Wright in the United States. Adopting a biometric, mathematical population approach, the three men defined evolution as the differential inheritance and propagation of gene frequencies over time. By assigning different gene alleles adaptive values with respect to fitness, it became possible to calculate how selection could fashion their respective frequencies in the population. Although the models of the three men emphasized different aspects of evolution and led to disagreements, they all used the gene, not the individual organism or the chromosome, as the functional unit of selection. Together with the quantifiable factors of mutation and migration rate, effective population size, and mating behavior, selection was shown to suffice as a mechanism for evolution. The force it exercised was powerful enough to render Lamarckian, and other forms of directed change, superfluous.[18]

The mathematical models were attractive to biologists both because they helped wed Darwin to Mendel in one grand scheme and because they lent an air of respectability to a science formerly nonquantifiable. But with little mathematical training of their own, most evolutionary biologists in the 1930s didn't actually comprehend them. Darlington stood firmly in the nonmathematical camp. (In a spirited exchange years later with the mathematical evolutionary biologist John Maynard Smith, Darlington quipped: "The trouble with you, Maynard Smith, is that you're illiterate." Maynard Smith fired back: "And the trouble with you, Darlington, is that you're innumerate.")[19] And yet, ironically, Darlington's genetic systems, with the mutual adaptiveness of all their different aspects, introduced too many dimensions for the mathematician to cope with. The gene was a one-dimensional variable. Whether or not Wright, Haldane, and Fisher took them to be so, their models were gross oversimplifications.[20] It was impossible for them to cope with all the effects of selection when the mechanism controlling selection and the unit that is being selected are themselves both being changed by selection. Unlike the gene in the mathematical population geneticist's scheme, Darlington's genetic systems had no single dynamic focus. They integrated several interacting incommensurables: the shape of the chromosomes, their number, the life cycles of organisms, their reproductive organizations, and their breeding behavior (outbreeding or inbreeding, sexually differentiated or hermaphrodite). All these were related in a system whose parts are mutually adapted and adaptively connected. De Vries's famous *Oenothera* proved the point. His mutant plants were trisomics (meaning they possessed three copies of a single chromosome), triploids (meaning they possessed three copies of all the chromosomes of the compliment), and products of segregation in a very unusual hybrid species. For this reason they were discredited in contrast to the gene mutants revealed in inbreeding *Drosophila* (and maize), which have more ordinary genetic systems. But the fate of *Oenothera* had important evolutionary consequences: it had properties of the building materials of evolution that single gene mutations cannot show. In the end, for Darlington, one type of analysis had served as an indispensable complement to the other.[21]

To be sure, Fisher realized that the hereditary mechanism itself could evolve under the pressure of selection. In his 1932 address to the Genetics Congress at Ithaca, Fisher offered: "Others have considered the bearing of the theory of heredity on evolution. I am going to consider the bearing of evolution on heredity."[22] His paper "Evolutionary Modification of Genetic Phenomena" reversed the accepted order manifested in Haldane's address, entitled "Can Evolution Be Explained in Terms of Present Known Genetical Causes?"[23] But the level at which Fisher posed the question—the *gene* level—was different from Darlington's—that of *the behavior of whole chromosomes.* Fisher was unaware at Ithaca that Darlington had just published an account arriving at the same principal conclusion, but when confronted by the young man with the notion of dynamic genetic systems, he "never appeared to hear."[24]

Haldane, who was closer to Darlington, worked beside him, and had acted as his scientific patron, seemed to acknowledge that present-day mathematical limitations were no reason to discount Darlington's complex new idea. In his praise-filled forward to *Recent Advances,* he remarked on the last chapter on the evolution of genetic systems:

> It is improbable that all the conclusions of Dr. Darlington's last chapter are correct. Probably comparative biochemistry will effect as great changes in our view of evolution as comparative morphology. Nevertheless I am convinced that the last chapter is a prolegomenon to every future theory of evolution. I do not pretend that I have yet fully digested it. Until I do so my opinions on evolution will be of somewhat restricted value. But I am already clear that no serious student of evolutionary biology can conceivably avoid reading it.[25]

Even if theoretical models could not yet be constructed to describe the full implications of the genetic system for evolution in quantified terms, it was clear that chromosomal inversions, recombination, polyploidy, balanced lethals, and ring formation play a more important role than simple point mutations, even if they constitute merely a shuffling of existing genetic materials instead of creating entirely new ones. Inversions, for example, affect the relation of any two genes on the same chromosome, a phenomenon known as "position effect." Their occurrence, along with gene mutations, enormously increases the role of variation in evolution. Just like with polyploidy and many other chromosome changes, they can have a great effect on breeding habit, and hence on the evolution of species. Darlington had grasped, more than anyone before him, that the truly important populations—as opposed to those studied by Fisher, Haldane, Wright, and countless geneticists and naturalists who focused exclusively on either the gene or the organism—are the little bits of chromosomes that are populations within which recombination cannot occur and that are therefore isolated from the rest of the species with chromosomes that are not inverted. It was within such populations that new species could arise. Those like Haldane, who understood biology, knew that the bell attached to this perspective must ring true in nature: the primary function of genetic systems is to generate, preserve, and recombine differences in the ways in which natural selection will most effectively be able to use them in furthering evolutionary change.

On purely scientific grounds, therefore, Darlington's theories met different fates. His cytological work initially encountered a number of specific reservations, but these were gradually overcome, until they were revisited in later years. Empirical scientific considerations cannot explain the immediate resistance to Darlington's cytology. Evolution was a different story. Here, Darlington's figuring of complex, dynamic genetic systems, subject to *and* responsible for heredity, encountered difficulties at a conceptual level, both from classical fly genetics and mathematical population genetics. Yet dynamism and complexity were not sufficient reasons for resistance in this case either. After all, a leading scientist like Haldane had paid great respect to the concept of the evolution of genetic systems, as did Fisher, even if he continued to focus his attention at the level of the gene. Something other than specific grievances, or even conceptual difficulties, was lurking behind the angry reaction to Darlington's work. Something more than mere empirical quibbles or technical disabilities was deeming it "dangerous" and, like Socrates' heresies, unfit for the consumption of young minds.

When Haldane predicted the improbability of the correctness of all of the conclusions of the last chapter of *Recent Advances*, he was making a claim based less on actual knowledge to the contrary and more on sound scientific intuition. In somewhat of a counterintuitive way, Darlington's conclusions seemed to matter less than his method of reaching them. Haldane could sense it: Darlington's method was what was at stake.

⁓ IN THE WINTER OF 1932, John Belling was engaged in preparing a scathing attack on *Recent Advances in Cytology*. Belling died of heart disease at the end of February, and in May of that year, "Critical Notes on Darlington's *Recent Advances in Cytology*" appeared posthumously. Its tone was scornful. "The method of the author," Belling wrote, "is to try to establish general propositions, and then to deduce from these. Unfortunately, to a general proposition there are sometimes enough exceptions to invalidate it. . . . The establisher of such general propositions tends to be somewhat impatient with the presenter of exceptions. . . . Hence, and for other reasons, his [Darlington's] writing savours somewhat of propaganda."[26]

"In my opinion," Belling continued,

> Darlington's book contains too many conjectures. A conjecture may be defined as a more or less plausible guess, stated and left, without work being done on it. No priority is due to such a bare conjecture, I think. If it is tested by all available relevant facts, and survives, it becomes a tested hypothesis. A tested hypothesis has some claim to priority. If the tested hypothesis is put to work in research for new facts, and succeeds in prediction, it becomes a theory, with strong right to priority.
>
> Now, if all cytologists printed all or most of their conjectures, the facts would be crowded out. Such literature would not be science. It is true that the discoverer of a new fact is allowed, I think, by general usage, a conjecture as to its origin. But the ratio of facts to conjectures should be kept in the ratio of food to wine. We know what happens when the wine is in excess.[27]

Belling suggested that Darlington cut out nine-tenths of his conjectures in the second edition. He could not see how "every 'fact' implies a hypothesis," and shuddered at the suggestion that "on several occasions where 'fact' and 'hypothesis' have recently come into conflict in cytology, hypothesis has won the day."[28] Before presenting his critique of caution, he added:

> Suppose an examiner questions a graduate student in cytology; is he to be answered with a conjecture from this book? We must remember that students are less sceptical than more experienced workers, and may actually believe in the conjectures they see in print. If all the conjectures were marked with a "(C)," little harm would be done.[29]

Belling's critique stands as a record of what an entirely nonspeculative cytologist thought of Darlington's bold methodology. Belling was a man deeply committed to the principle of inferring from observation nothing more than the eye had seen. Indeed, he had himself told Darlington emphatically that he would not dream of applying his own theory of segmental interchange as he had discovered it in *Datura* to *Oenothera*.[30] Here was a perfect demonstration of the conflict between the claims of general theory and particular observation, at the stage when neither could be supposed to have permanent validity.[31]

Darlington was unshaken. In a special appendix to *Recent Advances* on "Cytological Interpretation," he described how his methodology diverged from that which had become common practice in cytology. Whereas the cytologist investigating chromosomes had heretofore chiefly depended on observation of fixed and stained material, and used an inductive-deductive method of judging the value of his preparations, Darlington was now proposing a second, more empirical, statistically based method. This consisted of the comparison of observations applied in three ways: (1) observations compared with those of living material of the same organism at the same stage of development, (2) observations compared with those of related organisms at the same stage of development, and (3) observations compared with those of the same material at the stages immediately preceding and following. Direct inferences, Darlington claimed, had been admitted as "facts" and "observations" or denied as "artifacts" and "misinterpretations" when actually their whole weight depended upon their use for comparative inference. Comparative inference, on the other hand, had too often been deemed speculative and irrelevant when it was in fact the only means for any general advance in the understanding of the processes concerned.[32]

It seemed apparent to Darlington that the distinction between fact and theory (especially in cytology) was useless, or only useful so long as it was remembered that it was arbitrary. To his mind the two were in reality inseparable: facts had very little claim to existence without the theory that gave them meaning, as much as theory had little validity without the facts upon which it was built. In the elusive world of the chromosomes, where facts were difficult to establish with any degree of certainty, theory was crucial. Without it, little sense could be made of what had quickly become countless disputed details. Under these circumstances, it was imperative to be as open as possible about hypotheses, even when conjectural. In a reply to Belling tacked on to the posthumous "Critical Notes," he offered a forceful justification of his method:

> Belling has criticised my conjectures. We both make conjectures and describe them formally as hypotheses. The difference again arises from the opposition between the morphological and analytical points of view. . . . Belling maintains the morphological point of view, of which he is the ablest exponent. I maintain the analytical point of view, namely: every student should be aware of the conjectures that may reasonably be made in regard to the causes of the events he is studying. I maintain it partly, perhaps, because *to me the cause is more real than the event itself.* If the student is unaware of these conjectures he is apt to consider his present state of knowledge as unchangeable as the laws of the Medes and Persians. And he cannot be blamed for this error; it has been shared for over twenty years by some of our leading cytologists. But there is now no longer any excuse for it.[33] (emphasis added)

To which history was Darlington referring? What axe were both he and Belling grinding? A clue can be found in Darlington's own ascriptions. In his 1931 paper "Meiosis," an initial exposition of the theories that would become *Recent Advances*, Darlington quoted August Weismann's 1887 paper in which he conjectured that "There must be a form of nuclear division in which the ancestral germ plasms contained in the nucleus are distributed to the daughter nuclei in such a way that each of them receives only half the number contained in the original nucleus."[34] Darlington was consciously, although not yet fully, attributing his work to Weismann's earlier genius. Darlington's hero worship of Weismann was the source of much of his professional bravado.

⁓ LIKE DARWIN, WEISMANN HAD a remarkable facility for constructing hypotheses.[35] Like him, he was ridiculed for his speculations. Working on cells prior to the rediscovery of Mendel's laws, Weismann's grand conjecture of meiosis was attacked by outraged contemporaries such as Oscar Hertwig and Hans Driesch. But there was a logic behind the speculations. For Weismann had realized that the cause for the intellectual poverty with respect to understanding heredity lay in the inductivism then prevalent in German biology. Sensing that the accumulation of unexplained facts had begun to hinder progress, Weismann argued: "The time in which men believed that science could be advanced by the mere collection of facts has long passed away."[36] Biology, or so he believed, was now crying out for bold theory making. With a mass of unruly observations, Darlington had to hope that only brilliant deduction would come to the rescue.

Until the point Darwin, Haeckel, and he himself had begun to develop a conceptual framework of biology, Weismann wrote, "the investigation of mere details had led to a state of intellectual short-sightedness, interest being shown only for that that was immediately in view. Immense numbers of detailed facts were thus accumulated, but . . . the intellectual bond which should have bound them together was wanting."[37] Weismann constantly emphasized, therefore, that his hypotheses were to be viewed as heuristic devices, not as an ultimate or settled truth. "Even if it should be later necessary to abandon this theory, nevertheless it seems to me to be a necessary stepping-stone in the development of our understanding, it was absolutely necessary that it had to be proposed, and it must be carefully analysed, regardless of whether in the future it will be found to be correct or false."[38]

Forty years later, Darlington immediately recognized his own situation reflected in that of Weismann. Great changes had come upon cytology since the turn of the century. No better measure of these could be found than in a comparison of Wilson's first and third editions of *The Cell*. Wilson's 1896 book was a highly speculative and bold account of cytological knowledge on the eve of Mendel's rediscovery. Twenty-nine years later any sign of speculation had vanished without a trace: the initial 371-page deductive treatise had been replaced by a 1,232-page inductive collection of undigested observation. Wilson's third edition, full of hedging and prevarications on the laws of pairing and crossing over, was precisely what Weismann had been speaking of forty years earlier when he described the immense number of detailed facts accumulated, for which "the intellectual bond which should have bound them together was wanting."

Arriving on the scene shortly after the appearance of Wilson's book in 1925 and greatly confused by the "jumble of sound and unsound theory and observation"[39] around him, Darlington, like Weismann before him, felt that in an age of inductivism it was important that someone have the courage to speculate.[40] Weismann's holistic approach to biological questions, his theoretical disposition, and his grand, speculative style appealed to him. In part due to Haldane's influence, Darlington's disposition, like Weismann's, was theoretical, despite working in a descriptive field. Like Weismann, Darlington was not afraid to advance conjectures unsupported by observation, but rather deduced from genetic assumptions.[41] Like him, Darlington was ultimately using cytology as a tool with which to explore the problems of greatest interest to him—those of evolution. The result of all this was *Recent Advances in Cytology*. No less a critic than Haldane gave expression to this revolution, calling his young friend's research "the beginning of a new epoch, the transition from an essentially descriptive to a largely deductive science."[42]

When Belling attacked Darlington, he was representing an entire school of inductive, cautious cytology now under threat from what it saw as a young upstart. As Darlington's star quickly rose—a direct consequence of the ability of his bold generalizations to solve a number of outstanding problems in cytogenetics—he soon found himself leading a school methodologically opposed to Belling's camp. By this time, Wilson was too old to take part in the debate, and after Belling's death, the fight was taken up by younger men. Even after many of the specific grievances against Darlington's science, such as those held by Sax, had been put to rest, his method continued to ignite controversy. In a 1945 review of *Recent Advances* highlighting the methodological gulf separating that book and *Mitosis*, by Wilson's Columbia protégé Franz Schrader, Huskins wrote:

> At the opposite extreme is the school of cytology led by Darlington, which stresses deduction, particularly from genetic data, and asserts that in case of conflict greater reliance should be placed on logical theory than on direct microscopic observations.
>
> Most cytologists outside the Darlington school, particularly the maize and *Drosophila* workers, are Baconians, insisting on the importance of data, then on induction from the data. The function of induction to them is to suggest new experiments or tests, not to build up general theories. . . . Even those who accept Darlington's conclusions uncritically do not accept the tenets of his working principles for their own researches.[43]

Comparing the debate between the two schools to that which raged between the "Galileans" and the "neo-Aristotelians" over astrophysics, Huskins chose to examine Schrader's book first:[44] "In his *Introduction* he says, 'If a dispassionate discussion of the subject of mitosis is possible, it is perhaps chiefly due to the fact that our failure to solve most of its problems is so manifest. With rare exceptions we are filled with proper humility—the humility of the open mind.'"[45]

"To those who have been exalted and carried away by the brilliance and success of Darlington's approach to cytological problems and have not yet seen its flaws," Huskins continued, "Schrader's book will appear needlessly depressing, dull and lacking in imagination. What a contrast to *Recent Advances in Cytology.*" Quoting the injunction in Darlington's preface that "Those who refuse to go beyond the fact rarely get as far as the fact," and describing the bold manner in which generalizations and universals are later detailed, Huskins simply asked: "To which author should the young researcher turn for guidance?"

> Darlington is stimulating ("over-stimulating" wrote John Belling). His hypotheses do lead to discoveries. But he does usually state them as if they were facts; does . . . interpret data to fit his hypotheses; does work deductively from genetic facts to cytological observations, which are so often subjective, and prides himself on it. Are these demerits? Not necessarily. They are dangers and pitfalls for the unwary beginner, for the worker in allied fields, for the descriptive cytologist who is not experimentally or critically-minded. To the experienced worker, ably ready to distinguish fact from phantasy [sic], the book is invaluable. But one young research cytologist who keeps a copy of *Recent Advances* always handy on his desk has hit the nail on the head by labelling it "Poison for Students." Most scientists exercise the function of critic for their readers before they publish. (Schrader almost to excess!) Darlington leaves critical analysis to the reader. He continually publishes guesses—and very often guesses right! Unfortunately he rarely acknowledges it when he guesses wrong.

Concluding, Huskins could therefore only offer:

> Today there are not too many cytologists (and fewer geneticists) who know in what respect Darlington's hypotheses are still valid and useful and in what disproved. Schrader does! His book is therefore a current antidote, a "must" for every reader of Darlington.[46]

No wonder the young Penn graduate student was forced to hide *Recent Advances* in a drawer beneath his desk. Darlington was threatening to knock cytology from its foundations. Indeed, when his close friend from the John Innes, the Hungarian émigré mathematical geneticist Pio Koller, was visiting Cal Tech in 1937 on a Rockefeller Fellowship, he wrote back saying: "Many cytologists and geneticists do not take you seriously. They acknowledge that you have a brilliant mind, but think you are biased and disregard facts and evidences which are contradictory to your theories. . . . People here believe that we in England are getting too theoretical and instead of collecting facts, we build 'cobwebs'."[47]

The turbulant reception of Darlington's theories was all about method, about how science should be done. Darlington's challenge had caused a giant stir. But why was this so? Why was method so important? The answers were far from obvious, for hidden boundaries between cytology, genetics, and evolutionary theory were what lay at the heart of the controversy.

～ 7

Method, Discipline, and Character

T HE HISTORY OF CYTOLOGY FALLS rather naturally into quarter centuries, beginning with Oscar Hertwig's discovery of fertilization in 1875. From then until 1900, the mitotic apparatus, the chromosomes, gamete formation, and early embryology were all defined and recognized. It was an exciting period, the atmosphere charged with the conviction that science was on the verge of uncovering the secrets of the life processes and perhaps life itself. There was no longer any doubt that cytology represented an independent field of research at the forefront of progress in biology. Still, even though research in mitosis, fertilization, parthenogenesis, and embryology continued into the second quarter century, the feeling that answers to all the large questions lay almost within grasp gradually weakened.

When Mendel's laws were rediscovered at the turn of the century, cytology and genetics stood on almost equal footing as far as the new study of heredity was concerned, for the rediscovery was shortly followed by the Sutton-Boveri hypothesis, with its cytological demonstration of the parallel between chromosome and "factor" behavior. In the following decade, a close cooperation between geneticists and cytologists led to the acceptance of this paradigm and gave birth to the chromosomal theory of heredity. And while Janssens's chiasmatypy theory of 1909 marked the beginning of a trend that would redefine cytology for years to come, the price would be paid in autonomy. For when Morgan took up the notion that the observed chiasmata were in fact the locus of genetic crossing over, the mathematically based work of the geneticists suddenly became emancipated. The cytologist came to play a secondary role of supporter as the geneticist recognized that cytological verifications could be invaluable. The geneticist began to see that arguments could be made more convincing to biologists in general if they were buttressed by actual observation under the microscope, made by the cytologist. Morgan's use of Calvin Bridges's demonstration of nondisjunction to silence Bateson back in the early 1920s is the prime example.

A consequence of this new reality was that the cytologist's virtue now lay in his reliability. Proceeding with extreme caution, he would no longer make claims unless they could be verified. After John Ruskin, his motto now became: "In science you must not talk before you know." This environment quickly fostered an acceptance of Bateson's exhortation to "treasure your exceptions," and these soon amounted to hundreds of papers littered about the cytological literature. Not surprisingly, little of this was any use to the geneticist, for whom the accumulation of disconnected cytological observation hindered progress rather than spurred it on. In any event, the geneticist's theoretical framework and breeding experiments coupled with cytological verification were highly fecund, producing many successful results. The cytologist, therefore, gradually lost much of the status of path breaker he had enjoyed in the late nineteenth and early twentieth centuries. From a pioneer, he became a doubting Thomas and, in contrast to the budding geneticist, seemed to be doing little to help the cause along. Wilson's third edition of *The Cell*, with its presentation of a gargantuan mass of inapplicable and disparate observations, was a reliable gauge of the state to which cytology had sunk.

When Karl Belar spent a few months at the John Innes in 1928 helping young Darlington with his cytological technique, he succeeded in imprinting the novice with a sentiment that reflected the slough in which cytology now found itself. "Cytology," he told Darlington, "should not be the *ancilla* of genetics."[1] Belar himself had been working on establishing the universal character of mitosis and the chromosomes, a universality he believed must underlie the uniformity of development of plants, animals, and protista.[2] Just over two years later, however, he was dead, the sole fatality of a tragic car accident in the Mojave Desert involving the Dobzhanskys and Eileen Erlanson, Darlington's former John Innes colleague and lover.[3] Back at the John

Innes, Darlington now began to read the cytological and genetic literature. This is when he found the whole of cytology to be "a jumble of
sound and unsound theory and observation, inconsistent with one
another and with genetics; but its unfortunate derivation from histol-
ogy, and the technical subdivision of plants and animals had prevented
anyone from removing or even observing these inconsistencies."[4]
Indeed, plants and animals, chromosomes and breeding were usually
kept apart; wild and cultivated plants were not allowed to meet. It was
most fortunate then that Darlington had happened upon an institution
in which Bateson (through a private endowment) was able to keep
up Darwin's wide-angle biological approach. Now, under Sir Daniel
Hall's remarkable liberality, Darlington was free to set aside the
received academic orthodoxy and consider the subject for the first time
as a whole.

Belling had made an enormous contribution to cytology and genetics
by first recognizing the experimental value of supernumerary chromosome plants, or *polysomics*, in the analysis of meiotic phenomena, yet few
practitioners recognized the immediate importance of his insight.
By adamantly refusing to apply segmental interchange to *Oenothera*,

William Darlington, on left, Darlington, center, Nellie, above him,
and secretaries, Rectory House, 1931. It was during this period that
Darlington conceived and wrote the book that was to propel him to
world fame. Photo courtesy of Clare Passingham.

Belling was adhering to the traditional principles of cytology, which meant that his progress was painfully slow. There was no immediate prospect that he would attack the growing mass of cytological data to make it available to geneticists or noncytologists in general. Darlington thought otherwise. All the generalizations that had been made in his book were based on work with polysomics—they testified to his intuition that supernumerary chromosomes were crucial tools, and represented the step forward Belling was unwilling to make. Under Darlington, the unwieldy and undigested mass of cytological data began to give way. A usable system was beginning to emerge.

The immediate effect of *Recent Advances* was to re-empower the geneticist, for he could now once again use cytology as an employable, functional tool, as opposed to a mere *ancilla*, as Belar had complained in 1928. This was reflected in the favorable reception to Darlington that appeared in geneticists Sturtevant and George Beadle's 1939 book. "Cytological study of the chromosomes is older than the study of Mendelian heredity," they wrote. "Its development has been more or less apart from that of genetics, and it remained largely a descriptive subject until the work of Darlington, whose book in 1932 may be taken as marking the unification of chromosome cytology."[5] The new reunion was a positive step forward, and, after an initial period of skepticism, most geneticists were happy to give it their blessing.

Old-school cytologists like Belling, Sax, Schrader, and Huskins, were not nearly as pleased. To them, Darlington's new type of statistically based, bold, deductive, and often conjectural hypothesis-making represented a threat to their scientific autonomy and credentials. Cytology, after all, had come to be defined by its cautious, meticulous, "show me," Ruskinian style. Anything else was simply not science; it was not cytology! Darlington had in effect winnowed out the worthwhile material to arrive at a few very basic generalizations. Contrary to Ruskin, he consciously chose not to be afraid of making hypotheses public, however conjectural. But if caution were thus thrown to the wind, and faltered, what honor would remain in the profession? How could cytologists ever again be trusted? We now begin to understand the connection between methodological and disciplinary issues. More than simply scientific issues were at stake here; this was a matter of disciplinary pride and livelihood.

Making cytology useful again to geneticists and threatening old-school cytologists in the process were not the only consequences of Darlington's work. By giving shape to a formerly amorphous discipline, Darlington had succeeded in reinvigorating Sutton's science of cytogenetics. From the time of the publication of his book, nearly all younger geneticists, and some older ones, began to add cytological methods to those more traditionally genetic. At the Christmas meeting of the American Genetics Society in 1929, before this revolution, five out of thirty-nine papers presented, or 13 percent, involved some cytological work on the part of the author. By 1940 the papers employing cytological methods grew to 52 percent.[6] In just over a decade, the percentage had quadrupled. The chromosomes were no longer wrapped up in a spireme of unintelligible language quite beyond the interests of ordinary botanists, zoologists, and geneticists.[7] With Darlington's help, they were now once again becoming the focus of a living, relevant tongue.

Most important of all, evolutionist could now begin to speak cytogenetics. This had not always been the case. The early Mendelians were intensely interested in evolution. Models of heredity and mechanisms of evolution had been inextricably linked to the fierce biometrician-Mendelian debates in England at the beginning of the twentieth century. Bateson, after de Vries, had taken up the notion of spontaneous, internal, and discontinuous mutations as genetic mechanisms capable of explaining change in variation across generations— the genuine interest behind his (and de Vries's) inquiries. Across the Atlantic, William E. Castle in his *Genetics and Eugenics* of 1916 spoke for many when he claimed genetics to be "only a subdivision of evolution."[8] Indeed, almost the entire first generation of Mendelians were something other than what we now call geneticists: Bateson, Castle, Davenport, Morgan, and Wilson had all originally come from embryology, and a major reason for their investigations was evolutionary.

And yet beginning in 1910 two causes, one pragmatic and the other theoretical, began to drive a wedge between genetics and evolution. Upon embarking on his fly-room enterprise, Morgan had made a conscious decision to leave development alone for the time being, concentrating, at least experimentally if not in thought, on transmission genetics.[9] This practical decision resulted in a necessary divorce from evolutionary thinking. Coincidentally, and to a great degree consequent to this transmission-based line of inquiry, de Vries's mutation theory was increasingly called into question. For Morgan and his team had been showing how mutations could be responsible for slight phenotypic variations, a far cry from the Oenotheran mutants de Vries claimed represented new species. With the work of geneticists Nilsson-Ehle, Castle, Emerson, and East, Muller's hypothesis and description of "modifier" genes, and the beginnings of the solution to the *Oenothera* mystery, it soon became apparent that macro-mutations resulting in gross phenotypic changes could not be responsible for evolution. By World War I, de Vries's mutation theory was almost entirely extinct.

The next two decades saw the American genetics community withdraw from the evolutionary debate, in part due to the failure to resolve the dispute over the evolutionary mechanism, in part simply because with the incredible success of the transmission enterprise, there was too much to do, and too much to be gained by doing it. "Evolution was not forgotten," Sturtevant and Beadle wrote retrospectively in 1939, "but it seemed that more immediate and approachable problems were more profitable to attack."[10] So long as genetics generated fecund experimental results, evolution would become relegated to the realm of metaphysics and entered less and less into the geneticist's considerations. Indeed, by the 1920s Bateson felt that evolution had largely disappeared from the geneticist's agenda.[11] The experiences of Leslie Clarence Dunn and G. Ledyard Stebbins are a confirmation of Bateson's lamentation: as a Harvard graduate student during World War I, the former was encouraged to avoid evolutionary problems because they were too speculative; the latter was told outright that evolution was all right for Sunday newspaper supplements, but that real biology was biochemical.[12]

A dangerous schism in biology, with origins in the late nineteenth century, was now growing even wider. For as natural history (descriptive and speculative) was transformed in the twentieth century into a largely experimental, analytically rigorous, and integrative science, the gulf between naturalists and experimentalists widened.[13] Whereas the functional biologist focused on proximate causes in the form of the phenotype and how it came about through interaction between the genetic program and the environment, the naturalist-evolutionist tended to be interested in ultimate causes in the form of the origin of the genotype and historical reasons for antecedent adaptations and speciation. Bateson's 1922 piece in *Science* was in fact an impassioned plea for a closer collaboration between geneticists and systematists. For it began to be apparent that the true loser in this explainable yet ultimately arbitrary demarcation was the understanding of evolution itself, which Bateson considered "the salt of biology, the impulse which makes us biologists."[14] If such a gap were to be bridged, geneticists would have to come to terms with macro-evolutionary questions such as the multiplication of species, the origin of higher taxa, and the origin of evolutionary novelties. The naturalists, for their part, would have to become aware of new advances in genetics (for example, that Morgan's mutations were not those of de Vries), and learn to incorporate them into their own, population-based thinking.[15] Crucially, the power of natural selection to bring about gradual evolutionary change would have to be demonstrated. Many geneticists, Bateson, Morgan, and de Vries among them, thought in typological, essentialist terms, neglecting to appreciate the incredible variation readily observed within natural populations. For this reason they were unable to appreciate the power of selection. In contrast, many naturalists, seeing gradualism all around them, and erroneously assuming that Mendelism meant saltationism and hence a rejection of natural selection, turned to Lamarckism as the only possible explanation for speciation.[16] The story of the bridging of these gaps is the story of the "evolutionary synthesis."[17] The synthesis consisted of the work of the mathematical population geneticists Haldane, Fisher, and Wright in the form of a within-genetics synthesis of genetic gradualism with Darwinian natural selection, largely complete by 1932, and the further step of incorporating their models into the realm of natural biological populations for use as a workable methodology, in the service of explaining adaptation and speciation, largely completed by mid-century. The central character responsible for the initiation of this second step was Theodosius Dobzhansky, a Russian émigré who had arrived on a Rockefeller fellowship to Morgan's fly room and, after the death of his mentor, the field biologist Iurii Alexandrovich Philipchenko in 1930, decided to stay in the United States. A tireless worker, full of charm and charisma, Dobzhansky quickly became, both due to his personality and professional work, a leading figure in American science. With a strong background in both field biology *and* genetics and cytology, he was in a unique position to translate his friend Sewall Wright's mathematics to nature, chiefly by way of a close examination of chromosome behavior in natural populations of *Drosophila pseudoobscura*. His 1937 book *Genetics and the Origin of Species* is often referred to as the first account embodying the spirit and direction of the new synthesis.[18] In the following years, Huxley's *Evolution the Modern Synthesis* (1942) and the Columbia Series books—Mayr's *Systematics and the Origin of Species* (1942), Gaylord Simpson's *Tempo and Mode in Evolution* (1944), and G. Ledyard Stebbins's *Variation and Evolution in Plants* (1950)—contributed to the hardening of the synthesis between

Darwinian natural selection and Mendelian genetics and helped generalize it for both plants and animals, and for morphology, cytology, botany, paleontology, and systematics.[19]

There are exceptions to this narrative. The German preoccupation with evolutionary problems in contrast to the proximate/ultimate divide prevalent in the United States is one.[20] The ill-fated Russian school of population genetics with its emphasis on evolution is another.[21] The "Oxford School," in which a long-standing Darwinian tradition had been passed down from Edwin Ray Lankester through Goodrich to Huxley, Ford, Gavin de Beer, and Haldane, is yet a third. Despite the assault on Darwinism and relative strength of Lamarckian thinking in Britain, these men maintained a strong selectionist perspective in their evolutionary outlook, and the divorce between evolution and experimental work characteristic of the synthesis narrative was never true for them. Ford in particular had become a close friend of Fisher's and acted as a translator, much in the same way Dobzhansky had with Wright.[22] His 1931 book *Mendelism and Evolution* and Haldane's *The Causes of Evolution* one year later were early attempts at a synthesis prior to that of Dobzhansky.[23] Both books tried to show how micro- and macro-evolutionary phenomena could be placed in a continuum under the rubric of selection.

More than anyone in Britain, Julian Huxley played a leading role in the synthesis. He had cofounded the Society for Experimental Biology in 1925[24] and the Association for the Study of Systematics in Relation to General Biology in 1937,[25] both engaged in efforts to wed traditional naturalist methodologies with more modern, experimental practice, and unify increasingly specialized subdisciplines. His presidential address to the Zoological Section of the British Association in 1936 did much to help raise the status of natural selection to that of a theory (for Fisher it was only a *genetical* theory), and, with Ford and Haldane, represents a further, earlier origin of the synthesis.[26] Huxley's crowning *Evolution: The Modern Synthesis* finally gave the enterprise its name and, representing the efforts of many years, was both the most wide-ranging argument of all the synthetic works and the most accessible to the public.[27] Britain, therefore, was very much involved in the effort of bringing about the unification of Darwinism with modern classical genetics, and can be said to have produced the first real synthetic works of that enterprise. Perhaps this is not surprising despite the American domination in genetics, considering the strong theoretical tradition in Britain: where the United States took leadership in experimental, transmission genetics, Britain remained the home of a group of broad

Darwinian thinkers, led by the Oxford School. A claim can even be made that were it not for the war, Britain, and not the United States, would have been the home of the eventual synthesis.[28]

⁓ WITHIN THIS CONTEXT, Darlington represented a further exception. Despite holding a natural population perspective (as distinct from the geneticist's laboratory perspective), unlike Ford and Huxley, Darlington had no training as a natural field biologist. Unlike Haldane and Fisher, he was not applying mathematics to the problem of selection. Darlington, rather, was in the anomalous position of a cytologist extrapolating to evolutionary phenomena. There were disciplinary reasons responsible for making such extrapolations unpopular, and the resistance to Darlington's work within cytology can be understood to a great extent as stemming from his explicit evolutionary motivations.[29] It is telling that those cytologists working on cutting-edge problems, like Otto Renner and Belling, had not adopted an evolutionary framework, but rather a static view of genetics aimed at solving specific problems.[30] The lack of interest in evolution grew naturally out of the preoccupation with the exploitation of those breakthroughs that polyploidy and recombinational genetics were allowing. In the rush to apply new methods and to make experiments, few stopped to reflect on the bigger picture. And yet the lack of interest in evolutionary questions can also be understood as a logical consequence of the very nature of cytology at the time. Cytologists could see little to be gained from turning to a speculative field over which their discipline could exercise little control.

For Darlington as for his teacher Bateson, evolutionary questions were what biology was all about. Those geneticists, like Morgan and Haldane, who wrote about evolution tended to stress selection's dependence on mutation. Their books often read as if there was a choice to be made between mutation and recombination; true evolution, they argued, was spurred by the former. Indeed, before Darlington little had been written about the all importance of recombination as a source of genetic variation available for natural selection.[31] Now, in *Recent Advances*, he had shown how genetic systems regulate the amount of inbreeding and outbreeding—that is, the amount of recombination— and how crucial an evolutionary phenomenon this was. Years later, Mayr would comment that "no one made a greater contribution to the understanding of recombination and its evolutionary importance than Darlington."[32] Nevertheless, this approach would not be employed in a wider sense until Dobzhansky's 1937 book, and even then, not to the full extent it would later be accorded.

By 1939 Sturtevant and Beadle could once again declare evolution to be "back in style in genetics."[33] Cytogenetics, the science Darlington in no small measure helped resurrect in the 1930s, had ultimately become, through Dobzhansky, an important catalyst of the evolutionary synthesis. But even though Dobzhansky relied heavily on Darlington's work, citing him more than any other researcher beside Sturtevant and Sewall Wright, little credit would ever be afforded Darlington.[34] Ultimately it would take a geneticist to bring this point of view to a wider biological audience. Haldane had tried to use Darlington's dynamic genetic system framework in *The Causes of Evolution*, but because his technical ability to incorporate it into the mathematical models was limited, he did not consider its focus—the chromosomes—and their behavior in natural populations thoroughly. Dobzhansky did, and his book was immensely successful in placing the chromosome at the heart of evolution.[35] This, of course, had been Darlington's aim all along. Characteristically, he had done the piece of theorizing necessary before the next, empirical steps could be taken. Characteristically again, this theory first had to meet great resistance before it could finally gain wide acceptance.

Why did geneticists resist Darlington on evolutionary matters? "Perhaps it was the shock that *Recent Advances* served that prevented it from serving the synthesis in an immediate and direct way," the young Penn graduate, Hampton Carson, offered in recollection.[36] The controversy within cytology, creating an inadvertent smokescreen hiding the novelty and usefulness of the concept of the evolution of genetic systems, is a further reason. Another is that precisely because *The Evolution of Genetic Systems* (to be discussed later on) dealt with broad evolutionary trends, past changes, and relationships between uncrossable species, the conceptual system tended to share the untestable and unverifiable aspects of the earlier genetic models of the nineteenth century.[37] Still another is the conceptual difficulty of an evolving hereditary mechanism unextendable experimentally.

There is an element of truth in all these explanations: Darlington's evolutionary ideas were complex, controversial, conjectural, often not yet tested, or even worse, untestable. They had introduced not only the concept of genetic systems evolved by natural selection, but also the use of cytology to elucidate the mechanisms of recombination, and hence the production of heritable variation upon which selection could act. But, as we have seen, geneticists were glad to make use of the newly drafted cytology when its benefits became apparent to them. Their resistance to Darlington's evolutionary ideas, therefore, seems to stem less from any fear that Darlington was not sound than, on the contrary, from a fear that perhaps he was all too sound. No one likes an outsider telling him he is wrong.

An illustration proves the point. Darlington's theory stated that if two chromosomes must pair by chiasmata, and that if the chiasma is the result of crossing over, then the length of the crossing-over map must be 50 Morgan units (i.e., 50 percent of crossing over), according to Morgan's system of determination. But the Drosopholists claimed to have in the fly a very small chromosome only 0.4 units long, for which no one, as they told Darlington in 1932, had ever detected more than 0.4 percent of crossing over.[38] To be told that a cytologist could suppose, could *infer*, that there was more crossing over in one of his chromosomes than could ever be discovered by a breeding experiment was insulting to the last degree to the geneticists. That this inference had important evolutionary consequences was doubly unacceptable. Only when Dobzhansky made it legitimate to consider cytological mechanisms of chromosome behavior in relation to evolution could these originally Darlingtonian ideas be slowly understood and incorporated into the mainstream. In 1937 Pio Koller, the same man who had warned Darlington of his detractors in America, wrote worriedly about the appropriation of his friend's ideas. Sturtevant had become convinced through his experiments that structural changes such as inversions play important roles in wild populations, leading to genetic isolation and hence speciation. A loyal Koller felt outraged. "This idea is yours!" he wrote to his friend back in England.[39] And it was.[40]

⌒ DESPITE ALL THIS, there was in truth already a history of chromosomes being used as tools to investigate evolutionary problems preceding both Dobzhansky's learning of the English language and Darlington's first shave. This was true in particular in botany, whose professional preoccupation with evolutionary questions was long-standing.[41] Ernest Brown Babcock at Berkeley was one of those who appreciated both genetics and cytology as crucial instruments for the attack on evolutionary questions. Influenced by the work of Russian cytologist Sergey Gavrilovich Navashin, Babcock began in 1915 what would become a lifelong study of the genus *Crepis*, using chromosome number as a systematic tool.[42] For all its novelty, however, karyology was chiefly an extension of anatomy and comparative morphology, and by the early 1920s Babcock felt the need to go further. Once chromosome constancy was recognized, attention shifted to the nature and mechanism of the all-apparent chromosome alterations. At precisely this juncture taxonomy was undergoing a revolutionary reform, and it was from one of its leaders, Harvey Monroe Hall, that Babcock adopted the practice of applying genetic methods, in conjunction with comparative morphology and geographic distribution, to the solution of taxonomic problems.[43] By 1928 what had begun as an attempt to find a plant analogue to Morgan's *Drosophila* had already turned into a comprehensive phylogenetic and evolutionary study of the genus, with the objective of reaching "a clearer understanding of the evolutionary processes at work in this group of about 200 related species."[44]

It comes as no surprise that Darlington's Rockefeller application of 1932 was to Babcock's laboratory at Berkeley.[45] Where cytologists had failed to be dynamic, plant geneticists of a new ilk were now taking the lead, and Darlington was immediately attracted to their enterprise. After all, his work at the John Innes combined both cytology and plant breeding, and his perspective was always that of natural populations.[46] Nevertheless, it was principally from cytology that Darlington built his evolutionary base. For he had already independently realized that there were, in fact, good reasons for a student of the chromosomes to adopt an evolutionary perspective. While the experimental breeder could sort out linkage in one species, the chromosomes could reveal chiasmata in a hundred species and in every group of plants and animals. While the experimental breeder himself decided how his plant or animal should breed, the chromosome man had to pick up his cells and discover how nature had bred them, why, and with what effect. Even more importantly, to the naturalist and geneticist, the organism was an independent, discontinuous entity. To the cytologist it was part of a continuous process. Cell division was always a step between the past and the future; it was adapted to meet conditions that do not exist, to produce progeny irrelevant to the parents' success.

Darlington had already introduced another evolutionary-minded geneticist and plant breeder, Edgar Anderson, to his view of chromosomes in 1929 when he was a Rockefeller Fellow at the John Innes for the year. Anderson had taken up the importance of the effects of hybridization with his work on *Iris*, and the "ideographs" he developed became an important tool through which to accomplish population-based analysis with respect to plant evolution.[47] Back from Britain, Anderson now began a collaboration with Sax at Harvard on the cytology of the spiderwort *Tradescantia*, and wrote to Darlington in the spring of 1930: "Had it not been for you I would never have been able to study *Tradescantia* at all in the way in which I have done."[48] The resultant monograph, a combination of the cytological work and a taxonomical survey of the genus, was an important addition to evolutionary studies and was dedicated to Sax and Darlington, the American and British rivals.[49]

MORE THAN ANYTHING ELSE, Darlington was a cytologist interested in evolutionary theory. Evolutionary-minded plant geneticists like Babcock and Anderson were happy to collaborate with him as a cytologist, but were less likely to take his evolutionary theories seriously. After all, it was they who were doing the necessary fieldwork, carrying out laborious geographic surveys and *in situ* hybridization experiments. Darlington's theoretical hypotheses were suspect coming from one who had never stepped out into the field.[50] The notion that cytological extrapolations could be made to explain evolutionary phenomena without recourse to a broader set of data was anathema to the field men. Darlington's provocative form of expression, and the logical, deductive nature of his hypotheses, seemed too "clean" to be reliable descriptions of nature's ways.

An illustration makes the point. In 1939 Darlington published an expansion of the last chapter of *Recent Advances* in book form, called *The Evolution of Genetic Systems*.[51] The reaction from field botanists was immediate and harsh. Babcock's *Crepis* collaborator, G. Ledyard Stebbins, a botanist with a cytogenetic background, called *The Evolution of Genetic Systems* a "masterpiece of mythogenesis" and cautioned that its theories were "dangerous." The book, Stebbins concluded, not without a hint of malice, was "as informative on the content of Dr. Darlington's mind as it is uninformative on the subject of Evolution."[52] At the same time, strict laboratory geneticist Hermann J. Muller called the very same work the greatest single contribution to evolutionary thought since Darwin's *On the Origin of Species*.[53]

Just five years later the intransigent Stebbins had changed his mind. "I realise now," he wrote to Darlington meekly, "how little I appreciated this chapter when I first read it, and how unfair was the review which I wrote for *Chronica Botanica*. Please pardon me, and consider this review the product of an immature mind. . . . I realise now what a splendid bit of pioneering you did."[54] By this time Stebbins had already been asked by Ernst Mayr to contribute the botanical component to the Columbia Series of the evolutionary synthesis. That book,

Variation and Evolution in Plants, became the classic botanical text of the modern period, and its fifth chapter—"Genetic Systems as Factors in Evolution"—was widely reviewed as its most original contribution. Darlington replied to Stebbins's letter wryly: "I am not accustomed to people grasping the full meaning of what I write at the first glance, and I am very glad that you have been able to bestow a second glance on my little book."[55]

⌒ Two of Darlington's strongest supporters throughout had been Haldane and Muller. Through their own idiosyncrasies, the two men illustrate the degree to which Darlington fell between the disciplinary cracks characteristic of the biological landscape of his day. Haldane was both theoretically inclined and deeply interested in evolution. It is therefore not surprising that he would welcome theoretical models arrived at deductively, even if tinged by speculation. After all, his mathematical work was even farther removed from nature. This depended, however, on being able to recognize that single-gene frequency models oversimplified a natural complexity of far greater proportions. Haldane succeeded in achieving such an understanding where others had failed, as his foreword to *Recent Advances* humbly attests. The fact that Haldane moved so comfortably between disciplines, writing on biochemistry, evolution, mathematics, as well as philosophy and politics, may go some way in explaining this, and his backing of Darlington.

Muller, for his part, had followed a rigorous reductionism to as-yet-unheard-of theoretical conclusions, postulating the physical existence of the gene to a far greater degree than his teacher Morgan ever had. "One day we may be able to grind genes with a mortar in a beaker," he had famously prophesied in 1922. This was a speculation, in fact a very big speculation. Yet Muller continued to speak of genes as real, hard, physical entities, as if he had handled them individually with his own hands. It is therefore not surprising that a reductionist given to bold speculation would find much value in Darlington's approach. Muller's enthusiastic review of *The Evolution of Genetic Systems* testifies precisely to this point.

Muller and Darlington would become lifelong friends. The early and formative friendship with Haldane would sadly suffer a different fate. While this particular turn of events will be left alone for now, it brings us closer to the final consideration relevant to the reception of Darlington's science: his character.

IN 1923, WITH HIS SCIENTIFIC VIEWS not yet formulated, Darlington had arrived at a John Innes very much embroiled in biological controversy. Bateson did not believe in the chromosomes, Newton distrusted Darwin, and MacBride touted Lamarck. Darlington had sided with his two teachers against MacBride, but after their deaths was left to combat his detractors alone. A powerful force on the John Innes Council, MacBride had proclaimed in 1930: "Genes have played no part in evolution."[56] If this were not enough, Farmer and Gates, the two leading cytologists in Britain, both attacked Darlington for what they claimed to be his unsubstantiated and unscientific cytological work.[57] Some years later, Darlington would recall: "The principle that whatever MacBride and Gates wrote must be wrong, and dangerously so, gave my work direction and momentum until 1930. And the hostility that its application engendered provided—until the death of one and the disappearance of the other—a chorus of disapproval of great emotional value to me. For, next to friends, enemies, if well chosen, are the best stimulus to research. And I had no friends who knew the subject."[58] Darlington, it would seem, was conditioned into controversy. His first and formative experiences with science were contentious ones.

Darlington's preferred method of attack soon became ridicule. His cartoon of MacBride was a beginning. Next, when once Julian Huxley came to the John Innes to give a seminar on the chromosomes in 1929, the young Darlington, just twenty-six, said to MacBride's face that any intelligent schoolboy provided with the evidence could see that no other system could work.[59] As Darlington himself grew in prominence, the tone of his mockeries became even more unpleasant. Barbara McClintock's teacher at Cornell, the venerable Lester S. Sharp, published his *Introduction to Cytology* in 1934. Reviewing it, Darlington wrote:

> What will the student find? By the third page he will already be armed by a battery of words with which to confront his examiners. He will know that a living organism is a 'protoplasmic system.' Its transparent contents are always 'hyaline,' a network in a fixed nucleus is a 'reticulum.' He may later, if he reads Belar, learn to know it as an 'artefact,' but this is one of the words that had not got into the index (unless, indeed, Doubt refers to it). . . . Later the *Introduction to Cytology* will tell the student what an atelomotic isobrachial heteropycnotic idiochromosome is. It will tell him how an undefinable chromonema winds it way through an illimitable matrix. . . . Having learnt the passwords the adventurer will feel

free to enter the realm of constructive study. If he is used to the reasoning of chemistry and physics that is his misfortune. For he must now be inducted into the less rigorous methods of an inexact science. . . . There are many who delight in the study of facts compiled with impartiality and uttered with authority. There are some who find that a logical nexus is a hindrance rather than a help to this study. To such this book can be recommended. For amongst them it cannot fail to be useful: it will save the teacher teaching and the student thinking.[60]

Darlington was a young man fighting against what he took to be an archaic kind of cytology. His tendency to mock and deride those who disagreed with him won him few friends and allies among his colleagues. Some, especially old-school cytologists, seethed. Who was this young whippersnapper telling everyone what to do?

A comparison of the prefaces to the first and second editions of *Recent Advances* paints a telling picture of Darlington's growing cockiness. In the first version, Darlington began by stating modestly his plan to "attempt to describe one aspect of cytology." Promising "to be candid but not impartial," he forewarned his readers that the distinction between fact and hypothesis "as made in every day life becomes too naïve when applied to the data of a new study." Darlington openly admitted that some of his hypotheses may prove to be "will o' the wisps," but hoped that others, nevertheless, would lead to "tangible discoveries." Such candidness and modesty had all but disappeared in the 1937 edition. Now it was evident to the author that "General principles have been discovered of such wide validity that we can predict from them with considerable confidence, and on the rare occasion when the prediction is falsified, we are inclined to look for undetected causal agencies rather than to recast our first principles." Darlington now confidently claimed: "I have shown mitosis as giving rise to meiosis. . . . I have deduced the conditions of parthenogenesis from experimental observations of the breakdown of sexual reproduction." Even if they were true, contemporary cytologists were extremely sensitive to such immodest claims made in the first person. Huskins explicitly referred to the phenomenon in his 1945 review as the price that success, "The Bitch Goddess," exacts.[61]

A letter to his brother, Alfred, in 1934, though private, showed the same bombastic character: "A month ago I was acknowledging the toasts in my honour at a dinner of the Russian Academy of Sciences," Darlington wrote. "I was trying to be modest when the leading experts in America and Russia told me . . . that I was the greatest active cytologist in the world." Moving on to his own research, he added simply: "I have just made a stride in theory so important . . . that I can reasonably hope to be in the Royal Society in 3 or 4 years."[62] He was thirty-one at the time.

⟶ WEDDED TO DARLINGTON'S PROCLIVITY for derision and conceit was a penchant for mocking middle-class convention, as his acknowledgment of his marriage to Kate Pinsdorf to his parents in 1932 clearly shows. Whether predisposed by his upbringing or not, Darlington reacted to disapproval, both to his lifestyle and scientific ideas, by increasingly casting himself in the image of the iconoclast. The close friendship with Haldane, whom Darlington considered to be his "infallible mentor" for seven years since their first meeting in 1927, was also a factor. Haldane was not only quickly becoming a political radical, he was clearly invigorated, even titillated by his ability to scandalize and surprise.[63] His young protégé was now following suit. "Those who obey a general moral code," Darlington wrote in his diary, "will always resent the independence of those who adopt an independent and changeable one." And yet morality itself was but "a system of conduct in which one does what other people think right for one to do." Darlington thought he knew better. *"Vox populi,"* he wrote, "is always *vox diaboli.*" Realizing such a position would not make either professional or social life easy, he added: "I always want to preach what I practice: a dangerous course for a rebel." For Darlington the mind only worked by an appreciation of contrasts: day is good after night, water after thirst, company of women after that of men, solitude after society. He valued extremes and had no use for the boring middle ground. "The mean is safe," he wrote, "but it isn't interesting." To him it was plain that "The most important knowledge of any time is the knowledge whose truth is disputed." As he increasingly shaped his interests and crafted his actions in these terms, Darlington became not only to others, but also primarily to himself, a dissenter. Some years later, he would write in his diary: "No subject keeps my interest when I find my own view of it agrees with the accepted or majority view."[64]

As with Haldane, Darlington's boredom with the mean was often accompanied by an extreme and unattractive abrasiveness. As his star rose, so did the level of his arrogance, and fools, or rather those not in accordance with his own views, were by no means suffered gladly. Darlington's was a particularly acute case of the "I know better than you" syndrome. This was reflected not only in reviews of others' work, but also in more personal interactions. Tall and commanding, Darlington had a reputation for callously and pitilessly degrading colleagues. A single devastating remark, in a meeting or at a conference, could leave its unfortunate target shaken and humiliated.[65] Once, he refused to meet a visiting geneticist from Brazil who had come to the John Innes especially to see him, claiming that such a meeting would be useless because no work of any worth had ever come out of South America.[66]

⌒ Darlington's cutting character, dismissiveness, overt self-confidence, and conscious contrariness, no doubt contributed to the controversy surrounding his work. After all, the world of cytology and genetics of those days was no bigger than a pond; its denizens knew each other well. But to better comprehend why Darlington often did not get due credit for his own ideas, we must make a final observation. Put simply, the marriage of a theoretical bent, a synthetic mind, and a curious disposition often meant that Darlington was better at thinking up an idea than actually proving its truth experimentally. He was an "all at oncer," not a "creeper upper." This was fundamentally an aspect of character, of mind. Darlington was interested in general laws, in axioms that had the power to unify observed facts. The facts themselves cast a much smaller spell on his curiosity. For this reason, having theorized segmental interchange for *Oenothera*, Darlington did not pursue the experimental proof of the concept in relation to the ring formations (this was done by Emerson and Sturtevant, and by Cleland and Blakeslee). He also was not the cytologist to prove that genetic crossing over was accompanied by actual physical interchange of parts (McClintock and Stern were). Despite the fact that he had laid the foundations for the idea, he was not to be given credit either for elucidating the role of chromosomal changes in natural populations in evolution (Dobzhansky was). Yet again, Darlington didn't show how the evolution of genetic systems actually worked in nature in an experimental sense (Babcock and Stebbins did). There are more strictly disciplinary reasons for this: Darlington represented an anomaly as a cytologist interested in evolutionary theory. But there is also a basic quality of personal temperament responsible for this disposition.

"I already regret the early rapture at arriving at generalisations which I feel I can never attain to again," Darlington wrote to his parents from Cal Tech in 1933. "I am bored with the sordid details I now have to return to."[67] But boredom would exact a cost. Ultimately, the price Darlington paid would be measured in units of recognition.

⌐ MORE THAN ANYONE ELSE, Darlington recognized the centrality of the chromosomes in heredity, variation, and evolution. More than anyone, he saw that the chromosomes were not only the products of evolution, but also its guides; their structure and properties determined its course as much as the changes taking place at the gene level. Darlington grasped that the tiny nuclear constituents could be used not only as markers of evolutionary history, but also, and more importantly, as markers of evolutionary potential. Laying the foundations of the understanding of chromosome structure and behavior in cell division, he opened up infinite worlds going back into the distant past and forward into the future.

Many of Darlington's early bold deductions in cytology would have to be revisited following his first articulation of them, some even during his own lifetime. When Wilson's disciple and leading American cytologist Franz Schrader gave his address to the Zoologists' Dinner at the end of 1947, he described Darlington's immense impact on the field, saying: "Not often has a single man made such an impression on old established fields of science as has Darlington on cytology and genetics."[68] But the purpose of Schrader's speech was not to present an encomium to his clever British colleague, but rather an impassioned call for a return to precisely that kind of cytological practice Darlington had rebelled against. "The restless questioning of a Belling and the coldly analytical mind of a Belar," Schrader argued, "have been sorely missed during the past fifteen years." It was no longer affordable, he told his colleagues, "to hold our noses quite so high as we encounter the aberrant cases that trouble our beautiful generalisations." This was a direct attack on Darlington, for despite the immediate success of his law-governed cytological theory, and the strong spur it had served in the creation of cytogenetics, toward mid-century, cracks began to emerge in the edifice of Darlington's construction. To begin with, the universal law of pairing two by two ceased to be thought of as universal. Chromosomes do often pair in threes and fours, and evolution has found ways to safeguard orderly segregation nonetheless. Although not yet entirely understood or explained, it became apparent that other mechanisms besides chiasmata were playing important roles in keeping chromosomes paired in meiosis, and hence in their orderly segregation for gamete formation.[69] Darlington's precocity theory was also increasingly called into question as modifications in the interpretation of the timing of the split of the chromosomes into chromatids became necessary. In addition, the explanation of achiasmatypy in male Drosophila sex chromosomes was proven to be incorrect, and, as this anomalous exception remained unexplained, an ominous shadow was cast on Darlington's sine qua non deductions. For a time Darlington's dogmatic assertions, as Schrader had complained, may actually have had to have been tempered in order to make any sense of things. Darlington expressed this best himself, saying years later: "All great innovators become obstacles to further innovation when the period of innovation has lapsed,"[70] even if he was referring to Bateson and not to himself. Ultimately, it would turn out that many of the problems Darlington had tackled proved to be much more complex than he or anyone else could ever have imagined. The mechanism whereby homologous chromosomes find each other for pairing, and achiasmatypy in the fly, for example, remain unsolved mysteries; we are still far away from a complete understanding of the effects of all chromosome changes on the evolution of natural populations.[71]

Despite the necessary changes to his cytological tome, in the decade and a half following the publication of *Recent Advances in Cytology*, whether in his camp or not, cytologists were quite likely to be busy working on one aspect or another of Darlington's claims. His controversial work had created what no one in the twenty-five years before him had been able to create: a clear, well -articulated , evolution-encompassing cytological worldview. If it would prove in need of adjustments and modifications, it had nevertheless played a crucial role in the reunion of cytology and genetics, and in the solution of a number of outstanding and important problems in cytology, genetics, and evolution. When Darlington joined Bateson's John Innes in 1923 there were six permanent staff workers, who had produced nine papers the previous year. By 1939, when he would assume directorship, there were fourteen staff and forty-one papers; by 1953, nineteen staff and forty-eight papers. Among the staff, workers and students, eleven were elected F.R.S., eight became university professors, three directors of research institutions, and one vice chancellor during Darlington's tenure.[72]

More importantly, a new view of life was slowly emerging. Heredity, and the variation it produced, had brought about evolution, but the very mechanisms of heredity were continually evolving—arising first by chance, then being typed by selection, then changing yet again. The myriad kinds of recombinational devices produced by chromosomes created possibilities for evolution far exceeding those produced by deletions or mutations in single genes. Combined with these simpler phenomena, they painted a picture of nature far more complex, and far more dynamic, than anyone had imagined, and provided natural selection with a richer diet than was ever presumed. To be sure, Darlington had made bold and grand speculation the very totem of his science, but speculation is not always a dirty word, not even in science. While some of his cytological deductions would have to be revisited, even overturned, the overarching evolutionary theory that stemmed from them, and that had been their motivation all along, would ultimately last. In the end, this was largely due to Darlington's insistence that paying respect to the compartments of knowledge unnaturally created by the different disciplines of biology did more to obfuscate than to enlighten. Intellectual daring shepherded by methodological boldness may have led at times up mistaken alleyways, but they ultimately helped to forge a throughway in the direction of a more complete, more complex, and richer understanding of nature and its ways. Perhaps Italian economist Vilfredo Pareto said it best, referring to Kepler, "Give me a fruitful error any time, full of seeds, bursting with its own corrections. You can keep your sterile truth for yourself."[73]

Interlude

D ARLINGTON had grasped that the only way to make sense of the natural world was to combine the material basis from which it was built with the evolutionary framework providing the history of its future and past. Having begun with the first of these, he had been led to a grand theory of the second. After his return to England from India in the summer of 1933, he again became increasingly preoccupied with matter, focusing on the problems of the internal and external mechanics of the chromosomes. For the latter, he developed a theory of spindle formation based on timing relationships between the centromeres and factors outside the nucleus, and for the former, a theory of the dynamics of spirals, of which chromosomes were examples, to explain how they could become entangled with one another.[1] The study of spirals Darlington had begun at Kuwada's laboratory in Japan had persuaded him that all spiral relationships of chromosomes must be in some way connected, and, upon watching fellow John Innes researcher Margaret Upcott undo hanks of wool and asking her how it was done, he became convinced that observations of relational coiling could provide a mechanism for understanding the origin of chiasmata. (In those days, all wool was sold in twisted hanks, which knitters then unwound into smaller balls.) Spirals and spindles would both ultimately lead Darlington down false paths, but at the time, this seemed quite unimportant. Much more interesting was the bespectacled, somewhat timid young woman who was handling the hanks of wool. Margaret Upcott was a talented geneticist, and, slowly but surely, Darlington was falling in love with her.

Darlington was still a married man, though his legal wife was far away teaching history at Vassar. Flippant and capricious when the two had first met, Darlington was now a respected, if controversial, world-class scientist, and ready to take the family route. When it became clear to him in correspondence that Kate Pinsdorf did not share his wish for more than a few children, and would therefore not come to England to be with him, the two decided that it would be best to part ways. Shortly afterward, Kate secured a Reno divorce, and all contact between them came to an abrupt end. Back at the John Innes, in the meantime, Darlington's professional relationship with Upcott was becoming something more.

Six years Darlington's junior, Margaret Blanche Upcott was the daughter of Sir Gilbert Upcott, a top civil servant in the Treasury who later became comptroller and auditor general under King George. Growing up in upper-middle class Highgate, Margaret graduated in 1933 with a first-class degree in botany from Royal Holloway College. That same year she had applied to study for her Ph.D. at the John Innes, where the famous Cyril Darlington was to be her supervisor. Arriving in the fall, she began work under his tutelage.

These were good times at the John Innes. Under Daniel Hall's directorship, shared feelings of mutual purpose fostered a close camaraderie between the mostly young researchers. The Innes was no longer the isolated institution of Bateson's day. Haldane, Crane, Kenneth Mather, Irma Anderson, Dan Lewis, Geoffrey Beale, and Darlington were all scientists of world repute, blazing trails in their respective fields. The institution had built a name for itself not only as a leader in all horticultural research, but also for cytological and genetical investigations of the highest level. Its workers were a team, entertaining themselves with fun and games at least as much as laboring at serious work.

Treading a thin line between gravity and complete nonsensicalness, Haldane set the tone. "Dr. Darlington's book," he had written in an early draft of his forward to *Recent Advances in Cytology*, placed on Cyril's desk for consideration and approval,

> in which he covers the whole field of cytology, with special reference to boggly apparatus, mightn'tochondria, blepharoplasts, and ectoplasm, can be recommended with confidence as a textbook for elementary school children. There is not a word in it which could

They were young, smart, and leading the way, and they knew it. Darlington and Dobzhansky, right, enjoy themselves at a John Innes tea party in the grand Bateson tradition, 1936. Photo from the Darlington Papers, Bodleian Library. Reprinted with permission.

> shock the most delicate susceptibilities, and the importance of spiritual factors in the guidance of evolution is emphasized throughout. . . . Dr. Darlington's book should be in the hands of every progressive clergyman, and portions of it seem to be eminently suitable for broadcasting on Sundays.

(Cyril approved, but alas, had to defer to the publisher's better judgment.)

"It is well known that large numbers of scientific papers appear every year, but no classification seems to have been attempted in respect of the nature (apart from the subject) of the papers," declared one of the yearly Christmas feuilletons jointly written by the staff. *A darlington*, then, among some of the suggestions for definitions "put forward as a basis for criticism and suggestions," was "a paper calculated to raise the temperature of almost all other cytological laboratories by at least 5 degrees centigrade." *A haldane*, by contrast, was "a contribution on any subject to any journal from the Evening News to the Phil. Trans., preferably paid for;" *a macbride*, "a letter to Nature followed by at least five angry rejoinders." Whether staging productions of Shakespeare plays (Darlington always enjoyed playing Claudius in *Hamlet*), playing ping-pong with one another, or practical jokes on one another, the John Innes staff was having a ball. They were young, smart, and leading the way, and they knew it.

Darlington, Margaret, and Pio Koller had by now become an almost inseparable threesome. Koller's middle-European antics, his charm and good nature, had endeared him to his English colleagues. But true love, of the man and womanly kind, was already developing between teacher and student. At some point in 1934 Cyril and Margaret had begun a surreptitious affair, and, within the intimate John Innes confines, their secret was becoming increasingly difficult to hide. Despite a history of mental illness in the Upcott family—(he was always thinking as a geneticist)—Darlington was deeply in love and determined to make Margaret his wife. Mindful that his Reno divorce, while legally annulling his marriage to Kate in the United States, was not recognized by the laws in England, and fearful that his career prospects would be wrecked by the scandal of a public divorce, Darlington persuaded his (now) former student, and her respectable family, that they marry under Scottish law. Scottish marriage laws provided for "irregular marriages," including those of "habit and repute," and those where a promise was made to marry in the future *(per verba de futuro)*. This law famously allowed couples from England to marry without parental consent at Gretna Green, and was a godsend to many a young, rebellious pair, as any reader of Jane Austen knows. With the recent precedents of very much public, and ugly, divorce cases of both Haldane and Bertrand Russell, it seemed like a boon to the ambitious Darlington. As his common-law wife, Margaret would enjoy none of the rights of a wife under English law, but she put her trust in the man she loved. Following a small ceremony in Edinburgh on August 17, 1937, Margaret Upcott became Cyril Darlington's wife. Shortly afterward, she changed her name by deed poll.

"Much more interesting was the bespectacled, somewhat timid young woman who was handling the hanks of wool." Darlington and his second wife, Margaret Upcott, 1938. Photo courtesy of Clare Passingham.

DARLINGTON'S SCIENTIFIC STAR WAS now ascendant. A second edition of *Recent Advances* had appeared with great fanfare, and *The Evolution of Genetic Systems* was on its way. At the same time, the first tremors of a shake-up at the John Innes began to be felt. Increasingly, Darlington trained his sights on the directorship. Winning the post would not be easy, though, as a long history involving his good friend Haldane would quickly prove. Something, or somebody, it seemed, would have to give.

It all began in 1927, when Sir Daniel Hall had successfully lured Haldane to the John Innes with a verbal promise that he would succeed him in the relatively near future. By 1936 Hall had yet to show signs of retiring, and Haldane was beginning to become impatient. An intense dislike between the wives of the two men, and a barrage of Jewish refugee scientists from Germany Haldane was putting up at a cramped Merton, contributed to an increasingly strained, fractious environment. "At present this institution is in a state of hopeless indiscipline," Haldane complained in a serious but humorous memorandum to the John Innes Council. "I do not, for example, think it desirable that ping-pong should be played in a laboratory containing valuable micro-scopes." Nor was Haldane content with "the activity of Lady Hall, who has on at least one occasion eaten large quantities of fruit from plants of genetical importance."[2]

While Haldane knew much about genetics and nothing about gar-dening, Hall knew much about gardening and nothing about genetics. And because both were pluralists and mainly absentees, the institution had two heads who rarely met and then seldom to any purpose. As Hall approached the age of seventy-five, at which time he eventually retired, Haldane began to complain that he lacked advancement, that the John Innes needed a capable director, and that he himself was the only suitable candidate. But Arthur Tansley and Thomas Middleton, the two most serious members of an otherwise unsuitable governing body,[3] were not convinced by the new proposals for research Haldane submit-ted just before going off to Spain "to defend a foreign nation against German aggression and to prevent the conquest of Europe by a power hostile to my own country."[4] In Haldane's absence, Darlington saw an opportunity to organize his colleagues to make a proposal of their own for the reform of management.[5] The council spent a couple of years debating these proposals, but were quicker still in dismissing Haldane when, conscious that his bid for the directorship of the John Innes was all but over, he accepted the Weldon Chair of Biometry at University College London in 1937 upon returning from Spain. Meanwhile, the council accepted the proposal to create four departments—Genetics, Pomology, Biochemistry, and Cytology—and Darlington was named head of the last in 1937. Then, in September 1939, just a few weeks after the birth of his and Margaret's first son, Oliver Franklin, Darlington was chosen to become director of the John Innes.

⌒ WITH HALDANE'S DEPARTURE, Darlington had finally gained the stature of the ranking scientist in his home institution. Arriving just fifteen years before as a volunteer worker, he had now risen to the very top. Shortly before the birth of his second son, Andrew Jeremy, the British scientific establishment would recognize his accom-plishments when they elected him Fellow of the Royal Society in the winter of 1941. Two years later, he was elected president of the Geneti-cal Society. Shortly after that, in September 1944, Susan Clare was born, and after her, Deborah Jane, on July 12, 1946. Just a few months later, Darlington was awarded the Royal Society's top honor, the

Darwin Medal. "The importance of Darlington's work," Nobel Laureate and President of the Royal Society Robert Robinson told the audience on an uncharacteristically warm November London evening, "lies not so much in the discovery of new phenomena—although he has discovered many of these—but rather in the achievement of a synthesis which brings together a highly diversified body of apparently disconnected facts into an integrated system. . . . It is not too much to say that Darlington's results and theories are recognised as the basis of modern nuclear cytology."[6] Although Margaret still had to worry each time her children's births were registered—she feared that proof of marriage might be required to gain the respectable "long form" of birth certificate at a time when the "short form" was viewed with abhorrence as a mark of illegitimacy—Cyril always registered their children's births, and no questions were ever asked. With four healthy children, a loving wife, and world acclaim, Darlington's life seemed to have arrived at a coveted highland.

But not all was well. Margaret, who had quit the John Innes to devote all her energies to motherhood, began to suffer ill health after the birth of her second daughter. Gwendolen Adshead Harvey, who had been married to Darlington's cousin Jack Harvey and was a close family friend, often helped out by taking the children to her Leamington home for short vacations. But Gwendolen also began frequenting the Darlington's Merton Cottage home, with somewhat different intentions. She and Cyril had begun an affair, and, six weeks after the birth of their fifth child, Rachel Drew, in March 1949, Darlington informed Margaret that he was leaving. The John Innes was transferring its estate from Merton to Bayfordbury, and he and Gwendolen would be moving into the director's home the institution was providing.

Having spent the previous months searching for suitable schools for the children in preparation for the move to Bayfordbury, Margaret was taken by complete surprise. Suffering from illness and caring for a newborn, and four little children besides, she was stricken by grief, resentment, and anger. With nothing binding him legally to his wife, Darlington refused to pay alimony to Margaret, and at first refused to pay child support. His wife, Cyril claimed, had been lapsing over the past few years into bouts of depression, and he cited as grounds for divorce alienation of affection. It was she who had betrayed his trust. In a failed attempt to salvage some of his daughter's pride, and money, Gilbert Upcott made a visit to Darlington in a temporary London abode just after he had left. "When I gave my consent to this unfortunate arrangement," Upcott told Darlington, "I assumed that you were a man of honor and a gentleman." When the meeting dissolved in acrimony, all Margaret's furious father could say was " I don't believe you are sane. You are an unutterable cad, a terrible, terrible man," and slam the door on his way out.[7]

Margaret and the children moved to a cramped house in High-gate, and life thereafter became a financial struggle. The loss of the camaraderie of the John Innes staff, and the removal of any pros-pect of her eventual return to research, hit Margaret especially hard. Darlington, for his part, moved into the spacious Bayfordbury cottage with Gwendolen in June 1950, after a second fictitious wedding and subsequent change of name by deed poll. The children came to visit from time to time, and would later remember those occasions as forays into the lap of luxury. Their crowded city tenement was no match for the sprawling Bayfordbury grounds, where they could run amuck and collect nestlings and tadpoles, and other small animals, and plants. Gwendolen's cooking, too, was delicious. A pheasant or venison for dinner; cereal, eggs, and toast for breakfast; and treats like trifle pud-ding or egg custard with sherry within were a far cry from Margaret's measured meals back home.

But coming to Bayfordbury also meant seeing their father. Darling-ton could on occasion show great affection, rushing to the gate to kiss his children and swing them in the air when they arrived. He could spend hours explaining scientific subjects, like pollination of flowers, movement of the earth's crust, or the changing of the seasons. Once a week he would drive his children into Oxford, for a shopping spree that would always include a visit either to the Pitt Rivers Museum or to the Ashmolean. If the weather was good, he might take them to climb the tower at Magdalen, from which they could view the ancient town below: The neatly groomed, silent quads of the college, the magnificent dome of the Radcliffe Camera, the river gently curving beside the Botanical Garden, with its glittering greenhouses close by. At bedtime Darlington would read to his children from their favorite books, sup-plying voices for all the characters. To his children, Darlington must have been greater than life—tall, handsome, all knowing, and world famous. But despite their love for him, and despite their craving for his own love, their affection for their father would always be tinged with awe, even fear. At times he would dispense awful thrashings; at others he would force unwanted spinach down the throat of one or more of the kids. Invariably, bad report cards, or excessive noise-making, or a bro-ken dish, or an unwarranted interruption of his work would be cause for an awful display of temper. None of the children felt they could come to

Gwendolen started out as a baby-sitter for the Darlington
children, but soon she and Cyril were having an affair.
Darlington and Gwendolen Harvey, 1959. Photo courtesy of
Clare Passingham.

him with personal problems, and none did. "CDD wasn't a fellow you
talked to," recalled Clare years later. "You listened to him. If you were
sensible you kept your mouth shut."[8]

∼ IN THE SHORT SPAN of just thirteen years, Darlington had
managed to get married, father five children, leave his wife, and contract a third, less-than-legal marriage. He had gained the scorn of one

woman, the love of another, and an impossible mix of adoration and fear from his kids that would leave a lasting mark on their lives. Over the same period, he had become director of the John Innes Horticultural Institution, Fellow of the Royal Society, president of the Genetical Society, had won the Society's Darwin Medal, and been generally recognized in the world as cytologist extraordinaire. But this was not yet the half of it. Cyril Dean Darlington was just beginning to make his voice heard, his eyes, and mind, purposefully fixed on things to come.

~ *III*

Politics

8

The Lysenko Affair

As DARLINGTON WENT ABOUT HIS LIFE, the world around him was changing. In the Soviet Union, Stalin had consolidated his power, and in Germany, Hitler had led the Nazi Party to victory in the national elections. Europe, and Britain within it, was in the eye of a storm. Darlington had until now been relatively unpolitical, but it was becoming harder by the day to remain aloof. When politics finally invaded his own sanctuary—science—he felt the time had come to act. The match that lit the haystack was a controversy in the Soviet Union that would later become known as the Lysenko Affair. As Soviet science and Soviet politics became one and the same, and political meddling in scientific practice based on the party line came to be seen as a duty rather than an evil, scientists in the West were forced to ask themselves difficult questions about the social and political role of their profession. United in the service of knowledge, but bitterly divided politically, they fought over the principles of scientific freedom and planning, just as their profession was achieving previously unknown social, political, and economic prominence. Once again, Darlington would break ranks and follow his own path.

DURING THE 1930s, a group of "agrobiologists" led by a feisty plant breeder of peasant stock from the Ukraine, Trofim Denisovich Lysenko, began taking over genetics institutions in the Soviet Union. Charismatic, shrewd, but not formally schooled, Lysenko had risen to prominence in the late 1920s as a result of what were seen in Russia (and in the West to a degree) as revolutionary theories in plant physiology. In particular, Lysenko advocated what he called "vernalization," the treatment of plants with exposure to cold water to speed up their development and promote earlier flowering. Using this treatment, Lysenko claimed to have worked out a practical method for sowing winter wheat in the spring. The "miracle on the Lysenko farm" came just as Stalin was initiating his first five-year plan, at a time of massive forced collectivization and acute shortage of grain in the country. It also propelled Lysenko in 1929 to the head of a research program on vernalization at the All-Union Institute for Genetics and Plant Breeding in Odessa, one of the most important centers for agricultural plant science in the Soviet Union. Although the treatment of winter wheat quickly proved impractical, the "barefoot professor," as *Pravda* called him, was nonetheless incredibly successful in promoting his ideas, and his own person, to great prominence.[1]

Despite his limited knowledge of biology, Lysenko began to formulate genetic theories to complement what he claimed to be revolutionary practical successes. He linked them to the work of the famous Russian plant breeder Ivan Vladimirovitch Michurin, a national hero (similar to Luther Burbank) who had notoriously dismissed formal scientific methodology as sophistical. Academic plant physiologists and geneticists in Russia found the reports of Lysenko's results suspect, and the theory attached to them—a variant of the Lamarckian notion of inheritance of acquired traits—crude and vague at best, and otherwise utterly false and misleading. Lysenko showed a blatant disregard for the rules of scientific experimentation and testing of hypotheses. Seeing that his ideas would fare better if they were dressed in the garb of dialectical materialism, Lysenko enlisted the help of a young, zealous ideologist, Isaak I.—generally known as Isai—Prezent, and recast his biological theories in Marxist terms. Referring to the monk from Brno, Lysenko castigated Mendelian genetics as the "attempt of clerical reactionary elements to replace the materialistic biological system of Darwin by a veiled form of idealism"[2] and dismissed the chromosomal theory of heredity as determinist, antimaterialistic, metaphysical speculation. Lysenko argued that it is better to know less, but to know precisely that which is necessary for the production of tangible results. Practical use became the criterion for judging whether a science was "good" or "bad." Politically acute, Lysenko adopted the Marxist "Two Camps" philosophy of science—the one proletarian, materialist, practical, and Soviet, the other bourgeois, idealist, "pure," exploitative, and

Western—making it the base for his attacks on Western-originated genetics and its practitioners in Russia. Nikolai Vavilov, Director of the Genetics Institute of the Academy of Sciences in Leningrad, Bateson's former student, and now Darlington's friend, had led an enormous expansion of scientific agricultural research in the 1920s and made big promises at the start of collectivization in 1928. Now millions were dying of famine in the Ukraine.[3] What had happened to these promises, Lysenko asked? What benefit had all the theoretical research bestowed upon the starving masses?

The ideological fervor with which Lysenko surrounded his work, the constant declarations that he was transforming socialist agriculture for the benefit of the Soviet state and its people, and the terrible, pressing economic plight rendered replies to such questions difficult. They also led some scientists, Vavilov chief among them, to endorse Lysenko in the hope that despite the crude and unscientific nature of his genetical theories, some of the practical suggestions might yet prove valuable.[4] Criticism of Lysenko's claims could too easily be interpreted as a lack of enthusiasm for the goals of socialist agriculture; the consequences could be personally and professionally dire. Could sober academic geneticists, many scions of the old aristocracy, bending over trays of fruit flies in their laboratories, and constantly appearing to restrain progress by calling for more careful scientific procedures of control and verification, really be trusted? "We must remember," Lysenko warned, "that science is the enemy of chance."[5] Bourgeois scientists, propagating a genetics based on random genetic mutations and lacking the ability and the will to apply themselves to practical agriculture problems, were deliberately "wrecking" the socialist effort. They were "fly lovers and people haters,"[6] false prophets of a racist, determinist science, consciously advocated to serve their own class interests.

In such an atmosphere, a peasant agronomist who promised a revolution in agriculture had enormous political advantages. Quickly, a circle of sycophants clustered around Lysenko, claiming victory after victory for his methods, which were rapidly endorsed by the powerful Soviet scientific bureaucracy. In 1932 the Commissariate of Agriculture decided to implement Lysenko's vernalization techniques on a large scale, this time to speed up the ripening of spring grain to avoid some of the effects of summer drought. The area sown was expanded from forty-three thousand hectares in 1932 to 7 million in 1936.[7] In 1935 Vavilov was demoted to one of three vice presidents of his academy and replaced by a politician and agriculture official, Alexander I. Muralov, who promised to eliminate the gap between scientific research and practical agriculture. In an attempt to defend their science, and institutional positions, the geneticists initiated a public "discussion" with the Lysenkoists in 1936 and succeeded in convincing the presidium of the academy that Lysenko's genetics was not sound. But the resistance to Lysenko soon buckled under the weight of Stalin's terror. Muralov and his successor were arrested consecutively. Elsewhere, Solomon Levit, Director of the Medico-Genetical Institute, Isaak Agol, head of the People's Commissariat of Enlightenment's science administration, and Nikolai Gorbunov, a deputy head of the Organizing Committee for the International Genetics Congress, were arrested, accused of being "enemies of the people," and shot. Everywhere, scientists, philosophers, writers, artists, journalists, and politicians began to disappear. In February 1938 Lysenko was named President of the Academy of Agricultural Sciences, and a second "discussion" in 1939 was now dominated by the agronomists: "In order to obtain a certain result," the newly appointed director insisted, "you must want to obtain precisely that result; if you want to obtain a certain result, you will obtain it. . . . I need only such people as will obtain the results that I need."[8] Some months later, Vavilov disappeared, his fate unknown.

If genetics could not be applied to practical socialist needs, it would have to go. Soviet Ambassador to Britain Ivan Maisky expressed this position succinctly, declaring in 1941: "We in the Soviet Union never believed in pure science."[9]

⌒ THIS KIND OF ATTITUDE WAS actually rather new. In the 1920s Soviet geneticists had established close ties with Western scientific communities, and the science they practiced flourished. Many visited and worked in labs in Germany, America, and Britain, and in turn, many Western geneticists—including Julian Huxley, Sidney Harland, J. B. S. Haldane, Leslie Dunn, Richard Goldschmidt, and Calvin Bridges, among others, visited the Soviet Union. Some, like American Hermann J. Muller, even moved their labs there for periods of time. Soviet genetics in the late 1920s and early 1930s enjoyed considerable world prestige: in 1929 an estimated one-third of the world's population geneticists and approximately two-thirds of its Drosophilists were Russian.[10] Close personal ties and rich correspondence between scientists fostered in the international scientific community a sense of cooperation and collegiality transgressing politics and national borders.

With the signing of the Molotov-Ribbentrop nonaggression pact in August 1939, almost all contact between Soviet and Anglo-American genetics communities broke off abruptly. Western geneticists at the time generally assumed that the curtailment of contact had to do with Lysenko's rise to power, though it was actually a casualty of the Soviet Union's foreign policy and symptomatic of all aspects of Soviet-West relations under Stalin.[11] Accordingly, with the German invasion in the summer of 1941, the situation changed drastically, and the ensuing restoration of cooperation between the scientific communities buttressed a newly formed anti-fascist alliance between the U.S.S.R., the United States, and Great Britain. The Soviets were desperate for the opening of a second front in Europe, and employed every means to hasten it. With the renewal of international contacts with Western scientific communities, science became one such means.[12]

In 1942 the Anti-fascist Committee of Soviet Scientists was created by the Central Committee of the Communist Party, with the principle goal of spreading Soviet propaganda among Western scientists. The previously moribund All Union Society for Cultural Relations with Foreign Countries (VOKS) was revitalized, and served as a major channel for exchanges and correspondence. Summaries and reviews of the latest foreign research papers appeared in specially created columns in the Academy of Sciences Bulletin. From the West, large numbers of reprints, journals, books, and even *Drosophila* stocks were sent east, often through diplomatic channels. In the United States, an American Soviet Science Society and a Committee to Aid Geneticists Abroad were initiated and actively undertook to translate, edit, and review Russian manuscripts. By 1943 resumption of personal contacts was in full swing, and as the war approached its end, the Soviet government actively strove to expand its own, formal international science contacts. One major reason for this was the success of the Anglo-American atomic bomb program. Soviet officials recognized that great advances had been made in the West, and healthy international scientific contacts were essential if the U.S.S.R. were to catch up. Unlike physics or mathematics, genetics was not involved in military research, and its recent history of close personal relations made it a harmless vehicle for promoting scientific cooperation.[13]

At this time, agrobiology and academic genetics in Russia were in a compromised deadlock, with science entrenched in academic institutions of higher learning and aggressive Lysenkoism trying to expand from its agricultural base.[14] Soviet geneticists took advantage of their renewed contacts to make a plea to foreign colleagues to help them in this internal battle. When Anton Zhebrak visited the United States in May 1945, he made it clear to his colleagues that the support of

American geneticists was crucial if genetics was to survive in the Soviet Union. He was confident that with such support, "it will not be too long before Lysenko has enough rope to hang himself."[15] Back home, he argued that Lysenko's campaign against genetics was damaging the international reputation of the Soviet Union.[16]

Information about the fortunes of Soviet science and scientists, which had been rather scant and fragmented during the war, became more readily available at its close, as Western scientists were able to visit the Soviet Union and gain firsthand impressions. One such occasion presented itself at the 220-year jubilee of the Soviet Academy of Sciences in June 1945. Stalin personally invited delegations from the allied countries, and 122 delegates from eighteen countries took part in a grandiose celebration lasting two weeks. The foreign delegates witnessed a "parade of victory" with characteristic pomp and grandeur at Red Square, and heard Minister of Foreign Affairs Viacheslav Molotov propose a toast "for the development of close collaboration between Soviet and world science."[17] They also had an opportunity to visit laboratories, hear lectures, exchange ideas, and assess the overall health of Soviet genetics. Eric Ashby, a plant physiologist spending a year in Moscow as an Australian Scientific Attaché, and Julian Huxley both asked to meet Lysenko, and attended one of his lectures. *Science* and *Nature* carried enthusiastic reports on new developments in Soviet science by leading delegation members in biology, physiology, astronomy, physics, and mathematics. Huxley himself reported that Lysenko's position, both scientifically, as president of the Lenin All Union Academy of Agricultural Sciences, and politically, as Vice-Chairman of the Supreme Soviet, was strong. Nevertheless, he emphasized the solid research being done in genetics, and anticipated that the U.S.S.R. "will soon be leading the world in some fields, notably in the relation between ecology, field study, taxonomy, *genetics and evolution* " (emphasis added). The "certain spin of nationalism," which he observed in some branches of Russian science, was dismissed as of little consequence and in no way characteristic of the more general, optimistic picture.[18] Huxley's sanguine report was followed by a piece by Zhebrak in *Science*, arranged with the help of Western colleagues, in which the notion that biology in the U.S.S.R. was not free to develop was emphatically dispelled. In a reply to an earlier article expressing concern by Karl Sax of Harvard, Darlington's old chromosome rival, Zhebrak assured readers that Sax's beliefs "reflect a misunderstanding of the true facts, and . . . actually the science of genetics is making progress in the Soviet Union."[19] He blamed the misunderstanding on the impaired communication during the war, and painted a bright future for further research, development, and cooperation in "building up a common, world-wide biology."[20] Implicit in this hope was the rejection of Lysenko's "Two Camps" philosophy of science.

The concerted effort to coordinate a broad anti-Lysenko campaign continued in the West. In England, Huxley wrote to Darlington detailing a "plan of action" that sought to get Lysenko translated and have damning reviews published in the major scientific journals. He had gained "authoritative information" from Ashby that adverse reviews of Lysenko's work by Western scientists would be seriously considered in the U.S.S.R. and have a beneficial effect for genetics there.[21] This information was immediately disseminated through an elaborate communications network coordinated by Huxley in Britain, and Dunn, Muller, Dobzhansky, and Demerec in the United States. The plan came into effect immediately after P. S. Hudson and R. S. Richens, two British plant scientists, presented Lysenko's theories to the English-speaking public in a 1946 publication of the Imperial Bureau of Plant Genetics.[22] In America, Lysenko's own *Heredity and Its Variability*, published in 1943, was translated by Dobzhansky and was reviewed extensively.[23] A letter from Dunn to Russian émigré Michael Lerner at Berkeley clearly explained the underlying strategy:

> We believe the best way to deal with Lysenko's influence is to make known his ideas and evidence in the form in which he himself has published them. We have no doubt that the judgment of Americans will be adverse and that this will strengthen the hands of those in the Soviet Union who oppose him.[24]

~ BUT THE ORCHESTRATED CONDEMNATORY reviews were conspicuously and unanimously silent on one issue. American and British scientists alike were cautious in their criticisms not to attack the Soviet state and its role in the dictation of the development of science. Underlying the strategy was the assumption that any attempts to link Lysenko's ascent and biological views either to Marxist philosophy or to the Soviet government would compromise those Russian geneticists who had so far been spared and who were continuing their fight against Lysenko. Focusing instead on a purely scientific assault on Lysenko's claims and theory, Anglo-American scientists consciously stressed optimistic manifestations of Soviet genetics, and deliberately steered clear of politics. Accordingly, Dobzhansky laughed at the "preposterous" claim that when sex cells unite, they "assimilate each other"—Lysenko's answer to the chromosome theory of heredity. He attributed Lysenko's rejection of Mendel's laws of segregation to his ignorance of probability theory, and generally dismissed his philosophy as "a naive Lamarckist creed, reminiscent of some of its nineteenth-century versions, especially Spencer, yet alas, devoid of Spencerian finesse." Despite Lysenko, Dobzhansky was confident that "the brilliant and active group of geneticists now working there will keep the U.S.S.R. in the forefront of scientific progress."[25]

Leslie Dunn, Dobzhansky's colleague at Columbia, compared Lysenko's theory of heredity to Darwin's long disproved and discarded hypothesis of pangenesis, and judged it "an anachronism somewhat like the denial of the facts of evolution over large areas of a country as progressive as the United States. . . . In both cases the scientific position of the country is so strong that the heterodox views of small minority groups may safely be left to the judgment of time and progress."[26] Ashby claimed that although Lysenko's experiments sometimes worked in the field, they had "so far proved nothing" of his theoretical claims against Mendelian genetics.[27] Coming the closest to any kind of political comment, Ashby's popular book, *Scientist in Russia*, published in early 1947, was a romantic panegyric to a young and vibrant socialist state working to "build a much firmer bridge between practical application and pure science" than had been built in the West.[28] Regarding Lysenko's theory as wrong, but subordinate in importance to the principle of unification of theory and practice, Ashby had no doubt that the promise of science for the future of the fledgling Soviet nation could hardly be overestimated. The reviewers, and colleagues Sax, G. Ledyard Stebbins, Curt Stern, and Richard Goldschmidt, who wrote reviews as well, all carefully avoided any negative political criticism of the larger Soviet system.

WESTERN SCIENTISTS OF MARXIST PERSUASION likewise treated the Lysenko Affair as a purely scientific controversy, although they were as a whole more acutely attuned to the political dimensions of the science itself. In Britain, Haldane had long been combating the Lamarckism of anticommunist eugenists like E. W. MacBride, Darlington's longtime adversary.[29] "Reactionary biologists," he wrote in 1939, ". . . naturally use the theory of the transmission of acquired habits for political ends. It is silly, they say, to expect the children of manual workers to take up book-learning, or those long oppressed races to govern themselves." Fortunately, however, "laboratory experiments agree with social experience in proving that this theory was false."[30] Haldane was a committed geneticist and selectionist. As early as 1932 he had remarked: "The test of the Union of Soviet Socialist Republics to science will, I think, come when the accumulation of the results of human genetics, demonstrating what I believe to be the fact of innate human inequality, becomes important."[31] By 1938 he had grown closer to Marxism, and enjoyed appropriating Engels's assertion that "the real content of the proletarian demand for equality is the demand for the abolition of classes." Any demand for equality that goes beyond that, "of necessity passes into absurdity."[32] Haldane argued that capitalism ensured that the wealthy, who are genetically superior, will be outbred by the inferior masses. He was convinced that the mean IQ of the population was falling due to the differential birth rate. Socialism was therefore both morally and practically a necessary corrective. Marxist materialism, furthermore, was a vindication of classical, particulate genetics. Thus, Haldane's views ran counter to both the egalitarian *and* environmentalist emphasis of Lysenko's biology.[33] Accordingly, in 1940 he remarked that Lysenko's "attacks on the importance of chromomeres in heredity seem to me to be based on a misunderstanding," and that "this would be very serious if he were dictator of Soviet genetics."[34] Joseph Needham, a Cambridge biochemist and man of the Left, similarly attacked Lysenko's science.[35] The assaults on Lysenko continued after the war, as the communist *Modern Quarterly* printed three damning articles in the autumn and winter of 1947–48. One of these, by a Mendelian plant scientist named Fyfe, recommended that people read Lysenko's books "to satisfy themselves that Lysenko's own theories can be safely ignored."[36]

Lysenkoism during and after the war, and until the summer of 1948, posed little problem to the broader goals of the scientific left—those of a planned economy, equitable distribution of science-based goods and services, and the establishment of science itself as a model, vehicle, and determinant for the organization of all social activities and human needs. Since prominent geneticists like Haldane were in fact leading Mendelians, downplaying Lysenko's Lamarckian theories, and pointing to their divergence from Marxist materialism sufficed. Steering clear of the Scylla of state intervention in science and the Charybdis of rumors concerning the fate of Soviet colleagues, Haldane and others instead shifted the debate back to Britain, where, they claimed, backward national priorities were resulting in measly funding for the sciences.

Everyone knew, as Ashby put it, that "Russia has endowed science with the authority of religion."[37] The real issue was not the bad science of a Lysenko, but rather the good science being squelched by lack of funding at home.[38]

And so the political dimensions of Lysenko's rise to power, and the light with which they illuminated the problem of the subordination of science to politics in the Soviet Union, remained obscure. Both the politically mainstream Anglo-American genetics community and left-wing and Communist researchers in Britain publicly treated Lysenkoism as an isolated scientific aberration. Both were confident that the genetics controversy in Russia would be resolved scientifically.

⁓ As HIS CHROMOSOME BATTLES OF the 1930s gradually subsided, Darlington increasingly found himself drawn to his mentor Bateson's favorite pet—the cytoplasm. While arch-rival Morgan had declared in 1926 that "the cytoplasm may be ignored genetically," Bateson went to his grave in the same year believing that "cytology is providing some knowledge, however scanty, of the material composition of the cell, but of the nature of the control by which a series of orderly differentiations is governed we have no suggestion."[39] Bateson was convinced that the key to understanding differentiation lay in the cytoplasm, and spent much of his last days unsuccessfully trying to find it. Characteristically, as the chromosomal theory of heredity hardened, and became dogma, not least due to his own efforts, Darlington now turned once more to the controversial. "The first problem of heredity," he wrote in 1939, "is that of the parts played in it by the nucleus and the cytoplasm. It is also the last, for we can deal definitely with the indefinable cytoplasm only when the nucleus has been accurately defined."[40] In fact, the cytoplasm had been an interest of Darlington's since very early on, as his holiday talks with Chittenden and correspondence with Renner on plastids, dating from 1928, clearly show.[41] Having satisfied himself that the rudimentary workings of the nucleus were now sufficiently described to understand basic transmission, Darlington increasingly expended his energies on re-including development, and the cytoplasm, in the understanding of heredity.[42]

At the same time, rumors of Darlington's old friend Vavilov's ill fate slowly began reaching the West. In a bid to gain more certainty regarding his whereabouts, the Royal Society, having first sought and obtained the approval of the Soviet Academy, proposed the Russian scientist for election as a Foreign Member. Darlington, a newly elected member himself, was one of the outspoken supporters of the effort.[43] This was unacceptable. How could scientists be persecuted for espousing unwanted scientific theories? How indeed could it be that one of the world's greatest biologists, and a good friend, too, had gone missing under shady circumstances with not even a trace? Obviously, the Soviet authorities had been involved, and this would have to stop. Once Darlington's ire had been excited, there was little that could hold it back.

Before long, the rumors of Vavilov's death were confirmed. The great biologist and Russian hero had died, probably of starvation, in a gulag outpost called Saratov. Even the appeals for mercy voiced by Vavilov's despairing wife directly at the chief of the NKVD, Beria, himself, through Beria's own wife—a laboratory assistant of Vavilov's— had failed. The machine of terror put in place by Stalin was simply too powerful to stop. It had assumed a life of its own. Emotional and upset, Darlington was more than anything outraged. Together with Sidney Harland, a fellow horticultural geneticist and friend of Vavilov's, Darlington coauthored a *Nature* obituary—a careful and measured attack on the Soviet government.[44] But despite the general disgust with the circumstances of Vavilov's death, and sympathy for Darlington's attack, the vehemence with which he expressed himself was ill received by his colleagues. Karl Sax's "thanks but no thanks" letter in reply was representative: Darlington was commended for his obituary but reminded that Zhebrak's recent public criticism of Lysenko in *Science* was an indication of an encouraging trend in the fate of Soviet genetics.[45] In context, this was tantamount to saying: "Yes, things have been bad in the past, and shouldn't be forgotten. But now, things are changing, and we must support those changes to the best of our ability, by withholding some of our criticism concerning the political subordination of science by the Soviet regime."

Darlington's old friend Muller, despite, or perhaps because of, his own firsthand experiences in Russia, adopted this general strategy. "As I am told that negotiations are still afoot to induce the U.S.S.R. to invite the International Genetics Congress," he wrote to Darlington, new cofounder and coeditor with R. A. Fisher of *Heredity*, "it might be inappropriate for me to send you an article about genetics in that country in the near future." Instead, he was sending a "more technical" article, the title of which might be "A Comparison of the Mutational Potentialities of Individual Loci Under X-Ray and Radium Treatment."[46]

Darlington, who had purposely commissioned just such an attack from his socialist friend, flatly rejected this approach. "I think it necessary to deal with this thing at all levels, from the purely academic to the purely political," he wrote to physical chemist and philosopher Michael Polanyi, a Hungarian born German refugee.[47] It was now time to act. In December 1946 Darlington had sent a scathing article on "The Retreat from Science in Soviet Russia" to the editor of the *Fortnightly Review*, adding: "I have had to wait a long time for everyone of my friends mentioned in the article to be out of harm's way—all but one are dead."[48]

The story of the publication of the article is a telling one. John Armitage, the editor, declined the article, explaining that "at the present time we should not allow ill feeling to be inflamed unless we feel that all the arguments in an article are justified." Nevertheless, he added: "I must apologize for writing to you in this way but the fact is I am anxious to publish the article although fearful of being thought a stumbling block to better relations between the countries."[49] An irate Darlington tried placing the article with *Nature*, writing to its editor that "nothing will save fundamental science in Russia except an outright and unflinching exposure of the known facts." Here too the article was declined, its publication thought to be "undesirable."[50]

Finally, Darlington was able to place the article in *Discovery*, a somewhat marginal scientific magazine. Here he refuted Lysenko's claims on scientific grounds, as most Western critics had been doing. Going one step further, however, he asserted: "We see indeed the official overthrow of truth and reason and the men who stood by them in one branch of science. This overthrow is no less official than that in Hitler's Germany."[51] Just one month earlier American colleagues had helped Nikolai Dubinin publish a *Science* article in the vein of Zhebrak's 1945 piece.[52] Darlington was undermining their strategy, and his publication prospects were dimming.

Nevertheless, support came from an unexpected corner. A month after its publication, Darlington received a friendly letter from George Orwell, who had read his paper with great interest. Darlington replied: "I am so glad you were interested in this Russian persecution—I assumed you would be as the author of *Animal Farm*," and proceeded to send him a further article of his pen on the subject.[53] Precisely at that time news came from Muller that attempts to place Darlington's Russian articles in the *New York Times* had failed. They were rejected for being "too extreme."[54] In Britain, however, Orwell pushed for publication in *Polemic*, an anti-Soviet magazine to which he himself often contributed. When paper restrictions rendered its closure unavoidable, and publication impossible, he helped Darlington place the piece in *The Nineteenth Century and After.*

Here Darlington's attack was unsparing. "A government which relied on the absence of inborn class and race differences in man as the basis of its political theory, was naturally unhappy about a science of genetics which relies on the presence of such differences," Darlington wrote.[55] Granted, Hitler's was a distorted genetics, assuming the permanent, unconditional, and homogeneous genetic superiority of a particular group of people, the Germans. The easy retort, claimed Darlington, was to repudiate genetics and put in its place a "genuine," "Russian," "proletarian," and "Marxist" science. Lysenko was simply its prophet. Following with a collective eulogy for over ten persecuted Soviet biologists who had disappeared, and in contradistinction to other critical analyses of Lysenko, Darlington pointed an unwavering finger of blame at the Soviet regime, drawing a direct unwavering link between its official philosophy, Lysenko's teachings, and the political persecution of scientists. Addressing his Western colleagues no less than the Russian government, Darlington concluded: "No amount of glorification of science, will compensate for these evils either in making the good opinion of Western nations or, what is more important, in making the happiness and prosperity of the peoples who are ruled from Moscow."[56]

Dobzhansky expressed gratitude to Darlington for his article, but thought his rationalizations "over-shot the mark," claiming that the most salient feature of Lysenkoism, including its official support, was precisely the utter foolishness of the whole affair.[57] Lysenko was in every way—economic, political, and intellectual—a liability. Darlington's explanation of Lysenko's success stemming from the Soviet government's reliance on the absence of inborn class and race differences seemed to Dobzhansky to be counterproductive. It would undoubtedly be received by Lysenko as a most welcome gift; with Prezent's assistance, Lysenko himself had been arguing the incompatibility of Marxism and Mendelian genetics. In line with the basic strategy, Dobzhansky recommended that Darlington engage less in condemnation of official doctrines and more in personal attacks: "The most effective method of combating Lysenko is ridicule, not anger."[58] Indeed, just months earlier, in a tribute describing the circumstances of Vavilov's death, Dobzhansky himself expressed a hopeful optimism: "It is assuredly not true that all genetic research has been suppressed in the U.S.S.R. . . . Lysenko's influence was never sufficiently great outside the sphere of agriculture. . . . As the uninterrupted flow of publications demonstrates . . . the U.S.S.R. will retain one of the places of prominence which it has secured."[59]

Dobzhansky was not the only one to criticize Darlington. Some months earlier Huxley had reacted with caution to Darlington's piece in *Discovery*. The claim that "all those who are prepared to argue have been put away" was too sweeping. Huxley had been to Russia in 1945 and seen a number of excellent classical geneticists still doing good work in various labs. They just didn't argue, knowing that it doesn't pay.[60] By implication, Huxley was asking the same of his friend Darlington.

In addition to the resistance on the part of the mainstream genetic community, Darlington became the focus of communist sympathizers and fellow travelers' anger. Following the *Nineteenth Century* piece, he was attacked by readers of the *British Medical Journal* for his "wild charges," and for not producing "sufficient documentary authority" concerning his claims of Vavilov's fate.[61] Lay readers were not as polite, calling Darlington's claims "grotesque distortions" and promising that "in the long run empiricists will always make dogmatists like Darlington resemble a horse's kakoots."[62] Haldane, for his part, continued to ascribe reports of disappearances of scientists to vicious rumor. So contradictory were these reports to his Marxist preconceptions of Soviet society that he chose to view them as fabrications propagated by the enemies of the U.S.S.R. Dobzhansky's own sympathy-filled yet politically neutral description of Vavilov's demise seemed to Haldane completely unfounded.[63]

By the end of 1947, however, it was becoming all too obvious, at least to non-Stalinist Marxists, that the situation in Soviet genetics was grim. Ashby, previously exhibiting boundless optimism, could now no longer turn a blind eye: "Zhebrak's (1945) article in *Science* was a sign that normal geneticists in Russia were hopeful. Since then conditions have deteriorated, and my last information from Russia has been extremely disturbing. I feel now that what hope there was for development of intercourse between the Russians and ourselves on science has vanished."[64] Sax publicly acknowledged Lysenko's appeal to authority, castigating his "Two Camps" assault on classical genetics. Trying, nevertheless, to uphold a façade of optimism, he wrote: "The Russians are realists, and the political leaders as well as the farmers must eventually discover that Lysenko's claims have no basis in fact."[65] Few colleagues were convinced. Fewer still were prepared for what was to follow.

~ 9

Marxism and the Slaying
of a Mentor

The question is asked in one of the notes handed to me, What is the attitude
of the Central Committee of the Party to my report? I answer: The Central
Committee of the Party has examined my report and approved it.
(Stormy applause. Ovation. All rise.)[1]

With what has been called "the most chilling
passage in all the literature of twentieth-century science"[2] Lysenko's
biology became official Party biology. When Darlington obtained a
copy of the BBC summaries of the August 1948 session in Moscow in
which Mendelian genetics was denounced and its practice outlawed,
entire institutions officially closed, and geneticists made to recant, he
was right to gloat: "things are going as I expected and not as Huxley and
Ashby expected."[3] Now that Lysenkoist science was official, the bar to
criticism formerly withheld for tactical reasons was removed, and the
Western scientific community reacted with complete outrage—except
for the Marxists. The dismal story of the acceptance by many Marxist
scientists of the Kremlin's party line has been told elsewhere, and will
not occupy us here.[4] It will suffice to remark that when the scientists
were finally compelled to choose between their loyalty to their professional ethic and their loyalty to their political affiliation, some chose
the latter, sacrificing commitment to scientific truth in the process.
Mendelian breeder Fyfe, the same man who in 1947 had encouraged
readers to familiarize themselves with Lysenko's teachings so as to "satisfy themselves that [they] could be safely ignored," now, under Party
compulsion, wrote a book entitled *Lysenko Is Right*.[5] John Desmond
Bernal, a Marxist activist and world-leading physicist whose laboratory
was trying (with X-ray crystallography techniques he himself developed) to elucidate the structure of the DNA molecule, wrote of the
August conference triumphantly: "The importance of this decisive step

is that it marks for the first time the assertion of the independence of science in the Soviet union from the previously universal community of world science."[6] Haldane, for his part, knowingly obfuscated the political issues at stake by continuing to focus on the science. When it became pressingly clear that it was untenable for one of the giants of genetics to defend a man who denied the existence of the gene, Haldane quietly, and, if not pathetically then sadly, slipped away from the limelight.[7] He was the only prominent Marxist scientist in Britain to leave the Communist Party over Lysenko.[8]

The rest of the scientific world, however, discarded its former caution. Previously careful to confine their arguments to scientific refutations of Lysenko's claims, Western geneticists now argued that the issue was not about science at all, but was entirely political. Lysenko's biology was now Soviet biology, and there remained no other recourse but to attack the government directly. At a time of heightened political tensions, and the beginnings of vigorous expressions of anti-Soviet sentiment in Europe and the United States, Muller and Sir Henry Dale, a

The great geneticist J. B. S Haldane became for a time an apologist for the man who didn't believe in the gene, Trofim Lysenko. Ultimately, he left the Communist Party of Great Britain over the affair, but not before being ridiculed by colleagues and the press. A Low cartoon, from the Darlington Papers, Bodleian Library. Reprinted with permission.

former president of the Royal Society, both resigned from the Soviet Academy of Sciences, their letters published in *Science* and *Nature*. In his resignation, Muller compared Russia to Nazi Germany, arguing that "in both cases the attempt was made to set up a politically directed 'science,' separated from that of the world in general, in contravention of the fact that true science can know no boundaries."[9]

Huxley and Ashby, both previously sympathetic to the socialist experiment and hoping for its success, now joined the anti-Soviet campaign. Huxley attacked the Soviet distinction between "pure" and "practical" science, citing the claim of "useless" given for the closing of Dubinin's Cytogenetical Lab. "The Academy must have forgotten Faraday's answer to a questioner who asked him what was the use of his work: 'What is the use of a baby?'" he wrote.[10] Lysenko's ascent was attributed to a combination of his charisma and political clout and obscurantist ideas about theory and practice, as well as to Marxism's pro-Lamarckian bias and the rise of a patriotic nationalism in the U.S.S.R., which manifested itself in the adoption of a distinctively Soviet genetics. Lysenko's officially sanctioned terror reminded Huxley of that of Savonarola.[11] Ashby tried to explain Lysenko's rise as a function of the success of his practical methods, rather than as a consequence of any direct link between Marxism and Lamarckian notions of inheritance. Lysenko was playing a crucial social role as a bridge between the peasant and the government. Nevertheless, this situation had led to an unacceptable political intervention in science. "The significance of the August decree . . ." Ashby told his BBC audience, "is that formerly when biologists were liquidated, it was always given out that their liquidation was nothing to do with their biology; whereas now, the pretense is dropped."[12]

Dobzhansky too discarded his former caution, while continuing to reject attempts to rationalize Lysenko's victory. Dismissing Muller's and Darlington's argument that the Soviets felt genetics had to be destroyed because it proved the biological inequality of men, and Ashby's explanation that Lysenko's usefulness for improving the Soviet agriculture production made it expedient to sacrifice the theoretical science of genetics and ignore such damage, Dobzhansky argued that Lysenko was able to become dictator of biological science in the U.S.S.R. because he convinced the Communist Party that he was the heir and embodiment of an important tradition in Russian cultural history, a tradition that the Russian leaders regarded as infallible. In this context, rational thought had very little to do with anything. Indeed, Dobzhansky claimed, Dostoyevsky's old observation that what others regard as plausible speculation becomes indisputable dogma in Russia had turned into a grotesque prophecy. All that could be done now was to signal a warning: "Having placed a maniac in charge of their agriculture, they are bound to suffer grave loses in harvests, and this for a long time."[13]

As part of the now growing expression of outrage, Conway Zirkle, a genetical botanist from the University of Pennsylvania, edited a "white paper" in which translated minutes of the conference, *Pravda* editorials ("the most important principle in any science is the Party principle"), and scientists' recantations all told a sad story. Zhebrak, who had been hopeful that Western support for genetics would ultimately lead to Lysenko's downfall, was now quoted as saying: "since it has become clear to me that the fundamental aspects of Michurin's direction in Soviet genetics are approved by the Central Committee of the All-Union Communist Party, then I, as a member of the Party, do not consider it possible for me to retain these views which are recognized as erroneous by the Central Committee of the Party."[14]

The Governing Board of the American Institute of Biological Sciences (AIBS) issued an official statement, condemning the actions of the Soviet government and claiming that "the contention raised by Lysenko and his 'Michurinists' against genetics does not represent a controversy of two opposing schools of scientific thought. It is in reality a conflict between politics and science."[15] This had been Darlington's contention all along.

⁓ IN REALITY, HOWEVER, things were not that simple. A serious debate over cytoplasmic inheritance and its concomitant Lamarckian possibilities had in fact been raging in the West as a challenge to orthodox Morganist transmission genetics for quite some time.[16] One of the leaders of the debate, Indiana University geneticist Tracy Sonneborn, had produced experimental evidence for the inheritance of acquired traits. In fact, Sonneborn's work on *Paramecium* had been singled out both in Russia, and by Western communists, in support of Lysenko's theoretical claims. On September 6, 1948, *Newsweek* reported that such results were undermining the same classical theory of genetics that the "Soviet savants of Lysenko's group" had been attacking by dialectical argument. Sonneborn, of no Marxist inclinations whatsoever, naturally felt compelled to distance himself from Lysenko. He was, in fact, one of the instigators, and drafters, of the AIBS statement, in which the possibility of the inheritance of acquired traits was strongly denied.[17]

Darlington too, in his own work on cytoplasmic inheritance, had been advocating genetic mechanisms of a highly controversial kind. In fact, Darlington considered cytoplasmic inheritance, together with genetic crossing over, to be one of the two great exceptions to Mendel's law of segregation. Based on differences of size, structure, and complexity of genetic factors, Darlington postulated three systems of "government" within the cell, each with separate models of control. The first consisted of the nuclear genes, had a mechanical equilibrium, and predominated both in the government of heredity and of the cell. The second was corpuscular, had a physiological equilibrium, and could only be recognized in green plants with plastids. The final, molecular system constituted what Darlington called the "undefined residue of heredity" and was not associated with any visible bodies in the cell. This cytoplasmic mode of inheritance, of chemical rather than physiological or mechanical equilibrium, consisted in maternally transmitted genetic particles Darlington called "plasmagenes." It was here that environmental effects could direct developmental processes. Acting upon each other in complex ways, the three systems determined both the heredity and growth of the organism.[18]

Darlington did not believe Lamarckian mechanisms were the cause of forward evolutionary change, but he did consider the behavior of plasmagenes to be illustrative of a Lamarckian principle in evolution. Accordingly, the crystallographer William Thomas Astbury wrote to Darlington to warn him that his notions were "not far removed if not from Lamarckism then at least from something very deep underlying the superficial tripe that Lysenko has been getting off his chest."[19] Dobzhansky, more jokingly, had sent Darlington a piece from a Brazilian Communist intellectual journal, *Fundamentos*, in which Darlington's theories concerning the action of plasmagenes were praised in support of Lysenko's Lamarckian claims for grafting.[20]

Darlington then, like Sonneborn, actually had an *incentive* to attack Lysenko. Unlike Sonneborn, however, who came out against the Russian only when his own theories began to be used in his name, Darlington had been condemnatory all along. It seemed plain to him that this was principally a *political*, rather than a *scientific*, assault on accepted practice. Accordingly, Darlington refrained from publicly distinguishing his views on cytoplasmic inheritance from those of Lysenko, as Sonneborn had,[21] feeling that this would give the impression that an authentic scientific controversy was at issue. Darlington explained the difference instead to Huxley in a private letter: "I induced a theory having what I called 'a Lamarckian colour' on scientific evidence. They deduced a theory having a Lamarckian substance as well as colour from a political principle. This is obvious to the serious and non-communist student. With the ingenuity of a Haldane, or the naiveté of a Waddington, it is easily rendered confusing to the layman. Plasmagenes do not account for the fakes or fancies of Lysenko."[22]

Despite the debate over cytoplasmic inheritance, Darlington thought, things were clear-cut and simple, after all. Back in Britain, he joined a saddened Fisher and Harland and an evasive Haldane in a BBC broadcast to the nation about the Lysenko Affair. "Many people in our country have been shocked by these events," he commented dryly. "I have several times predicted them as the inevitable end of the long continued persecution of science in Russia. I am therefore more shocked at the indifference to this persecution on the part of our own scientists."[23]

Why had Darlington's reaction to Lysenko been so divergent from the rest of his non-Marxist scientific colleagues? Did he know something they didn't? Had his personal and institutional experiences and his scientific work led him to other conclusions about the state of genetics in Russia?

Darlington had grown up in a patriotic, middle-class, working family; his domineering father was a Socialist. Young Cyril's own sensibilities were cast early on in the mold of both youth and of his progenitor, to which early diary entries, somewhat obsessively concerned with the coal strikes, testify.[24] When he heard Julian Huxley's 1926 speech about the genetic inferiority of the multiplying masses, Darlington reacted with anger and disdain. Was it really all that surprising that the leaders of the eugenics movement were all well-to-do men, an irate Darlington had asked not so facetiously?

At twenty-three, Darlington gravitated toward the mentorship of Haldane when the latter joined the staff of the John Innes in the spring of 1927. Haldane was not then yet a Marxist, but, as this mischievous note to his secretary anecdotally, but also instructively, shows, he was headed that way:

Wanted:

1) Insignia of the Order of the Holy Ghost
2) Mr Ford's bank balance
3) Clara Bow
4) Info on all trains bet. 11 A.M.—5 P.M. from London-Paris
5) £5 (cheque cashed)
6) safety razor blades
7) If you can manage, a max of botches
8) Karl Marx's Kapital[25]

Despite the fact that Haldane shared Huxley's views on the genetic underpinnings of the class system, and was as privileged as they come, he nonetheless espoused social, political, and scientific views that appealed greatly to Darlington. At thirty-four, Haldane was Olympian in form and manner, the idol of the forward-looking, emancipated, protesting intellectual youth of the time, and the model of the scientist of the future. "For about seven years," Darlington would remember, "I regarded him as my infallible mentor. He took the place of my father and of Newton in my still immature mind."[26] Gargantua and Pantagruel took holidays together in Austria, stayed late after work discussing science, politics, and the philosophies of Marx and Engels, and developed a close friendship.[27] Haldane was gradually reconciling his firm belief in the innate inequality of man with his growing attraction to revolutionary politics. Concomitantly, he was moving further away from the Kantian idealism, anti-mechanism, and organicism he had imbibed at home through his father, the physiologist John Scott Haldane.[28] This was having an effect on his protégé. Struggling to unite moral and mental engagements with both the political and scientific, Darlington embraced Haldane's scientism and his confident pronouncements that "the progress of society depends . . . on the progressive application of science."[29] From this sentiment sprang an equally fervent antireligious bias. "Scientific faith consists in belief in a uniformity derived from evidence of one's own senses," Darlington wrote to himself. "Religious faith consists in a belief in a uniformity derived from evidence of someone else's senses. Scientific faith, therefore, implies belief in oneself and doubt with others, religious faith—belief in others and doubt in oneself."[30] As science increasingly came to be defined for Haldane by the metaphysics of dialectical materialism, Darlington's own reasoning followed suit.

Unlike Haldane, Darlington did not engage in political writing until later in life, but his early leftist tendencies are unquestionable: "It is even stranger than usual to hear an Englishman refer to the lower classes as 'they' after living in the U.S.A.," he had written in his diary in 1933, upon return from America.[31] Neither are the strong friendships he was developing with Russian colleagues doubtful. In September 1934, just as *Recent Advances in Cytology* was successfully appearing in its Russian edition, Karpechenko and Vavilov invited Darlington to lecture in Leningrad and Moscow.[32] The foreword to this translation is instructive. Describing cytological relationships as "the focus of a dialectic field," Darlington wrote:

> The internal contradiction in the theory of chromosome permanence is found to consist in genotypic control of chromosome behaviour; the chromosomes are subordinate to the integrated action of their own genes. A quantitative change in the timing of division qualitatively changes mitosis into meiosis and a quantitative change in the genes, polyploidy, gives a qualitative change in the genotype. In parthenogenesis there is a negation of the sexual process which is itself a negation of the asexual process. An old form is returned to at a higher level. In crossing over (or chiasma formation) genetical and mechanical functions interlock, for the same event which conditions the pairing and segregation of the chromosomes also determines their internal reconstitution. In the quantitative properties of crossing over we therefore see the struggle between its opposite and incompatible advantages.[33]

Darlington went on to list the shortcomings of the undialectical approach in cytology, which, he claimed, examines the relation between facts by "idealistic processes of teleology and analogy." Meiosis is thus explained as the physiological completion of the sexual act; polyploidy as the means by which the chromosomes of hybrids get mates suitable to their otherwise thwarted desires; and crossing over as a logical consequence resulting from the need for four unique tetrad cells. Due to such muddled thinking, Belling, Sax, Morgan, and Sturtevant, who did not think dialectically, had been unable to appreciate Darlington's cytological work and the manner in which it led him to the study of the evolution of genetic systems. "It is on account of the sharp cleavage that exists between my treatment of cytological problems and the orthodox or academic treatment," he wrote in the foreword to the Moscow edition, "that I find it difficult to explain my views to a well-trained cytologist. I hope on the other hand that my colleagues in the Soviet Union who are accustomed to approach scientific problems with a conscious and consistent philosophic outlook will find none of the snares and pitfalls that have troubled readers of this book in other countries."

On one level this was simply catered rhetoric, conscious lip service to his Russian-speaking, Communist audience. After all, Darlington had never couched his claims against his detractors in such unequivocal terms in English.[34] This was a foreign-language, easy, unassailable cheap shot at his critics. It was also a requisite for the success of any scientific book in Russia. On another level, however, it is evident that although not quite a Marxist, Darlington, like his mentor Haldane, was becoming convinced that Engels's metaphysics of dialectical materialism, short of being a method of discovery, was a valuable framework for scientific research and exposition. This conviction was enduring, as Engels's *Ludwig Feuerbach* quotation on the opening page of the 1939 *Evolution of Genetic Systems* shows. After all, that book was an attempt to express evolution in terms of dialectic change. Founded on the generalization that crossing over was coextensive with sexual reproduction, species formation and apomixis were shown to be alternative contradictions to a process of expanding hybridization. Indeed, Darlington had been such an attentive student that his Russian foreword appeared more than two years before Haldane's first public attempt at a dialectical interpretation of his own research![35] It was, in fact, possible to be a dialectical materialist without subscribing to Marxism by adopting it as a heuristic, or tool, in the scientific realm alone, without extrapolating to social phenomena. Haldane, for one, was, years before he finally joined the Party in 1942.[36] In 1935 he had in fact told the mathematician Hyman Levy and Barnet Woolf, both Communist Party members, just that.[37] Conrad Hal Waddington, with his discussion of the importance of dialectical materialism for understanding ontogenetic development in *Organisers and Genes* was another example.[38] Darlington, it would seem, viewed things in a similar way, although he was never as explicit in English as either his geneticist or embryologist friends.

But Darlington and Haldane, alas, were like two travelers approaching each other only to continue walking in opposite directions. Darlington's enthusiasm for Soviet Marxism gradually began to wane just as Haldane's was waxing. The seeds of separation were sown at the John Innes, and would ultimately grow an insuperable divide between the close friends.

⌁ WHEN IN 1936 THE BID FOR DIRECTORSHIP OF the John Innes began, Darlington found himself Haldane's rival. By now he had matured enough to see his mentor's intellectual shortcomings, even his character flaws. Darlington was fighting to become his own man. But more than just growing pains began to drive a wedge between the two men. The directorship bid came, after all, precisely at the moment Haldane was moving closer to Communism, and Darlington, of the same obstinate, obdurate, and ruthlessly critical nature as his teacher, was becoming increasingly disillusioned: disillusioned with Haldane, his science, and his politics.[39] Haldane, Darlington now believed, not only lacked the manual and visual skills requisite for experimental work. He also lacked the ability to transcend a formalist, mathematical approach to evolution and genetics which utterly failed to understand the dynamics of which Darlington was speaking in *Evolution of Genetic Systems*.[40] In addition, he failed to conceal a persnickety rudeness and irritability that went far beyond the failure to suffer fools gladly, and an utter unsuitability for administration. Most of all, Haldane lacked the ability to form true friendships, and Darlington began to see that for that same reason, more than most men, he craved renown. It was this need, Darlington thought, that was now being cloaked in the bogus emotions of the Communist cult: expressing a violent love for mankind combined with indifference to every individual man or woman.[41]

Of course, Darlington himself suffered from precisely the same flaws as his mentor. Egotistic, prickly, determined, knowing it all, and with a particular proclivity for the hysteric, Darlington too found it hard to sustain close relations. Even his own children could place little trust in his friendship. And, after all, Darlington had in effect betrayed the man who had shepherded his early, and shaky, scientific career. Haldane had consulted him *as a friend* in the drafting of his own memorandum to the John Innes Council, in which he pointed out that despite the complete disarray of the institution and unreliability of its director, Darlington was "a first rate man" who could be relied upon to direct cytological research. To win the directorship, Darlington had gone behind the back of his benefactor to produce a proposal contingent on Haldane's resignation.

The similarities between the two men help explain Darlington's particular sensitivity to Haldane's flaws. The difference was that as Haldane's loneliness, dramatics, and vengeful feelings toward the establishment led him to embrace Communism, in Darlington the same traits, combined with a highly developed antitotalitarian streak, led him to reject Soviet-style Marxism. Nothing places this opposition in such stark relief as the Lysenko Affair, for just as Darlington was becoming disillusioned with Haldane's science and psychology, Haldane was beginning obstinately and systematically to ignore the terrible news from Russia. Although reports of arrests and purges of scientists were often contradicted by the denials signed by the purged themselves in *Nature* and *Science*, Soviet tyranny was becoming increasingly apparent. Geneticists in close contact with Soviet counterparts were particularly privy to information that put the lie to the party line. When a letter from Hermann J. Muller arrived in March 1937, Haldane's stubbornness became increasingly difficult to endure.

⌒ DEEPLY UNHAPPY IN TEXAS, feeling betrayed by his former teacher Morgan, and suffering from a rocky marriage, Muller recuperated from an attempted suicide, and followed his twin enthusiasms for socialism and eugenics to Leningrad, where in September 1933 he joined Vavilov's institute.[42] A firm conviction that a truly viable and effective eugenic program could only be implemented successfully in a classless society—where the absence of social conventions and barriers would allow scientific specialists to identify true talent, intelligence, and ability—led Muller to view the Soviet Union as the ultimate locus for the creation of a genuinely equitable and progressive society.[43] Unfortunately for him, his timing could not have been worse. The eugenics movement in the Soviet Union had a decided class bias, and after the Nazi ascendancy in Germany, "eugenics" became to an even greater extent a dirty and despised word.[44] Muller was present at the 1936 "discussion" between the geneticists and Lysenkoites, and was well aware of Lysenko's growing power. All around him, institutes were being shut down and fellow researchers were disappearing. In what was therefore perhaps a final and desperate bid, Muller sent his book *Out of Night* directly to Stalin himself in the hope that he might take to his progressive eugenic ideas. The gamble failed miserably, however, and Muller soon realized that his Soviet dream would have to come to an abrupt end. In March 1937, confused and frightened, practically fleeing on a train destined for Spain, he penned a letter to Huxley, which he asked be passed on to Darlington and to Edinburgh geneticist Francis Albert Eley Crew. Here, for the first time, he spoke of the disappearances and of the terror:

> I have been asked to write private letters to my geneticist friends abroad, telling them that things are going well again for genetics in the U.S.S.R. and asking them to use their influence with the international committee, to have the congress held there. . . . While I will not do that, neither will I do the opposite—tell the truth to the world about the situation there. It would be too damaging to the opinion of scientists about the U.S.S.R. . . . I do not want to become an agent of anti-Soviet propaganda. While what I have told you are only facts, they cannot be appraised without taking them in connection with *favorable* facts concerning the U.S.S.R. and its system. I know you are familiar with these, and so I can tell you the above facts, but the *mass of people can hardly see two facts at a time*, and so these facts might have a dangerous effect on them. When they are finally given out it must be in just the right setting.[45]

Underlining the last sentence with a red pencil, Darlington wrote in the margin "the Marxist trapped!" Indeed Muller, who argued with force for a rigorously materialist and reductionist particulate theory of heredity, showing how Lenin's doctrines actually *supported* such an interpretation,[46] had now witnessed firsthand the cruel suppression of this science and its practitioners and could do little to help save run. Muller wrote that he would return to Russia after his time in Spain with the medical unit he was to work with, for he feared Vavilov would "suffer considerably" otherwise, "for he's always supported me." He planned to return for a few short months of organization and farewells, before leaving for good. By December 1937 Muller was already in Edinburgh in a post Crew had secured for him, and wrote to Darlington: "I abhor dictation from either Hitler or Stalin, and regard the system known as democratic as a much better basis for the building of a more ideal society than totalitarian systems run by lower, hungry, and vindictive ignoramuses."[47] Darlington's proposal to the John Innes Council suggesting that the institution do its utmost to secure Muller as Haldane's replacement had failed.

⁓ IT INFURIATED Darlington to think that Haldane remained

singularly unable both publicly and privately to renounce Soviet tyranny when a source as reliable as Muller reported the extent of the
horror.[48] Solomon Levit, another close friend of Haldane's, had been
shot. What more evidence did he need? As Haldane increasingly
dodged the political aspects of Lysenkoism, shifting the issue instead
onto Britain, Darlington's disillusion with his onetime mentor, and
with Soviet Marxism, became complete. Following the Russian withdrawal of its delegation to the 1939 Seventh International Congress of
Genetics in Edinburgh, in reply to an invitation to chair a talk on
the recent 1939 "discussion" in Moscow at the Society for Cultural
Relations etween the Peoples of the British Commonwealth and the
U.S.S.R. in September 1940, Darlington wrote curtly, "in my opinion
the discussion has been going on there for five years too long, and I can
see no purpose in any continuation of it here."[49]

For Darlington, Haldane's fall mirrored the Soviet Union's fall, as it
signed a pact with the German devil in 1939. In 1936 Haldane had
gallantly defended his profession against the anti-genetic MacBride,
reminding him that "since the time of Galileo authority has counted for
less in science than experiment."[50] Now, like Muller before him, he was
the Communist, trapped, except without the integrity, or the insight to
know. Darlington felt he knew better. And his reasons were sound.

When Sidney Harland sent him a copy of a letter Vavilov had written
him in the fall of 1938, Darlington could read distress between the
lines. Vavilov was cautiously hinting at the increasingly repressive conditions. "The problem of interspecific hybridization in cotton [a
Lysenko claim with no theoretical justification]," he wrote, "seems to
be of much more interest now than it did before. At Tashkent there are
many changes in personnel."[51] Still, Darlington could not have known
the full truth; that Vavilov was in fact already a condemned man, that
the NKVD (People's Commissariat for Internal Affairs) was simply
waiting for an opportune moment to pounce.[52] By early 1940, when
correspondence between Russian and Western scientists had all but
ceased, Darlington wrote to Vavilov expressing his interest to have his
great book, *Theoretical Bases of Plant Breeding*, translated into English
and published in Britain, indicating that he understood the noose was
quickly tightening around his Russian friend's neck.[53] Vavilov's reply,
assuring Darlington that he would "finish it [the translation] soon," was
written less than six weeks before his arrest in the Ukraine, on the
trumped-up charge of divulging state secrets to Western scientists.[54]

At the end of October 1945, an International Genetics Conference
organized by the Genetical Society and the British Council met in

London. Darlington, since 1943 the president of the Genetical Society, welcomed the foreign visitors, proudly proclaiming that it had taken a full nine years after World War I to arrange an international gathering; on this occasion it had been possible in nine weeks. The war was over. Life, and scientific research, would finally resume its regular course. The conference was held with great feelings of relief and optimism for the future.[55]

As the organizer, however, Darlington knew not all was well. "In September 1945 I had the privilege as President of the Genetical Society of inviting seven distinguished Russian geneticists to a conference in London," he wrote to the editors of *Nature*. "I received this reply from one of them, N. P. Dubinin:

> Unfortunately it reached me on the 27th of May 1946 so that I couldn't possibly avail myself of the opportunity. I do not lose hope of meeting you in the near future at the next session of your society.

Since I find communication with my Russian colleagues so difficult, especially with my old friends Lewitsky and Karpechenko, may I be permitted to make use of these columns to say how much I, and my colleagues in this country, look forward to the renewal of our scientific collaboration?"[56]

Just months before publishing his piece in *The Nineteenth Century and After* Darlington, whose name was known in Russia ("Pravda have painted a vivid portrait of my personal character"),[57] and who consequently became an address for those in the East eager to get information out, received a letter from a Polish refugee in a recalcitrant camp near Thurso on the northern coast of Scotland. J. S. Alexandrowicz had shared for a number of years a tent in Iraq with a Polish lieutenant who had been in prison with Vavilov, and related reliable information on his worsening predicament.[58] Darlington had already written Vavilov's obituary, and was under no illusions whatsoever as to what was happening in the U.S.S.R.

⌒ IN JANUARY 1948 DARLINGTON ACCEPTED an invitation to become president of the Rationalist Press Association (RPA), which had emerged from its nineteenth-century roots as an organization of free-thinkers and reformers. By now, Soviet Communism had become to his mind such an evil, that just over a year later he felt obliged to resign. Explaining his untimely departure, Darlington wrote:

> God is no longer the war cry. It is from non-Christian sources that superstition is being most potently organised for most potent political ends. The perversion of culture and suppression of science which has been attempted by one political dictator after another is now being attempted on a larger scale than ever by Communism. In these circumstances, for the RPA to confine itself to attacking religious agencies while pretending to be unaware to the existence and purposes of Communism is simply to serve communist ends. Instead of being means of liberating humanity we are in the process of becoming tolls of those who would enslave it.[59]

By this time it had already been years since Darlington and Haldane had exchanged any words. Their great friendship had long since died.

MANY INITIALLY SYMPATHETIC TO the Soviet socialist experiment (though some later than others) lost faith in the enterprise as the tyrannies of the Russian regime became apparent.[60] Darlington was not unique among scientists in that respect. In the small genetics community with a rapid network of dissemination, it is clear that he did not have more information than his colleagues, his contacts and official positions notwithstanding. So while it may be true that men such as Huxley and Ashby held out, if not in their hopes then in their actions, longer than Darlington, it has no bearing on the question of the response to Lysenko before the August 1948 congress in Moscow. Many turned a blind eye to Soviet abuses in the hope that they would ease the conditions for Russian colleagues at risk under the Lysenko regime. The reaction of the Western genetics community before that time was about *strategy*, not *ideology*. Dobzhansky, to be sure, was just as irate as Darlington at the Soviet suppression of science. Within the non-Marxist scientific community, *how* the West was to react to the Soviet reality was at stake, *not* the reality itself. Darlington's path of disillusion, along which his mentor lay slain, has yet, therefore, to explain the divergence of his reaction. Upon closer examination, a much larger set of issues can be seen to inform his actions. For Darlington, nothing less than the role of science in the shaping of humanity's future was at stake.

～ 10

Science in a Changing World

ONE INITIAL AND IMPORTANT SOLUTION TO the problem of Darlington's nonconformity lay deep in the idiosyncrasies and vagaries of his character. In his youth he had formed a fierce aversion to orthodoxy; what began at St. Paul's and at Wye as a rebellion against the establishment—the hatred for games, indoctrination, and the tyranny of "fair play"—became a philosophy and *raison d'etre* with the orthodox reaction to *Recent Advances in Cytology*. If there was one thing Darlington felt was worth fighting for, it was freedom of expression, the right to buck the system, to defend unpopular views. He despised all forms of totalitarianism. Bedtime readings for the children were invariably either *Uncle Remus* or *Animal Farm*, with Darlington mimicking the different voices, and singing the revolutionary song "Beasts of England."[1] Darlington's strong antitotalitarian streak, wedded to a potent blend of dismissiveness, impatience, and pugnacity, goes a long way to explain his seemingly reckless outburst against Lysenko after the war. Colleagues, too, were well aware of Darlington's short-fused temper: in his 1937 letter to Huxley, Muller made a point of entreating Darlington not to air his grievances publicly. This is not, however, the whole story, for the rational did play a part, alongside the impetuous, in Darlington's reaction. What, then, were these potently organized "superstitions" of which Darlington spoke in his resignation letter to the RPA, and how had they contributed to his disillusion with the Soviet Union? To answer these questions, more than just the internal strategies of the Western genetics community and the lost friendship with Haldane must be considered. In fact, despite what has been presented thus far, Darlington was not the only, nor even the first, British scientist to attack the Russian government over the Lysenko Affair.

⌒ FOLLOWING THE NOVEMBER 1948 BBC BROADCAST, eugenicist Carlos Paton Blacker wrote Darlington a letter objecting to the comment he had made concerning the indifference of British scientists to the genetics controversy:

> But have our scientists been indifferent? Public attention was first drawn to the full significance of the Soviet genetics controversy in February 1943, by Professor Michael Polanyi and Dr. J. R. Baker, at a meeting of the Institute of Physics. Baker was most bitterly attacked by impassioned communists in the audience who accused him, among other things, of base ingratitude to an ally who was winning the war for us. Shortly afterwards Baker reopened the question in a public discussion with Haldane. . . . It required courage to speak thus at that period. Baker's wife is Russian and it was she who translated the documents on which Baker founded his case. . . . The matter was kept before the public through the activities of the Society for Freedom in Science.[2]

Darlington quickly replied that his reference was to those who had attacked Baker, and who were now attacking him. Polanyi and Baker deserved all the credit for their actions. "I did not raise it earlier myself," Darlington explained, "because I felt it dangerous to speak until I knew that all my friends in Soviet Russia had already been murdered."[3]

⌒ THE SOCIETY FOR FREEDOM IN SCIENCE (SFS) was an international organization, founded by Oxford zoologist John Randal Baker, Oxford ecologist and botanist Arthur George Tansley, and Michael Polanyi in Britain in 1940. Its constitution outlined a belief in science as an autonomous, not necessarily utilitarian, pursuit that can only flourish under conditions of freedom—especially freedom from any political ideology.

This seemingly prudent and uncontroversial position had been formulated as a "counterblast" to a movement the leaders of the SFS saw as endangering the very fabric of science and society.[4] In the 1930s, economic depression, the rise of fascism, and an increasing call for the mobilization of the nation's defenses for the possibility of war conjoined in Britain to raise the political consciousness of many previously apolitical scientists. As in the United States and France, the failure of science to stage an economic recovery through effective utilization of technology was blatant. But peculiarly, in Britain government remained wedded to aristocratic cultural forms that were often hostile to industrial and scientific aims. Scientists were denied access to the inner circles of political life, and were often demeaned, discounted, or demonized in political speeches, humanist discourse, and in more popular mainstream works. There was a significant lack of interest shown by most professional politicians in the social ramifications of scientific research.[5] The Hunger Marches at home, the Nazi perversion of biology abroad, and the long history of ruling-class disdain for science and scientists had together brought the profession William Whewell had coined almost a century earlier under severe attack.

In these adverse circumstances, compounded by the meager funding allocated for scientific research and development, a "social relations of science movement" arose, becoming a dominant force on the British intellectual landscape from 1932 to 1945.[6] It is not difficult to see how, with its stress on the importance of science to society, its call for a shift in national funding priorities, and demand for a greater and more careful utilization of science for the good of the nation, the movement became attractive to many practicing scientists. Indeed, in the early 1930s many British scientists became attracted to the Soviet planning doctrine, predicated as it was on the harnessing of science for the good of the Union and its people. The first Soviet five-year plan, Julian Huxley observed in 1932, was "a symptom of a new spirit, the spirit of science introduced into politics and industry."[7] If science was not a disembodied activity—a detached body of knowledge to which great minds added and improved upon—but was rather intimately linked to human history and destiny, the argument went, then Britain would be committing a gross injustice were she not to capitalize on the manifest opportunities to harness it to improve the welfare of her citizens. Centralized planning based on scientific principles and knowledge was not only prudent and humane; it was vital for national rebuilding and fortification. It was the way toward a more prosperous, brighter, and more secure future.

Increasingly, Stalin's Russia became the model for inspiration. "Our present rulers and those who support them," Haldane wrote in 1927, "will be well advised explicitly to imitate the extremely capable

Bolshevik leaders, and adopt an experimental method."[8] Ten years later the Webbs, who had visited the U.S.S.R. like Haldane and like other Fabians such as George Bernard Shaw and H. G. Wells, proclaimed: "Unlike groups of landed proprietors, lawyers, merchants, bureaucrats, soldiers, and journalists in command of most other states, the administrators in the Moscow Kremlin genuinely believe in their professed faith; and their professed faith is science."[9] Indeed, no left-wing movement in the West ever became so obsessed about the scientific road to socialism as the one in Britain.[10] Twice an "outsider"—on one account due to its politics, on the other because of its profession—the scientist-led Left adopted science as its own distinctive cultural symbol. And, when in the 1930s the Communist Party of Great Britain loosened its sectarian attitudes and began actively recruiting students, artists, academics, and professionals, Marxism became a vehicle through which the scientific Left sought to advance its professional agenda. In that decade and in the 1940s, over half of the editorial board of the *Modern Quarterly*, Britain's most distinguished journal of Marxist thought, were practicing scientists. Haldane, physicist P. M. S. Blackett, Needham, mathematician Hyman Levi, Lancelot Hogben, and J. D. Bernal were interested in Marxism *as a science*; they were concerned with the social relations of science, and saw the reconstruction of science and the reconstruction of society as interdependent tasks.[11] J. G. Crowther summed up this position in his 1936 book *Soviet Science*: "The social philosophy of Western Europe has roots deep in a pre-technical era. The social philosophy of Soviet Russia, dialectical materialism, is founded on modern physical and biological investigations. . . . [A] social system established according to the principles of this philosophy must be founded on technology and science, and the scientific mode of thought must permeate the intellectual activity of its governors."[12]

〜 THE SOCIAL RELATIONS of science movement, however, were far from monolithic. Reformists—notably represented by Huxley, Cambridge biochemist Frederick Gowland Hopkins, Sir Daniel Hall, and *Nature* editor Sir Richard Gregory—sought to secure a greater voice for science in the existing political order. Radicals, on the other hand—led by Bernal, Levy, Hogben, Haldane, and Blackett— fought to revolutionize the political order, believing that the fullest and most humane use of science could only be carried out in a socialist polity. Nowhere were the divisions between the two groups more apparent than in their respective responses to Nazi Germany and the Soviet

Union. For while the Radicals championed a fruitful interaction between science and society, the Reformists, predisposed to pessimism about the effects of systematic social controls on their profession, were much less enthusiastic. To them, the integration of German and Russian scientists into the new political regimes in those countries, and the zealous nationalism characteristic of their attitude, was frightening. Such sentiment, Reformists believed, stood diametrically opposed to the spirit of the international scientific community.[13] Radicals, on the other hand, had attacked German scientists for their apolitical stance before 1933. Since scientists could never fully insulate themselves from social pressures or concerns, they felt pressure, internal as well as external, to align themselves with those political forces best equipped and most committed to the advancement of science for the benefit of the entire community. German scientists had only themselves to blame, for had they stood up to Hitler, Nazism would have been stifled and their subjugation to it, avoided.[14] Russia, by contrast, was a model of the prudent application of science for the betterment of mankind. Beholden to her scientists and dedicated to the use of scientific resources to solve important social and economic problems she was bound to succeed.[15] In 1939 Bernal published *The Social Function of Science*, a book that became the Left's manifesto for the reorganization of society along scientific lines.[16] Here he argued that the socialist transformation of Britain had become a scientific necessity and virtually a historic inevitability. As new scientific knowledge led to new technological advances, the economic field would be altered, and the distribution of wealth and labor, and consequently social values, transformed. The pursuit of science itself not only stood above the class war, but also was actually opposed, in the long run, to the preservation of a class system. The march of science, not the actions of the working class, would ultimately ensure the downfall of capitalism, whose structural limitations precluded the effective and humane employment of science.

Just as this book was being championed by the British Left, the Reformist section of the social relations of science movement was hardening its opposition. The growing awareness of the "extent to which political organizations can affect the direction of scientific research, and even frustrate its efforts"[17] served only to diminish their support for, and augment their suspicion of, state intervention in scientific practice. Science and politics, they argued, shouldn't mix.

Then came the Nazi invasion of Poland in 1939. At a time of national crisis, government and science were thrown, the one reluctantly but voraciously, the other excitedly but cautiously, into each other's arms.

Bernal's statistic—that only 0.1 percent of the GNP in Britain in 1939 was devoted to scientific research and development—could no longer be ignored with either equanimity or impunity.[18] Science, and her technological children—bombs, radar, encryption, and war machinery—were desperately needed, ruling-class hostilities aside. A scientific advisory committee was quickly set up by the British Government. Scientists of every description were recruited en masse and served as a cornerstone of the British war effort.[19] Under such conditions, political squabbling was duly set aside. "Obviously, the war overshadowed everything else," remembered primatologist and expert on the effects of bombs on people Solly Zuckerman.[20] The celebrated social dining club Zuckerman founded in 1931 and resurrected in 1939 met expressly to mobilize scientists and the country's scientific resources in the fight against Hitler's Germany, *regardless* of the rife ideological divisions among its members. The club's very name—Tots and Quots—expressed this political variation, taken as it was from the Latin *'quot homines, tot sententiae.'*[21]

Wartime scientists coalesced around the consensus, as Zuckerman put it, that "a general scientific point of view is foreign to the direction of the country, and in particular to its administrators. If this defect is to be remedied, men of science themselves must show the way."[22] One way to do this was to begin by correcting the popular view of scientists as magicians at best—untidy, absent-minded, slightly comic miracle workers with "magnetic eyes"—and at worst Mephistopheles; to establish an everyday relationship with the public, and convince it of the need to organize a scientific approach to current problems. *Science in War*, anonymously written by members of the club and published in the summer of 1940, was designed to serve that end. The establishment of scientific advisory groups to the government, and the division of "Operational Research" was yet another way, and one in which the club left a significant mark.[23]

As war continued, *Nature*, which had previously stressed the need for freedom in science as a crucial check to the inherent dangers resulting from its social function, became an unhedged ally of the movement for the social relations of science. The journal's new editor, Lionel Brimble, wrote a number of editorials in which he rejected the distinction between pure and applied science, and argued that to function at all, science must be properly planned.[24] Under the Bernalist influence, a Marxist view of science inspired the activities of the Association of

Scientific Workers and informed its journal, *The Scientific Worker*. Here socialist scientist Blackett declared: "The main lines of its [science's] growth are broadly governed by urgent human needs of the time, arising out of the way in which men gain or seek to gain their living."[25] Compelled by circumstance to unite, Radicals and Reformists came together to champion the cause of planned science. "How is it that in recent years science has come to the front in such strength?" Nobel laureate crystallographer Sir William Bragg asked his BBC audience in 1942. Because "Knowledge is power," he answered, and government had finally caught on.[26] It was high time for the scientist-led Left in Britain.

THESE WERE THE DEVELOPMENTS that prompted Baker, a conservative, even reactionary eugenist, to found the Society for Freedom in Science, the constitution of which was described above.[27] In his 1945 book *Science and the Planned State*, Baker argued that science does not exist solely to serve man's material wants, that radical central planning was damaging to scientific progress, and that totalitarian forms of government were least in accord with scientific principles.[28] Polanyi, a liberal who had left the Kaiser Wilhelm Institute in 1933 in protest against the Nazis, likewise argued for the mutual dependence of scientific progress and liberal democracy.[29] On a visit to Russia in 1935, he had been shocked to hear Nikolai Bukharin argue that science under socialism must cease to be carried out for its own sake, and be firmly harnessed to the needs of the five-year plan.[30] He was now committed to expending his energies on combating the trend to define science in terms of its social utility. Together, the conservative eugenist and liberal émigré fought to counteract the movement for state oversight of university research, insisting that researchers should be provided with adequate funds and modern facilities, and be given free rein to pursue their investigations.

The SFS's seemingly prudent and innocuous constitution was therefore far from innocent. Socialist planning and liberal democracy represented two irreconcilable worldviews, vying over the control of a political landscape in upheaval as science assumed a greater social role. Significantly, the SFS seized on the uncritical admiration for the Soviet Union voiced by leftist leaders of the social relations of science movement. Here, they argued, was a prime example of the dangers to freedom resulting from government involvement and direction of science. "It might seem churlish to criticize the institutions of our ally," Baker wrote in 1945, referring to Soviet science and the Lysenko Affair,

> when we all know that her action in self defence has made our task in the war so much lighter. . . . Nevertheless, scientists who are supporters of the Soviet political regime have created a situation that makes it desirable, indeed necessary, that criticism should be offered. . . . Truth cannot come from one-sided argument. Nothing but freedom of speech and publication can reveal it. It cannot be right to praise the science of another country simply because that country is our ally. Good science is to be respected wherever it comes from; bad science, neglected and condemned.[31]

During the war, Baker and Polanyi's critical voice was singular. Commitment to freedom at the expense of detachment, and condemnation of the West's greatest ally, as Blacker had noted in his letter to Darlington, were not popular positions at a time of war. Brimble spoke for many when he said: "There are those who deliberately hold themselves aloof from any effect their science may have on human society; they are, to say the least of it, selfish, though how often has one heard them claim that they are the only champions of scientific freedom."[32] J. G. Crowther, then secretary of the Scientific Section of the British Council, who had himself defined science as the system of behavior by which man acquires mastery over his environment, intimated that Baker's philosophy would have denied Britain victory in the war.[33] These were harsh accusations in war times.

But Baker and Polanyi were political outsiders. At a time when science was enjoying newly earned power and respect, their calls for caution were drowned out. Compounding this was the willingness to overlook certain rumors of Soviet breaching of democratic process in relation to science so as not to threaten East-West cooperation. To some degree, the hints of purges were only rumors after all: no one knew for sure. By the time the war ended, the membership in the Parliamentary and Scientific Committee had grown from fourteen in 1938 to 146 in 1946.[34] Labour was firmly in power, the membership of both the Communist Party and the Association of Scientific Workers was at an all-time high, and a new, widely shared consensus about the importance of science in national life had been born. Most importantly, however, the government had adopted a new outlook on science and scientists: expenditures for research and the number of practitioners in the corridors of power increased in tandem. The Division for the Social and International Relations of Science (DSIRS), which the British Association had created in 1938, gradually began to lose its appeal, and eventually withered away in the 1950s. This was not for lack of relevance, however. Indeed, Bernalist ideas continued to have influence precisely because modern science was not just a reflection of free-market activity, as Polanyi and Baker had claimed.[35] In later years, Harold Wilson's government was to take up many of Bernal's ideas, and to create a Ministry of Technology that would give birth to the British computer industry. For now, though, the crusading days of the movement for the social relations of science were over.

Some of the effects of the government's adoption of science were less encouraging. Science had been militarized to a great extent, and a top-bottom administrative approach was now led by a select group of scientists "in" with the military, including Zuckerman and other members of the Tots and Quots Club. Atomic weapons test programs were beginning by the late 1940s to generate political fallout and a further wave of distrust for science. Humanists buckled down to fight what they took to be the dangerous cultural and technical imperialism of the Bernalist creed. Nevertheless, the in-fight over freedom was over. In July 1947, the SFS reported that "very little is heard nowadays about the movement against the freedom of science."[36] This was indeed true, for it quickly became evident after the war that the freedom of scientists was in no real danger. In a publicized piece in *Nature*, the Association of Scientific Workers heartily accepted the SFS's five principles. It was at this point that the debate shifted more to the plight of fellow scientific workers in the Soviet Union.[37]

⌒ ON HIS VISIT to Russia in 1934, Darlington had been greatly impressed by the role the Soviet government had taken in encouraging scientific research and education. He especially noted the part played by Soviet radio in the dissemination of scientific knowledge, and began himself to participate regularly in science education programs for the masses on the BBC and in other fora.[38] Through his friendship with Haldane, his personal experiences, and scientific work, Darlington had grown to reject the demarcation between "pure" and "applied" science.[39] In fact, he became resolved to champion their inherent unity and necessary unification. Back in England, the short-tempered controversialist became increasingly frustrated with the disparate and disconnected nature of the scientific enterprise. In modern times, the use of science for the solution of social problems was at base a matter of political will and education.[40] The Russians had understood this. "The Soviet Government," Darlington told his radio listeners, "has instilled into its people the need for approaching all the problems of life in a scientific way without regard to tradition, that is without regard to the beliefs handed down from ancestors living in the world before science. To them the increase of food supply and the control of the birth-rate are equally scientific problems which must be considered together and resolved together."[41] Not so in England. This "novel" method was to the English "repugnant." In stark contrast to Russia, Darlington complained, "the administrator, the editor, the judge, the bishop, the diplomat, the military commander, the governor of the school or corporation (such as the BBC), the cabinet minister, the company director (even in the chemical industry), and the headmaster of the public school, came to be chosen for their high responsibilities in about ninety-nine cases out of a hundred by a process which excluded the chance of their having been given an insight into scientific discovery, an interest in its possibilities, or a respect for its use."[42]

"Research must be seen not as an end unto itself," Darlington told an audience at Caxton Hall, Westminster in January 1943, in the presence of P. M. S. Blackett and Hyman Levy, "but as providing the universal raw material for the growth of industry and government."[43] Two months later, at a conference organized by the DSIRS in London, Darlington offered a further exposition of his views:

> Scientists used to consider that science was their private concern, just as landlords considered that land was their private concern. But the world is much less private today, and this is particularly true of science. The scope and purpose of science have grown as its power has increased, and we are now compelled to fit it with a larger size of definition. *Science as we now have it, is nothing less than the arrangement of knowledge for the benefit of mankind.* This view is based not on a generousity of sentiment but rather on an economy of sentiment, on the good old principle of Occam that you should not waste your ideas, or any one else's ideas either. . . . The man of science and the man of government are each largely ignorant of the other's job. In recent years the scientist has shown a laudable amendment. He has begun to take an interest in politics and has made some progress with his lessons. But what of the other side? The picture is not hopeful (emphasis added).[44]

Darlington's affiliation was not in question: he was a strong friend and active member of the movement for the social relations of science. When it came to exhorting the Russian adoption of science as a tool for social reconstruction, Darlington's praise equaled that of Hogben, Huxley, Haldane, Crowther, Levy, and Bernal. He was a staunch advocate of planning.

Like these men, Darlington was a scientific humanist, a champion of the worldview that science, *pursued properly*, offers the greatest hope for mankind.[45] If a premium were put on progress—on maximizing the efforts of schools, universities, all branches of government, including agriculture, health, industry, and defense—as it well should be, scientific humanists argued, science and the scientific enterprise would need to be harnessed to ensure that planning be prudent and effective. This necessarily entailed the reformulation of government's, and the general public's, relationship to science and to scientists. There was no other viable path to tread: rational planning was necessary for the making and maintenance of a civil society. This was the reason why Darlington spoke so often and so forcefully to the nation on radio, and at open public talks. It was the reason why, usually allergic to clubs and soirées, he joined Zuckerman's Tots and Quots, and was one of the more involved contributors to *Science in War*. Significantly, however, it underlay Darlington's violent reaction to Lysenkoism and his contempt for his colleagues' deliberate strategy after the war. Where Darlington diverged from both liberal and radical scientific humanism was in his understanding of "proper utilization." What did it mean for government to use science properly?

Darlington's disillusion with Haldane, his friendship with Muller, and the conclusions deduced from information from Russian colleagues and informants, both personally and officially, had all led him to reject Soviet Marxism. But Darlington was also beginning to reach *scientific* conclusions that were distancing him from the Marxist interpretation of history. A growing belief in the biological diversity of man, and disbelief in the possibility of changing his nature by anything other than biological means were coming into direct conflict with both Marxist egalitarianism and environmentalism. Slowly but surely, Darlington had become convinced that socialism—Soviet style—was in fact incompatible with genetics. That was indeed the meaning of the rationalization to which Dobzhansky had objected in Darlington's anti-Soviet piece in the *Nineteenth Century and After*. For Darlington had become a *racialist* scientific humanist, convinced both that genes determine

Darlington, circa 1952. Photo courtesy of Barry Juniper.

behavior, skill, and intelligence, *and* that if scientifically guided social planning were to ignore the inherent biological differences in capacity and function between people and races, it was bound to fail.

Haldane's *scientific* position, like that of Muller, had in fact been very similar. He enjoyed reminding his audiences that "The dogma of human equality is no part of communism. . . . [T]he formula of communism: 'from each according to his ability, to each according to his needs' would be nonsense if abilities were equal."[46] Like Darlington, Haldane believed that social classes represent distinctive biological units. But whereas the mentor increasingly understood class to be a dysgenic social construct, because environment did not necessarily mirror genetics, his student took it to be a consequence of the invisible hand of nature. People made it to the upper classes because their genes brought them there; others remained faltering because they lacked the abilities requisite for advance. Ironically, the irate youth reacting against Huxley's 1926 eugenic talk was now convinced of the biological basis of class, and as Haldane increasingly moderated his eugenic views, Darlington intensified his own, vehemently rejecting Marxism in the process.

By 1945 it became impossible for Darlington to accept what he viewed as either short-term liberal policies or the splendid Marxist utopian solutions of the scientific Communists. Marxism taught how classes of society, by pursuing their interests politically, intellectually, and economically, and in conflict with one another, were responsible for the evolution of society. Like enlightened scientific humanism, it remained silent, however, about the genetic processes responsible for their survival or decay, achievement or failure. Darlington would not. "For the fundamental problem of government," he told his audience at the April 20, 1948 Conway Memorial Lecture in London, "is one that can be treated by exact biological methods. *It is the problem of the character and causation of the differences that exist among men, among the races, classes, and individuals which compose mankind*" (emphasis added).[47] Marx preached love between nations and hatred between classes. But nations and classes equally owed their existence, Darlington argued, by virtue of genetic differences among races. If governments were to prescribe social policy prudently, they would have to admit these biological truths.

Like Lord Acton before him, Darlington came to believe that equality was the foundation of tyranny, a mental placebo of poisonous social effect. The great irony, to his mind, was that in the first serious materialist social theory, the very materiality of man had been left out. Marx and Engels had dreamt of establishing a materialistic account of the whole of nature, but could not: genetics had not yet been developed. When Western science finally filled up the yawning gap that existed in the knowledge at the time of Darwin and Marx, the theory of equality had already been ensconced in the foundations of Soviet Marxism, in direct contradiction to the materialism it had set out to expose. Although Darwin's theory of evolution by natural selection depended on the existence of inborn differences between individuals, the application of the theory to man became inadmissible in the Soviet Union. The closing of the Medico-Biological Institute in Moscow and suppression of its 1,000-pair monozygotic twin study (in which heredity was shown to be decisively more determinative than environment), the persecution of geneticists, and the appointment of a quack to head agricultural research were the latest manifestations of the conscious and official subjugation of fundamental biological research to a petrified Marxism.[48]

Like many others, Darlington was concerned that the adoption of an absolute science by an absolute state was a recipe for the return to the backwaters of the Middle Ages reign of Albertus Magnus; Soviet totalitarianism spelled disaster for free inquiry and prosperity. Above and beyond this, Darlington was concerned with a transgression that would prove harder to remedy. There was nothing wrong with the principle of planning; there was everything wrong, however, with the assumptions the Soviets had adopted with which to direct it. For even within a totalitarian regime, if the Russian people had sound plant breeding, they would have more wheat. If they had a sound investigation of human heredity, their medical and educational services would profit.

What was true for the Kremlin was true for Whitehall, too. Darlington surely did not awaken any argument from his fellow scientific humanists as he exposed and ridiculed the continued disregard for the lessons of fundamental scientific research perpetrated by government, industry, and the academy in Britain. The failure to employ modern statistics in the formulation of policy, the separation of medical from fundamental research (especially in cancer), the failure of the Agriculture Ministry to use artificial insemination techniques to improve livestock and milk produce, the resistance of the Forestry Commission to the use of genetic methods for seed production, and the failure of the universities to introduce cellular biology and genetics into a new departmental structure might have been grievances particular to Darlington, but they would no doubt have enjoyed the support of many activist scientists.[49] Where Darlington parted with such company was in a growing, and increasingly intrepid, determinism when it came to genetics and man. English popular sentiment and social policy, he bemoaned, inconsistently ascribed success to the law of heredity, and failure to the accident of environment. This was not so far as might seem from those assumptions steering the Soviet Union into the mire of Lysenkoism. If they were to expect any progress, the people and government of both countries would have to internalize genetics—the materialistic biological science that taught that different organisms resulted from different heredities and were selected by nature for different tasks—and use it as a guide.

The SFS had been warning that he who pays the piper calls the tune, but Darlington was convinced that ultimately it was the tune, that genuine reality the discovery and description of which was the call of the scientist, that determines success or failure, progress or stagnation, happiness or misery. He applauded the scientific point of view adopted by the Russians, embraced dialectical materialism as a valuable heuristic, and championed the doctrine of planning as a necessary way forward. Indeed, Darlington's politics were in no way reactionary, not *yet* at least: in 1944 Sir William Beveridge had personally written to him asking that he run as the independent progressive candidate for the University of London seat in parliament (a Mrs. Mary Stocks was ultimately chosen in his stead).[50] Nevertheless, because Darlington had become convinced that "We can now control and predict over a wide range the ineradicable inborn differences which determine the varying qualities of the individuals, families, classes, and races of mankind"[51] it became imperative for him to attack the Soviet Union.

Why was this so? The political landscape of the time included radicals and liberals—both supporters, to varying degrees, of the principle of scientific planning—and the "liberal" alternative, the SFS, staunchly opposed. During the war Polanyi and Baker's attack on the Soviet restriction of freedom in science, bound as it was to their calls for the disentanglement of research from immediate utility, was looked upon as dangerously detached and self-interested. In the excitement of war and growth of science ensuing therefrom, attacks on the U.S.S.R. were marginalized. And yet after 1945, within the scientific community, the SFS continued to argue for the incompatibility of socialism and free scientific inquiry, and criticize the Soviet purges. Although the movement for the social relations of science was now waning, this remained the case *as a direct consequence of the strategy of silence* adopted by the scientific mainstream, which had aimed to help the struggling geneticists in Russia. The overall effect of this situation, however, was that all champions of planning publicly ignored the Soviet subordination of science to politics, while the outspoken critics of Russia were the leaders of the movement to prevent government control of research. To make things worse, such visible champions of the leftist scientific cause as Bernal and Haldane had in fact bent over backward to defend Lysenko and the rise of his totalitarian, profoundly *anti-genetic* scientific program! This was ominous, Darlington thought, for a nation still negotiating the character of its relationship to science; still fighting over whether, how much, and in what way science could be applied to problems of government and social welfare.[52] "I am doing my best to discredit Communism amongst scientists," he told a London audience in December 1948, "because I feel it is a deadly danger to the country at the moment."[53]

Like George Orwell, Darlington was probably too self-willed and too egotistic ever to be a convinced Communist. Nevertheless, it is indeed an irony deserving of his own pen that Orwell, a leader in the humanist fight against the perceived Bernalist technocratic plan for the future of Britain, had first helped Darlington publish his attack on the Soviets.[54] For Orwell also believed that a new kind of neo-liberalism, a blend of social conservatism and technophilia, with the second lending an air of respectability to the first, was just as bad, if not worse, than the pure radical leftist strain.[55] It was precisely for these reasons that Darlington, an advocate of scientific planning, but one predicated on the principle of inequality between men and therefore giving rise to radically different social solutions than those proposed by the Left, felt compelled to raise his voice.

~ PUBLICLY LASHING OUT AT THE PRESIDENT of the Royal Society shows the extent to which Darlington thought science needed to inform politics, and the irascible manner in which he went about making his point. In December 1941 Sir Henry Dale had commented to the press that British scientists might not preserve their freedom of research (a freedom denied them elsewhere) if they were to urge scientific evidence in favor of "any special political doctrine." Darlington took Dale, and the *Manchester Guardian* and *Times*, which applauded his words of council, to task in his public 1948 Conway Memorial Lecture, intimating that by "any special political doctrine" Dale meant one not held by the government of the time. "In other words," Darlington lamented, "In England scientific research was to be regarded as on sufferance to-day just as much as it had been in the past. Those who wished to pursue it must deny themselves the political privileges of the manual worker."[56] Dale immediately wrote Darlingon an apologetic letter, in which he claimed to have been misunderstood. "My warning was against the danger of allowing scientific truth to be compromised by any kind of extraneous influence—political, religious, social etc." Darlington replied—instructively—that he had understood Dale's meaning *all along*, but that since it was expressed in such a way that it was applauded by non-scientists, he had to react. "In the fields in which I work, scientific discovery and its application are hopelessly frustrated by the incompetence, ignorance, and downright hostility of civil servants and politicians," he wrote. "This commits and perhaps distorts my point of view!"[57]

In his obstinate and absolute way, Darlington held strongly to the belief that social planning based on scientific research would become an imperative for the future of mankind. "It is no longer sufficient," he told his London audience, "to have our systems of research controlled by authorities, some of which are indifferent to the result and some anxious to avoid any result at all. . . . How are we to get anyone to notice it? We need a Ministry of Disturbance, a regulated source of annoyance; a destroyer of routine, an underminer of complacency, an *enfante terrible*."[58] Because Darlington was convinced that not only Russia and Britain, and their respective scientific establishments and governments, but also humanity as a whole were going to have to face up in the coming years to the lessons of genetics, which would no doubt be difficult to swallow,[59] and because he believed this battle to be absolutely crucial for the welfare of mankind, he took it upon himself to be this *enfant terrible*. Given that it was a controversy concerning planning, politics, and genetics, the Lysenko Affair was as good a place as any to start.

⌣ IN THE SUMMER OF 1950 Darlington wrote to the honorary treasurer of the Association of Scientific Workers, reminding him that he had discontinued his subscription the previous year. "I found some slight divergence between the course you set towards scientific research—as represented by your official statements, publications, and conference resolutions—and the one I should find suitable," he remarked with characteristic terseness and sarcasm. "If I have overlooked a public condemnation by the Association of the dismissal of Soviet geneticists I shall be prepared to reconsider my decision."[60] Darlington's break with the scientific Left was now complete, and the next big battle, which would occupy him for the rest of his days, was under way. By the time the second edition of *Evolution of Genetic Systems* came out, the Engels citation had been scrapped. In its place were the following two quotations, affirmations of Darlington's now entrenched determinism and antienvironmentalism:

> For if each organism had not its own genetic bodies, how could we with certainty asign to each its mother?
>
> For nothing is born in the body in order that we may be able to use it, but rather, having been born, it begets a use.[61]

The author was Lucretius, the Latin poet-philosopher Darlington's father had taught him to love as a boy.

∼ THE LYSENKO AFFAIR HAD COME AT the height of the Cold War. The Berlin Blockade, the Polish and Hungarian elections, the Chinese Communists' drive toward Shanghai, the emergencies in Malaya and Indo-China, and the increasing anti-Communist feeling in Britain and the United States were all manifestations of the icy divide that was to separate East from West in the years to come. It also came, however, at a time in which science was metamorphosing, and a cultural and political battle was being waged to negotiate its social role. Here politics and science were hopelessly enmeshed. At precisely this point, Darlington became convinced not only that such a reality was unavoidable, but that, on the contrary, it was necessary. In parallel, his views about the *kind* of scientific knowledge that would have to be harnessed by government sharpened. As he quickly noticed, these ideas would lead him down a dangerous and difficult road, but one that he would make his home for the rest of his days.

Like others of his generation, Darlington's path from Marxist sympathizer to castigator cost him not only friendships but innocence, too. Unlike some, however, who found themselves relinquishing faith in a particular political program, Darlington's loss of innocence sunk directly to the depths of human nature itself, and stemmed first and foremost from scientific considerations. The young man walking down the streets of Baghdad and marveling at the glories of human variety had by now reached conclusions about man that were to excite the anger of many. Reacting to his claim that the fundamental problems of government can all be solved by biological methods, Lancelot Hogben suggested that we may recall, with profit, the remarks of Einstein in a letter to Freud:

> How is it possible to control man's mental evolution so as to make him proof against the psychoses of hate and destructiveness? Here I am thinking by no means only of the so-called uncultured masses. Experience proves that it is rather the so-called *intelligentzia* that is most apt to yield to these disasterous collective suggestions, since the intellectual has no direct contact with life in the raw but encounters it in its easiest synthetic form—the printed page.[62]

How was it that Darlington became convinced that the genetic determinism he observed under the microscope in his plants was not only applicable to mankind, but in the long run its only hope? Why did he embrace genetics precisely as Hitler was making eugenics the recipe of the Devil, and all other practitioners were distancing themselves from it like from a criminal half-brother? What connections did Darlington draw between history, biology, and the evolution of society, and its moral order? We turn now to the final chapter of Darlington's story, in search of answers.

~ IV
Man

~ 11

The Conflict of Science and Society

Darlington's 1948 Conway Memorial Lecture, *The Conflict of Science and Society*, was a characteristically brash assault—this time on society and its institutions at large. "In the present lecture," Sir Richard Gregory introduced the speaker,

> he [Darlington] deals with the reactions of academic and other social groups to knowledge of any kind which requires readjustments of old looms to produce acceptable patterns with the warp and woof of newly discovered threads. He is as outspoken and fearless in his denunciation of such obstacles to progressive thought . . . as he has always been to the suppressive attitude often presented to interpretations of new observations in his own biological field. He stands for freedom of scientific thought and expression without control by any authority, as is represented in the spirit of the motto of the Royal Society, *Nullius in verba*.[1]

With a soothing, almost petting prosody, replete with soft S's and punctuated T's, Darlington's gentle, confident voice could have almost masked the rupturous message he stepped up to deliver. "In speaking of the Conflict between Science and Society," he began, "I mean the conflict between Discovery, which I take to be the active principle of science, and Continuity, which is in some measure the necessary condition of society."[2] Scientific discovery, he explained, is often carelessly looked upon as the creation of new knowledge that can be added to the great body of old knowledge. This was true of the strictly trivial discovery. "It is not true of the fundamental discoveries, such as those of the laws of mechanics, of chemical combination, and of evolution, on which scientific advance ultimately depends. These always entail the destruction or disintegration of old knowledge before the new can be created. And it is this destruction, or the fear of it, which arouses the opposition of the well-trained and well-established scientist, as well as of those outside of science whose beliefs the new ideas threaten to disintegrate."[3]

The "conflict of science and society" was not a conflict between the scientist and the layman, Darlington told his audience. It was not even a conflict between any particular material interests or between any particular classes. The "conflict of science and society" was at bottom a conflict between the rival powers of tradition, belief, and security—represented by all the established organs of society—on the one hand, and innovation, doubt, and discovery, on the other. The history of science from Copernicus, Giordano Bruno, and Galileo through Harvey, from Dalton, Waterson, and Joule through Fourier and Darwin, was nothing but a narrative of this age-old discord. New knowledge was always disruptive. It always demanded a price, always commanded change. It was therefore always resisted—first in the mind of the discoverer himself who rightly fears the reaction; then by his colleagues who stand most to lose from the revolutionary knowledge; then by academies, universities, and schools, where it is necessary for those who teach to understand and believe what they teach, and in the interest of the pupils to learn what they are taught, and where new knowledge must break old, and interested divisions; and finally by industry, ministries, and governments, controlled by administrators who secure their offices precisely because they appease a system that resists change. At base, all such relations were manifestations of the same conflict. The purpose of teaching, of tradition, of precedent, and of belief, Darlington argued, is to cover up what ignorance cannot remedy. Yet ignorance is the very foundation of discovery, the very fountain of all that is new.

"New scientific research arises out of the teaching in the Universities," Darlington told a BBC audience following the lecture.

> If it is trivial it arises in loyal discipline. If it is fundamental it arises by outrageous rebellion. But that it arises from what has been taught none can gainsay. And since what is taught has been planned—largely by the generations of the dead—it follows that

our research is proportioned to this same outmoded plan. Freedom to follow this plan is what, for some reason, is usually described as freedom of research. . . . The freedom to teach what has not been taught before has never existed anywhere. It has always had to be fought for, in every place and in every age—and today more than ever before.[4]

Advocacy of mainstream scientific humanism, then, was not enough. Neither was promotion of its handmaiden, the doctrine of planning, sufficient. Darlington was readying himself for something bigger, and far more radical. This would be a battle that he knew full well would draw fierce resistance at all levels of the conflict of science and society. All levels, that is, save one. The discoverer no longer needed any convincing, nor did he suffer from any fears, or doubts. Like a master chess player, he was now setting the board, maneuvering the pieces for attack. "It is no accident," he told the crowd, harking back to the past in service of the future, "that bacteria were first seen under the microscope by a draper, that stratigraphy was first understood by a canal engineer, that oxygen was first isolated by a Unitarian priest, that the theory of infection was first established by a chemist, the theory of heredity by a monastic school teacher, and the theory of evolution by a man who was unfitted to be a university instructor in either botany or zoology."[5] No accident indeed, for the inherent conflict between tradition and innovation meant that interlopers, men with little or no training in the field they have invaded, were often the true agents of novelty, discovery, and change. Darlington, self-fashioned iconoclast and plant nuclear cytologist, was ready to assume his role. Invading sociology, history, philosophy, education, criminal justice, linguistics, anthropology, and economics all at once, he did not seem to blink. "The fundamental problem of government," he told his London audience, "can be treated by exact biological methods. It is the problem of the character and causation of the differences that exist among men, among races, classes, and individuals which compose mankind."[6]

᠕ As the half-century mark neared, Darlington had been peering less and less down the ocular lens of his light microscope, instead focusing his energies and sharpening his pen on a subject infinitely larger, but, to his mind, intimately connected to the minuscule chromosomes he had made a career out of observing. Concerned with the development of a new kind of social genetics, which should interpret language, class, race, society, and their history in genetic terms, Darlington, in each of his remaining decades, produced one-third of what was to become his trilogy on man. Although semipopular in form and manner, *The Facts of Life* (1953), *The Evolution of Man and Society* (1969), and *The Little Universe of Man* (1978) were considered by Darlington to be the intellectual pinnacle of his career. They were the fruit of a lifelong intimacy with what their author took to be the most important determinative factors in the unfolding of the past, present, and future of mankind. In them were contained those ideas entailing the "destruction and disintegration" of old knowledge, of all that society held dear and cherished.

⌒ THE LINK BETWEEN MAN, CULTURE, and biology had captured Darlington's imagination even before he had set foot on the cobblestones of ancient Middle Eastern alleyways during his 1929 journey in search of tulips. Even before he had scrutinized the myriad professional, religious, and ethnic groups, and carefully noted their linguistic and marital behavior, the young botanist had embarked on a quest, a struggle really, to discover the links between the movements of his tiny chromosomes and those of mankind. It all began during a trip to London with the aging Bateson in 1924. On that occasion, the twenty-one-year-old Darlington had heard a talk by Edward Murray East on inbreeding and outbreeding at the Royal Botanical Conference, noting in his diary the impression the American's argument against the crossing of human races with markedly different characteristics had made on him.[7] Man was no doubt a fascinating topic for study, he thought, but soon returned, struggling young scientist that he was, to Bateson's breeding tasks, and to those magical moments when Newton showed him plant chromosomes under the microscope.

Almost three years later, in February 1927, Darlington received a letter inviting him to join the Eugenics Society. "It is widely appreciated among biologists in general," it read, "that the neglect of biological and genetical knowledge in modern legislation may lead to incalculable damage to our own people and those of other civilized communities." The signatories, among others, were Huxley, MacBride, and Fisher.[8] But Darlington was cautious. To his brother, Alfred, he wrote: "I shall join when I know that my influence will be felt but not until then. In the meanwhile, I hope you will do your utmost where you have any opportunity—and it is only medical men who have any opportunity—to

watch any possibility of defects or susceptibilities being hereditary."
Showing an early interest in human heredity, Darlington went on to
suggest to his brother that he pay close attention to any occurrences of
the inheritance of deviations from symmetry, and of mixed eye colors,
and asked that he record instances of color blindness and exceptions to
Mendelian inheritance wherever he could.[9]

Darlington's initial scientific debates were largely waged in his own
mind, and were there becoming inextricably linked to human problems.
The solution he was beginning to work out for the *Oenothera* puzzle he
had first encountered in Berlin, and for the mechanics of meiosis, were
closely paralleled in diary entries on man. "Cultural evolution of which
the family is the unit," he wrote in 1927, "proceeds by the mingling of
families (marriage), and the selection of their joint resources (by the
children) in the same way as organic evolution proceeds by the mingling of its units (chromosomes) and the selection of their products
(segregation)." Going one step further, and translating directly from
what he had been finding in experimental breeding of plants, he added,
"The future of the world will be determined and peopled by descendants of only 1/10 of the living world. On the question of which 1/10
depends the future of the species . . . Man's great hybridity is the key to
his evolutionary progress. Only illogical pride of race prevents me
taking the logical step of marrying a Jewess or a Turk."[10] Indeed, when
Vavilov, on a visit to his old friend Bateson's institution, asked the young
man, "Darlington, what is your philosophy?" he was astounded.[11] For
the handsome Russian had realized what would take Darlington several
years to discover, that his interest in the future of the human species
and the origin of sexual reproduction were twin obsessions, reflections of the organization of the nucleus and its chromosomes. In
Huxley, Haldane, and Vavilov, Darlington had been meeting men for
whom such integrations came naturally, and he quickly began to follow
their leads.

Over the next five years, as he painstakingly observed chromosome
movement and worked out its laws, the idea that the same deterministic
processes that govern outcomes in the breeding of *Oenothera* and
Drosophila could bear too on the breeding of men, and hence on the
shape of human culture, began fermenting in his mind. But how could
one compare a primrose, or a fly, to man? How could free will and
consciousness be accounted for in genetic and cytological terms? Could
class, race, religion, and language have anything at all to do with chiasma formation, recombination, and apomixis? Still unsure, or rather
lacking the confidence to express his certainty, Darlington kept his

developing thoughts to himself. But a burning intellectual desire would not let go of him. He *needed* to make connections. He knew they were there. They must be there! He pressed ahead. Echoing Goethe's dictum that "volition is nought but willing what we have to will," Darlington wrote in his diary in 1933: "A man can surely do what he wills to do, but he cannot determine what he wills." "Racial uniformity means racial stagnation; cultural uniformity means cultural stagnation," he added the following year. By 1937, upon return from a second visit to India, a balanced mix of fatalism and opportunity had come to describe his thoughts. "Life is like a game of chess," Darlington wrote, "chance playing one side, fate playing the other."[12]

Darlington had been reading heavily: histories of peoples and empires long gone, theories of language, archaeological tomes, ethnographies, philosophies, codes of religion, new and old studies on the genetics and behavior of man. He had been talking to his mentor and friend Haldane, who thought that "When the truth about human behavior is discovered, it will probably appear that philosophers of all schools had failed to predict it as completely as they failed to predict Heisenberg's uncertainty principle. Human behavior is a subject for scientific investigation rather than *a priori* pronouncements."[13] All these had informed his thoughts, buried between hundreds of small pages in crimson pocket notebooks, yearly rubber-banded, and discretely stashed away.

Darlington finally made his private views public in 1941. Shortly after his election to the Royal Society, he published a barely noticeable, half-page piece in *Nature* that was to signal a major refocusing of his intellectual energies in the years to come. In his first publication on man, he wrote simply: "Science is concerned with what a man (or a thing) must do, ethics with what he thinks he should do. Until the contrary is proved, therefore, we must suppose ethics to be derivable from science." How this was to be done precisely was a question on which scientific men could not yet be expected to agree. But the historical relationship between ethics and the subject matter of science, the material conditions of society, was surely commonplace. The irreversible succession of changes characteristic of all integrated systems— that which is termed "evolution"—was common to both. Referring to society as a "system," Darlington argued that already during the present war political expediency had led to violent changes in the relations of the individual to society, changes which, to his mind, scientific method could have directed long ago and without any compulsion. These were not merely political, as opposed to ethical, issues; the distinction lapsed, said Darlington, as soon as both are subjected to scientific treatment. "What a man must do and what a man should do are always the same for the man himself at the moment he does it. . . . Science is therefore bound to be the foundation of ethics of the future."[14]

If this somewhat obfuscated message of biological determinism and its necessary political correlates failed to be picked up, or make an impression, it was not long before Darlington detailed, with characteristic audacity, precisely what it was he meant. He had already determined privately both that "the important distinction is not between useful and useless knowledge, but between integrated and disconnected knowledge," and that "the microscopist must believe that by examining what is very small he will come to understand what is very large."[15] Taken together, these statements distilled all that Darlington stood for.

⌒ IN HIS WORK ON THE GENETIC SYSTEMS OF PLANTS, fungi, and animals, Darlington had developed the notion of hybridity equilibrium (HE), a measurable quantity of genetic homogeneity within a species. When organisms self-fertilize, their HE is low; when they out-cross, it is high. HE, therefore, is a function of (1) the rate of the recovery of homozygosity, which depends on the type of breeding, (2) the frequency of lapses from inbreeding, and (3) the amount of heterozygosity produced by lapses, which in turn depends on (4) the amount of genetical variation in the breeding group, itself influenced by mutation. In nature, many examples existed of switches from one form of breeding to another, often triggered by environmental factors such as change in temperature or the arrival of a pollinator. Wild tomatoes growing in Peru, for example, are regularly cross-pollinated. In England, however, where no pollinating insect exists for the equivalent wild tomato, the tomato quickly becomes self-pollinating in the modern glasshouse, inbreeding having replaced outbreeding. The salient point was that such switches affecting the HE, important for the evolution of the species, are genetically controlled. In plants, dioecy (sexual differentiation—with sexes borne on different plants), monoecy (sexes borne on the same plant), protandry and protogyny (timing differences in the production of male and female flowers), the various forms of incompatability (e.g., heterostyly—see page 219 of Chapter 12), apomixis (the suppression of sexual reproduction), and complex hybrid systems (such as that in *Oenothera*) are all genetically controlled mechanisms devised to manage the inbreeding and outbreeding of the group. Such variety of mechanisms is not matched in animals, for which sexual differentiation is the basic device. Unlike plants, many animals can exercise mate discrimination, elaborately testified by the courtship behavior of fishes, birds, and mammals. The important point, however, was that, just as in plants, this discrimination is under genetic control. All such genetic systems, whether in plant or in animal, achieve the same end: the control of the mating system. The means adopted reflected the circumstances and capacities of the species. Plants make use especially of the diploid style as a "sieve" for sorting the pollen delivered to it by a pollinator, over which they can exercise no direct control. Animals have powers of perception that they use in accepting or rejecting mates.[16] Darlington had considered all such phenomena in the formulation of his evolutionary views. But what about *man?* How did they apply to him?

Darlington's answer was not long in coming. When D. P. Riley of the Association of Scientific Workers wrote to the new president of the Genetical Society asking that he give a talk, "perhaps on the pseudo-science of racial theory," Darlington jumped at the opportunity.[17] "Race, Class and Mating in the Evolution of Man" was delivered at Cambridge on July 16, 1943, and appeared a few months later in *Nature.*[18] It was quite a bit more than poor Riley had bargained for.

〜 DARLINGTON AIMED TO SETTLE the long-standing dispute between those, following Rousseau and Marx, who emphasize the importance of differences in environment rather than in heredity concerning both class and race, and those, following Darwin, who do the opposite. While the former were in line with the New Testament, the latter echoed the worldview of the Old Testament. Since, in their extreme forms, both were being politically applied, Darlington suggested in his article an examination of the scientific value of the claims of each. Thankfully, decisive conclusions could be reached on the basis of the postulates of, and by the method of *analogy* from, experimental genetics. As in his work on chromosomes, Darlington would begin with first principles.

Darlington's point of departure, like Darwin before him, was breeds and races of domesticated plants and animals. When a plant is self-fertilized, he explained, its progeny constitute a group of indistinguishable individuals, termed collectively a variety, or race. With cross-fertilization, however, comes diversity, and any unitary character possessed by the group is now owed to the collective chromosome pool of its ancestry. Since it is a property of germ-cell formation that half the chromosomes of the parent are rejected in the formation of each germ cell, chance will determine the regrouping of that pool in each generation. Still, something more than chance is at work. Different groups are invariably differentiated by selection, whether natural or artificial. The conditions of survival and reproduction of the recombinants produced by cross-breeding differ in different geographical locations, climates, and for different ways of life. Through its effect on differential survival and fertility, selection continually changes the character of the chromosome pool. Furthermore, it gets rid of unsuitable genes, and combinations of genes, thereby balancing the genes contained in the chromosomes against all the others with which recombination from the pool brings them into relationship. As Fisher's student, and Darlington's John Innes colleague, Kenneth Mather had recently shown, this internal selection produces a "relational balance" characteristic of a particular mating group.[19] Thus, at the environmental, the genetic, and the jointly environmental and genetic, or cultural, level, any mating group possesses a unified selective response that differentiates it from all others.

In order to differentiate domesticated varieties, breeders and geneticists use certain clear genetic "markers." Because more minor "polygenes" are not considered for this purpose, artificial cross-fertilized breeds are invariably more heterogeneous than they seem. Although marker genes are used to distinguish among wild plants and animal as well, they are not as useful for separating natural subgroups in animals or man. Such conditions are recognized by the use of geographical and historical criteria in addition to the strictly morphological. Because it is now known, Darlington argued, that natural groups have often become inter-sterile without any difference having been noticed by the morphologist—and because if invisible polygenic differences separate races, then their internal homogeneity due to their inbreeding must also be invisible—it therefore follows that natural races are more homogeneous internally than they seem, by comparison with artificial breeds. Due to this crypsis, Darlington concluded, the only measure of the convergence or divergence between races, apart from experiment, is the breeding history of the group.

Having established the invisible homogeneity of groups within natural populations, Darlington turned to consider man. The evolution of man, derived from a small but genetically diverse group of ten thousand to a hundred thousand ancestors, has from the outset been limited by three kinds of variables, he explained. First, his numbers have set the *genetical* limit to his inter-breeding group. Second, the effective breeding groups have been limited by *physical* barriers. Mobility depended on geography, and on invention, either of which could hinder—as in the case of agriculture—or help—as with navigation. Finally, *social* limits to the mating groups further circumscribed choice of mate from the sum total of those physically accessible. These depended, in turn, on man's intellectual discrimination—itself evolving—and have come to be applied with greater rigor as the density of the human population has increased. "When the cultural barrier arises from the physical limitation," Darlington told his readers, "we have the characteristic origin of race. When it arises from a social differentiation we have the origins of class. When the two are combined and stabilized we get the formal climax of caste in India." The "invisible divisions of civilization," as he called them, could be detected by the study of the frequency of blood group genes, recessive diseases, and marriage patterns.[20]

Part of the subdivision of races into smaller mating groups, Darlington continued, was inherently adaptive. The close social intercourse within groups following the same trade favors the mating of the similarly adapted. Since groups tended to lose those individuals least suited to them, the principle that "first a man selects his mode of life, and then his mode of life selects him" governed their racial evolution. The saying that it takes three generations to make a cotton spinner, Darlington surmised, was more literally true than imagined.

"In monogamous societies based on hereditary wealth," Darlington continued,

> economic differences cut across adaptive differences and establish a still finer division of mating groups. This effect will always be strongest in the layers constituting the governing class. This class is of paramount importance in determining the real and apparent 'national character' and hence in controlling racial evolution. In its origin, like other groups, a governing class must in a sense be adaptive, that is to say selected for its capacity to govern under the conditions prevailing at the time. Is it capable of maintaining its adaptability?[21]

Darlington's answer was that there were two conditions operating to prevent a governing class from doing so. The first was that groups within the governing class are not highly inbred and specialized for their function. Only 25 percent of eminent churchmen in the past fifty years, *Who's Who* informed, had been sons of churchmen. "Segregation of unsuitable types" was therefore frequent. Since the inheritance of wealth, or power, gives the governing class the means of releasing itself from selection pressures, this meant the prevention of a large proportion of its members from assuming their proper adaptive level.

The second condition preventing the ruling class from maintaining its adaptability was a consequence of the mechanism of social diffusion. On the one hand, diffusion between classes could be prevented. In that case, Darlington claimed, not only are human individuals wasted, but genetic materials, useful to the race, are apt to be thrown away, since they are tested in the wrong environment. Such materials normally accumulate until disruptions like the English, French, or Russian revolutions set them free and make use of them. The counterexample is the Indian caste system, in which having been stored for long periods of time, the genetic materials are forever lost before they can be used. In such a system, the social tension will be released, and internal stability achieved. On the other hand, where social promotion occurs, it is to be expected that merit is always the chief agent at work. This expectation, however, Darlington informed his readers, is false, because just as a governing class in its origin is always mixed, so it will always be in its reinforcement. This is due to the fact that social promotion in men is favored by the reduced fertility of their parents, which has been shown to be heritable, and to the fact that social promotion of women will always be based on qualities of a different kind from those favored in the promotion of men, and often conflicting with them. "Thus class division combined with diffusion are necessary for the efficient utilization of the race where the differentiation of the culture is stable; but the conditions of diffusion, unless suitably controlled, may make it little better than no diffusion at all."[22]

Having explained how to genetically maximize ruling classes, Darlington turned to the problems of culture and language. "If mankind had long been broken up into genetically isolated fragments," he wrote,

> it would at once be obvious to us that race made culture and language. Any group, however heterogeneous, would have to be held genetically responsible for its aggregate cultural activities. Now both race and culture are continually being mixed, but culture is a more important obstacle to this mixing than race. Hence the relationship becomes reciprocal. Culture and language often make race just as much as race makes culture and language. It is

this argument, combining genetic premises with a dialectical change, which has hitherto defeated historians and philosophers.[23]

Because of the dominant cultural influences of certain Western languages, English, French, German, Spanish, Swedish, Russian, Turkish, Persian, minor languages, and dialects have at various times become socially and culturally depressed. Where these groups have held on to their language in spite of this, they have by social diffusion lost genetically useful elements to the dominant group. Darlington claimed that in recent times such depressed and genetically extracted groups have reemerged as nations, with disappointing results. The same kind of process had benefited the town at the expense of the country, and the capital at the expense of the provinces.

Darlington then turned to demonstrate how genetics could be usefully employed as a handmaiden to history. When conquest breaks down mating barriers and races fuse, he explained, outbreeding momentarily succeeds inbreeding. Spanning a thousand years in time, and from Marakesh to Delhi in place, the spread of Islam was just such an example. For Islamic conquerors offered the vanquished a simple choice: outbreeding—often against the tenets of their respective religions—or death. The result, Darlington claimed, was always the same. New genetic combinations were able to take advantage of new cultural combinations and a striking cultural development ensued, although limited, for the reasons explained earlier, to seven or eight generations. The effect was due to the breakdown of class as well as race barriers assisted by polygamy. The forced conversion favored by Christianity, on the other hand, had no such striking cultural consequences, because it did not destroy the existing barriers established by class or by race.

Next, Darlington turned to the practice of exogamy, the forbidding of marriage and sexual relations within certain degrees of relationship. When such rules break down, he claimed, the result is an instructive human genetical experiment. Where the Ptolemies and royal Incas engaged in sister-brother marriages, few unions were fruitful, and none that lasted more than a generation. But successful incest had frequently been maintained by half-sister marriages, as in the cases of the Eighteenth Dynasty in Egypt, ancient Athens, among the Mongols, at Ur, and in the royal house of Siam. This practice was helped by half-sisters being very numerous, and by the matrilineal incest taboo. This taboo, in turn, could be made sense of by the evidence from experimental breeding. "Only in five generations of self-fertilization and in ten of brother-sister mating will sterility usually wipe out the inbred stock," Darlington explained.

> The effect will be most drastic when inbreeding is suddenly intro-
> duced in an outbred stock, and the inbreeding must be continuous
> to be cumulative. The exogamy rules of man, like most moral laws
> seem, therefore, to have been derived by an extreme extrapolation.
> Incest is directly dangerous to the progeny only under special
> conditions which have nothing to do with those governing the
> indirect effects of inbreeding on a whole group.[24]

Inbreeding, then, is a conservative agent, for it produces homogeneity, and hence predictability of offspring from parent. As such, it allows
for easy cultural transmission. When applied to specialized classes, it
conserves their differences and increases their fitness. But inbreeding,
while increasing temporary fitness, reduces flexibility, or the means of
adapting to new conditions. Homogeneity provides the optimum conditions for epidemics. Heterogeneity, on the other hand, permits selective survival and recovery. This advantage shows how inbreeding, in
effect, frustrates the long-term function of sexual reproduction and its
resulting genetic recombination.

But outbreeding, too, has its advantages and disadvantages. Hybrid
vigor, which produces positive effects in the short run, is an
example. Here Darlington quoted a study by the Swedish human
geneticist Gunnar Dahlberg in which increased average height of
recruits in Sweden was attributed to the increased outbreeding following increased mobility.[25] Since the change was steady over a hundred
years, it could not be due entirely to improved nutrition, which had not
been steady. While such effects of hybrid vigor are usually thought of as
advantageous, it is possible that if the previous range of heights was an
optimum, secured by adaptation to conditions which had not changed,
those Swedish recruits were now in fact *too* tall. Thus a conflict resides
between the advantages and disadvantages of outbreeding at both the
genotypic and phenotypic levels, and between the short and long view.
How could it be resolved?

"In general," Darlington offered,

> the combination of inbreeding and outbreeding in parallel rather
> than in sequence gives the greatest efficiency in the utilization and
> selection of the available variation of mankind, and consequently
> the most rapid evolution. A subdivision of mankind into races and

classes is, therefore, highly advantageous provided that we can assure its instability.[26]

∼ THE CASE HAD NOW BEEN MADE, albeit in outline. On the one hand, man could not, like plants and animals, be experimentally bred. On the other hand, however, the prodigious refinements of historical, cultural, linguistic, and medical studies could provide the most exact and prolonged account available of variation in any plant or animal—and not just variation, but mating habit, too. Just like in plants and animals, the hybridity equilibrium in man was a crucial measure for his evolution. As in the fly, or the fish, mating discrimination was responsible for the creation of varieties and races. Man's intellectual and cultural evolution had led not to an *absence* of the races found in other species, but merely to a *special character* in these races, resulting from the unique combination of artificial and natural selection. The intellectual and cultural discrimination at man's disposal invariably accelerated his evolution. Migration and conquest, on the other hand, continually created new groups of hybrid origin, groups that could not be classified by marker genes, but that by working, *not breeding*, together, had made advanced societies possible. For all these reasons, human races and classes are more homogeneous than they seem; animal breeds, by contrast, less so. Darlington concluded:

> We must, therefore, regard those who would have us shut our eyes to the genetic differences between races and classes, lest the recognition of unlikeness should generate antagonism, as offering us the council of timidity and escape. Let us rather consider that all the races and classes of men, however distinct, are likely in the end to have their posterity in common and to the common advantage; and that this posterity will still be classified and subdivided. Meanwhile, let us use the methods at our disposal for evaluating the different genetical and cultural prospects of different races and classes and systems of mating. In doing so we shall recognize the reciprocal connexions that exist between changes in the mating system, the economic system, and the political system. For in changing one we are likely to change all three, and the systems which are functionally the most advantageous have a prospect of being eugenically the most desirable.[27]

⟶ DARLINGTON'S ARGUMENT FOR THE GENETIC BASIS of race had several elements. Reasoning upward from first genetic principles was one; reliance on analogy from experimental genetics yet another. So too were the use of direct human data, mainly from the work of others; the assumption of strict genetic determinism; recourse to history as natural experiment; rejection of the single gene; and embrace of the genetic system approach. Darlington's argument also emphasized the importance of adaptiveness and variability of groups, and the message of connectedness. Its tone and tenor reproached maudlin, uncritical, and cowardly social sensibilities. Finally, the argument comprised of suggestive yet nonspecific eugenic undertones. In outline and miniature form, these were the scaffolds with which all of Darlington's claims in the remaining thirty-eight years of his intellectual career would be built. Together, they were assembled to challenge what Darlington saw as the terrible conflict between science and society. As with his more strictly cytogenetic work, they would get him into a lot of trouble still.

～ 12

On the Determination
of Uncertainty

B Y THE LATE 1940s, life at the John Innes was beginning to lose its appeal. This had much to do with the increasingly
overbearing administrative responsibilities of directorship at a growing
institution, a calling Darlington had never particularly enjoyed. A
chance to reach a wider audience, to teach the new ideas he had been
developing about man, was what he craved more than anything else.
But how, from within a horticultural institution, could this be done?

Darlington's preoccupation in the 1930s and 1940s with spirals and
spindles had both increasingly brought him to the physicochemical
study of the chromosomes, and to a theory of the material basis of
heredity. In fact, he had been one of the first to pursue a physicochemical elucidation of the gene,[1] and was a key participant in a secret
Rockefeller-sponsored workshop in 1938 at Klampenborg, Denmark,
at which cytologists, geneticists, chemists, and physicists were brought
together for the first time for the express purpose of cracking the
mystery of the gene.[2] In the years that followed, inspired by the work of
Torbjorn Oskar Caspersson and Jean Brachet, Darlington had
embarked on a collaboration with Len La Cour at the John Innes on the
study of nucleic acid starvation. Like many of his colleagues around the
world, he was convinced that DNA could not be the genetic material,
but that protein most probably was.[3] Starving nucleic acids wrapped
within proteins in the chromosomes might help to understand how
those proteins went about conducting the business of transmission.
Darlington knew this was still just a guess. In line with his normative

scientific strategy, he believed that in rewriting nuclear cytology on a chemical basis "we must be prepared to make great mistakes . . . to embrace and reject hypotheses with appropriate carelessness."[4] But Darlington did not seem prepared to relinquish his protein-based theory of heredity so easily, even as the evidence began to mount against it. Not only after Avery, MacLeod, and McCarty's transformation experiments at the Rockefeller Institute in New York in 1944, which proved that DNA was the hereditary material, but also immediately after Watson and Crick's elucidation of the structure of DNA in 1953, Darlington continued to argue for a cooperative nucleoprotein-DNA symbiosis in which DNA served as a "midwife molecule" helping the chromosomes coil and assisting in replication.[5]

A molecular revolution was sweeping across the biological world, and Darlington, the chromosome man, could sense it. Still adamant about the importance of studying chromosomes, Darlington knew what staying at a horticultural institute would mean for his career. As bacteria and viruses replaced plants (and to a degree flies) as the model organisms of the new molecular age, great changes would be inevitable. Thirty years after his innocent plea to join the organization, Darlington now found himself in a similar bind to that of the chromosome-rejecting Bateson. As the biological world began to catch the new wave of the molecular era, just as it had the new wave of the chromosome a generation before, life at the John Innes felt increasingly isolated and stale.

⌒ ONE WAY TO ESCAPE THE SECLUSION WAS to establish an independent journal. Darlington had in the past encountered difficulties placing his scientific papers in peer-reviewed journals. Haldane, at the *Journal of Genetics*, had rejected some of them on account of their being cytological (and not genetical enough). James Grey, of the *Journal of Experimental Botany*, had rejected them as not experimental enough. The *Proceedings of the Royal Society* had refused to publish others, having sent them for review to Darlington's arch-rivals John Farmer and Ruggles Gates. In consequence, Darlington often published his work in foreign journals, chief among them a German journal whose name, *Chromosoma*, he himself had suggested in 1939.[6] Elsewhere, Darlington's attempts to attack the Soviet Union in print were rejected by *Nature*, bringing about the Orwellian alliance and the article in *The Nineteenth Century and After*.

If strong resistance had been mounted to his scientific work and to his attacks on a scientist generally considered a crackpot, a vehicle of independence—a mouthpiece—for Darlington's social genetics would prove indispensable. In 1947 Darlington cofounded *Heredity: An International Journal of Genetics* with the biological statistician Ronald Aylmer Fisher.

Fisher had already secured himself as the greatest statistician of his age. A bearded genius with Coke-bottle glasses, he was also painfully insecure, and like Darlington, had a fierce temper. Once, in a fit of fury against a laboratory assistant (the story went), Fisher crushed the mouse he was holding in his hand and then exclaimed: 'Look what you've made me do!' Following the death of his son George in a plane crash in 1943, Fisher had moved out of the home he shared with his wife and remaining seven children, and taken up residence in the bachelor's quarters at Gonville and Caius College, Cambridge. Darlington and Fisher, it seemed, were a match made in hell.

Based on the belief that genetics needed to become the causal framework not only of biology, but of the social and medical sciences too, *Heredity* was addressed

> "To the botanist and zoologist by way of evolution and systematics. To the physiologist by way of cytology and experimental technique. To the medical research worker on account of both diagnosis and treatment; on account, too, of both man and his enemies. To the social scientist since the study of nature and of nurture is the foundation of his work. To the agriculturalist since plant and animal breeding is the key to his future. And finally to the physicist and the chemist since here the bridge is being built which joins their sciences with biology."[7]

Darlington and Fisher's wide-angle approach to genetics at the time seemed to them unique: the traditional policy of the *Journal of Genetics* of keeping within the narrow limits established at its inception by Bateson seemed to them "dangerously restrictive." *Heredity* would break new ground.

Unlike all other respectable scientific journals, *Heredity* would eschew the blind peer-review system of assessment. Though widely considered to be a fundamental mechanism to ensure the quality of scientific publication, and hence of the advance of scientific knowledge, Darlington saw the process as nothing but an obstruction. He and Fisher would decide what would be published, and what would be sent back to the

Darlington and R. A. Fisher, 1947. Together the two men cofounded *Heredity*, in which Darlington was free to express his thoughts on a wide range of subjects uncensored. Photo courtesy of Clare Passingham.

author. More importantly, perhaps, Darlington himself would be free to pass uncensored judgment on a wide range of publications, from investigations in demography, linguistics, history, and moral philosophy to studies of human genetics, behavior, and evolution.[8] Fisher was deeply interested in the application of genetics to man. A conservative eugenicist, he had already understood, like Darlington, that the fundamental integral insight in genetics was that the mechanism of heredity itself underwent evolution. Sympathetic politically and intellectually, however, Fisher never really became a friend. Sensing that their prickly natures would draw blood if housed in close quarters, Darlington effectively became sole editor, and, despite the joint venture, Fisher let it be. He was content that Darlington was providing in *Heredity* a ready platform for the selectionist crusade of his protégé E.B. Ford and Ford's students Bernard Kettlewell, Arthur

Cain, and Philip Sheppard, as well as the occasional eugenic essay by Fisher himself. With Fisher's death in 1962, Darlington became sole owner of the journal. With 1,350 subscribers in eighty-two countries, he turned over the journal to the Genetics Society in 1970, a profitable gift in return for which he asked only that he be sent copies of each part and supplement during his lifetime.[9]

~ As MID-CENTURY NEARED, man continued to preoccupy Darlington's thoughts. If helping somewhat to extricate an aging scientist from the gushing, difficult-to-master molecular wave, the study of man was, regardless, an infatuation that would not tire its grip on Darlington's mind. The more time spent trying to understand, the more difficult the questions became. In "Race Class and Mating" he had argued that by exercising his discriminative breeding choice, man had affected profound genetic and cultural consequences for his species. But was man's discrimination in the choosing of mates itself under genetic control, as in the lily and frog? Darlington's assumption was that in man natural selection operates on breeding systems that are culturally determined, and not therefore *directly* under genetic control. Human races resembled the races of animals and plants except in this one important particular: they were consciously made by man himself and not by external or genetic accidents. Conquest, slavery, conversion, and migration had in man produced an alternation in cycles of inbreeding and outbreeding of a rapidity unknown in any other organism, successively favoring fitness and flexibility, the rises and falls of races and classes. Owing to the rapid changes in humankind's cultural capacity and economic power, the equilibria of races and classes in man had always been unstable throughout human history. Genetic principles, combined with simple observations of mating practice, could thus explain the chief characteristics of human societies and make sense of their social, cultural, and economic evolutions.

But there was a further consequence to man's conscious manipulation of his own environment. The basic principle, whereby natural selection in other organisms depends on the action of varying environments on varying genotypes, became distorted in man by the predominant long-range effects of the genotype and of reactions between different genotypes whose aggregates man calls culture (this was the "system" Darlington referred to in 1941). This was due to the fact that human innovations, of theory or practice, increase the chance of survival not merely of the innovators (discoverers, inventors, rulers—"Great

Men"), but of their imitators who may be of unlike race and class. Since Darlington assumed a strict genetic determinism affecting ability, this meant that man's control of his environment had given an exponentially increased value to the genotype. Just like in his work on the control of development, where he had theorized a nucleus acting on a cytoplasm which it has earlier modified, so becoming the predominant partner,[10] Darlington surmised that in man the genotype works on an environment that it has earlier modified, becoming, like the nucleus, the dominant of the two. "Man's social and cultural organization," he wrote in 1949,

> depending on the integrated effects of the genotypes of groups of individuals, often with a few of predominant importance, is responsible for the weighted, or exaggerated, or indeed catastrophic, fluctuations which are so evident in human history or evolution. The outstanding individual, that is the individual having an outstanding reaction with his environment, is exceedingly important in human history even though his emergence can be roughly predicted from the flexibility of the mass. The adaptations and fluctuations are, in turn, responsible for the increasing tempo of human change and the increasing contrasts of genetic capacity for culture between different genetic groups.[11]

ONCE SUCH GENERAL PROPOSITIONS concerning the genetic basis of culture were stated, practical examples could be given. Darlington chose to begin with language.[12] Since the precise genetic control over the development of all bodily structures extends to the organs of speech, it must limit, Darlington argued, the ease with which races and individuals are capable of uttering various sounds.[13] While education affected to obliterate such differences, the unlettered and infants provided the evidence that they were indeed genetic. In his study, Darlington showed how the contour map of Europe for O blood groups not only manifested a significant relation with the known history of migration in the last three thousand years, but also a significant agreement with the present distribution of the phonetic "th" sound in European languages. Not knowing what would be discovered in later years—that blood group and disease susceptibility are in fact closely related—Darlington concluded that since both of these characters are nearly neutral in regard to natural selection, they reflect the long-standing genetic properties of the population. Students of language had often held the view that some inherent property of peoples influences the sounds they use and the changes this use undergoes. Darlington was now waving his hands at them, claiming that the undefined "substratum" assumed in linguistics was in fact the aggregate genotypic action of the community. Languages changed over time because conquering peoples, such as the Danubian Aryans, and later the Huns, and the Mongols, could suppress the aboriginal languages so far as words and arrangements of words were concerned. But they could not, except by their slight genetic contribution, ultimately modify the limited capacity for sound production of the conquered peoples. Hence the peripheral similarities of certain sounds would seem to depend on their aboriginal origin.

With his study of phonetic preference and its historical changes and geographical distribution, Darlington was trying to show how a new, genetic approach could be used as a key to understanding the history of culture.[14] Underlying all his arguments was an assumed genetic determinism—a tacit understanding that genes were responsible for differential ability in humans. In order to progress any further, therefore, Darlington would need to produce a complete philosophy of genetic determinism, to make his implicit assumption explicit. This he began to do in a talk at Oxford in May 1950, "The Coming of Genetics," which soon ballooned into a book with a simple, somewhat presumptuous title.

⌒ IF DARLINGTON WAS NOT PREPARED at fifty to jump on the new wave of the molecular age, unlike Bateson before him he had many battles in him still. Until lately, forsaking his cozy directorship to become one of many department heads in a university famous for being a capricious mistress seemed the height of folly. Now Oxford's advances actually began to sound inviting.[15] The thick medieval walls that once had seemed erected simply to keep him and his ideas out were now offering Darlington professional and intellectual sanctuary. By teaching undergraduates, appointing new staff, coordinating with other departments, establishing international conferences, and writing books, Darlington hoped that his assumption of the Sherardian Chair of Botany at Oxford University would allow him "to make genetics in my own broad sense into the central framework of biology, and biology itself integrated this way into the central framework of education."[16] This had been his goal for quite some time, and never before had such a fitting opportunity presented itself before him.[17] In the summer of 1953, Gwendolen and Cyril moved out of Bayfordbury and settled in an old farm cottage in Woodside, just outside of Oxford, where Darlington would assume his responsibilities at the university beginning in Michaelmas term in the fall.

⌒ IT WAS THEN THAT THE BOOK WITH the presumptuous title appeared. *The Facts of Life* was published in 1953, just as Darlington and Gwendolen were settling in at Oxford, and immediately became a popular success. Here Darlington's determinism was finally afforded an uncensored, unabashed performance. Part history, part biology, part philosophy, and part method, the book walked its reader through the gripping story of man's search for self-knowledge. The history of man's understanding of heredity, Darlington began, was nothing but a path of fallacies and superstitions in which, from the time of the bible, natural and moral law had been confused. The great culmination of this quest finally arrived with Mendel's proof for the theory of particulate heredity, in which he spoke, like Lucretius two thousand years before him, of "elements which determine." Unlike his Greek predecessors, however, Mendel had by inductive experiment succeeded in deducing the basic laws of heredity. Although his champion, William Bateson "the immaterialist," had thought that Mendel's central discovery was not the elements-which-determine but the fact of segregation, the first fifty years of the twentieth century had overcome his skepticism, firmly ensconcing the material gene—strewn along the chromosome in the cell nucleus—as the fundamental unit of heredity.

Nevertheless, whereas the one-celled bacterium, which merely seems to propagate itself, did not arouse suspicion or confusion, in higher organisms, and man, the mystery of development, lying between a minute, apparently structureless germ cell and a large and elaborate adult organism, continued to disguise the fact of determination. With the understanding of the relations between genetic particles of different origin: nuclear, cytoplasmic, and viral, however—an understanding to which Darlington himself felt he had contributed—development, heredity, and disease had finally been put on one footing. They were all consequences of the complex interaction of Mendel's genetic, determining factors. From his origins, man had sought after the self-knowledge that comes with the understanding of inheritance; now, finally, he could be taught the "facts of life."[18]

With transmission, and particulate determination firmly in place, evolution could be viewed in a new light. For it did not arise from a property of progress inherent in life, but rather from the occurrence of changes in heredity due to chance. Such changes, affected by mutation and recombination at the genetic level, represented the element of uncertainty in heredity. The organization, and the suppression of this uncertainty, was itself subject to heredity, and where uncertainty was completely suppressed, evolution ceased to be. The great paradox of genetics, and therefore of evolution, was that so much determination, in the form of the laws of mitosis and meiosis, chromosome pairing, and crossing over, should be organized to express the results of uncertainty.[19]

Darlington wanted to make things clear. Although chromosomes recombine by pure chance, every step in this process was determined at a lower mechanical level. Uncertainty—a situation in which it is impossible to predict the future of a system on the basis of all the *available* physical data—was not the same as indeterminacy, a situation in which it is impossible to predict the future of a system on the basis of all the *relevant* physical data. Indeed, Darlington's friend Hermann J. Muller had used X-rays to show that mutations in a gene and the breakage of chromosomes were determined, at nonbiological levels, just like chemical reactions. Although it was impossible to exclude the possibility that the residual unexplored uncertainty of mutation arises from an ultimate subatomic physical uncertainty, such an assumption could only lead to sterile contemplation, rather than to experiment and new knowledge. The truly interesting questions pertained to the *results* of the determination of uncertainty. Could such a system have been introduced in evolution because it was advantageous?

⌒ DARLINGTON'S EVOLUTIONARY UNIT WAS the genetic system— not the gene, the chromosome, or even the individual. By using fertilization, crossing over, and recombination, the *system*, defined by the genetic control and adaptive connections between hereditary structures, the reproductive process, and the breeding habit of a population, could exploit uncertainty to bring about rapid and efficient adaptation of an unlimited lineage of sexually reproducing individuals. This process, Darlington argued, was in itself deterministic, because it allowed organisms, including man, to respond deterministically to changes in their environment, that is to say, changes outside themselves too large in amplitude for the individual to react to, or even to perceive. Echoing his hero Weismann's logic, Darlington cautioned that if this situation were mistaken for the operation of indeterminacy, the point of the evolutionary mechanism would be lost. Thus, the first principle of heredity was not chance, as it had become common to believe. Rather certainty comes first, *then* chance, and finally certainty again.[20]

⌒ Now that the fundamentals of his biological philosophy of determinism had been described, Darlington could turn to man. His central claim was that social scientists had forsaken biology. Much of this was related to the fact that they had learned their Darwinism from a man of whom Darwin himself had said: "If he had trained himself to observe more . . . he would have been a wonderful man."[21] Herbert Spencer's stylized ideas of evolution, his indefatigable belief in progress, based on the principle of the inheritance of acquired traits, did not seem to work, however, when applied to the brief, intense intellectual transformations, to the cycles of advance and decay, to the diversity of culture that characterizes human history.[22] Disappointed, the social scientist satisfied himself that evolution does not work in Spencer's way, and, imagining that nothing fundamental in biology had changed since his day, felt justified in deeming evolution and natural selection irrelevant to his own inquiries. The biological interpretation of man was thus repudiated; it was not man's nature that distinguished him from other men, but his habits.

Surely, there were good reasons for this. The notions of progress and Lamarckism, translated to society, were infinitely more pleasing than a random, deterministic Darwinism. But were they credible? Arnold Toynbee, in his *Study of History*, expressed the idea that the greatness of nations is not due to any permanent genetic or racial quality, but rather arises out of the challenge of the environment.[23] The Dutch were challenged by having so little land to take some more from the sea. The English were challenged by having invaded so small an island, with such bad climate, and therefore set out to conquer empires, bring about the scientific revolution and industrialization, breed many poets, develop their own idiom in landscape gardening, and so on.

An agreeable idea! But was it not the challenge of the environment which stimulated Lamarck's giraffe to grow his neck so long? And did not Lysenko's wheat change to rye in response to the same provocation?[24]

The theory of the challenge and stimulus of the environment,

Darlington claimed, is a deterministic theory of limited explanatory power. It stood on a genetic vacuum. Darwin himself had understood this, writing in his book *The Variation in Plants and Animals under Domestication:* "Some writers [referring to Mr. Buckle, the historian, who had recently published his *Civilization*] who have not attended to natural history, have attempted to show that the force of inheritance has been much exaggerated. The breeders of animals would smile at such simplicity; and if they condescended to make any answer, might ask what would be the chance of winning a prize if two inferior animals were paired together."[25] But Darwin got the mechanism of heredity wrong. His first cousin, and the father of eugenics, Francis Galton, on the other hand, knew how to use the evidence to make an argument about heredity.[26] Galton's *Hereditary Genius* appeared in the same year as Marx's *Das Kapital.* Both books were concerned with improving the lot of mankind, but with opposite interpretations. To Darlington's mind, it was his fellow Englishman who had come closer to the truth.

⌒ ALMOST A CENTURY FURTHER ON, the time had come for reintroducing biology into the study of man, to wed the tools of modern genetics to the Galtonian cast of mind. Citing the twin studies of Otmar von Verschuer and Johann Lange,[27] Darlington argued that not only physical characters, but also behavioral ones, such as criminality, were inherited genetically. In fact, he claimed, "sexual capacity and drive," "degree of imagination and reason," "understanding of truth and beauty," educability, temperament, susceptibility to disease, and musical ability were all genetically controlled and determined. Success and failure in school and in marriage, criminal and sexual proclivity, genius and disease could all be explained by genetics. Economic factors, Darlington conceded, could *alter* the expression of vices as well as virtues, but in themselves they played little part in *creating* them. Conversely, poverty or social depression, as countless examples from Dickens to Dostoyevski, Lawrence Sterne to Charlie Chaplin, and Benjamin Franklin to Jean-Jacques Rousseau had shown, could not stifle the expression of true genius. Indeed, the role of the environment was continually overestimated because it was not grasped that an environment is partly, and always, genetic: no two distinct genotypes can ever share the same one, no matter how uniform. In addition, the genotype of the group is the environment of the individual, and individual genotypes seek out their own environments. Taken together, these three axioms meant that genes were more determinative than even Galton had thought. While for "timid souls" this meant a sentence of predestination—"scientific Calvinism" as one soldier had once remarked to Bateson—for Darlington, and "bolder minds," it was a charter of individuality.

Genetic variety at the group level, moreover, was the scientific justification for political tolerance. Referring to the advantage of the white-plus-Negro society of the southern United States over a pure Negro society, and over the previously displaced Indian society, Darlington explained:

> The American Indian has been found not to be able and willing to help the other two. Their capacity and his capacity are both, of course, racial and genetic. They are determined and limited by heredity. . . . Thus we arrive at a paradox which timidity usually hides from us. The assumption of a genetic basis for race and class differences provides the evidence, the only scientific evidence, in favour of racial tolerance and co-operation.[28]

That the American Indians had been forcibly resettled seemed to Darlington almost irrelevant. On the contrary, it merely proved his point: had the Indians' genetic makeup been different than what Darlington supposed it to be, the kind of displacement that had occurred would not have been possible.

For Darlington mutual aid between genetically diverse groups could not be assisted in the long run by make-believe of any kind, certainly not by a make-believe of equality in the physical, intellectual, and cultural capacities of such groups. Tolerance was crucial because division was a built-in human attribute. However stimulating individual and group variety may be, Darlington explained, there is a point at which it produces friction. Such friction is relieved by the agencies of social and mating groups and of family.[29]

Finally, Darlington could come full circle and link these ideas to his work on genetic systems of plants and animals. Man's breeding habit may not be as directly determined by genetics as in the lily or fly, he now argued, but those beliefs, and social behaviors that affect his choice of mate, result from his fixed capacities and proclivities, *themselves* directly determined genetically. In turn, the mating group, such as the social class, or professional guild, selects and concentrates the genetic capacities of its individual members, acting as a mechanism shaping the breeding habit. Between groups, language, just like heterostyly in plants, also becomes a powerful agent in the determination of breeding habit. By making intermarriage less likely, language determines the size and shape of the genetic pools within which the genes and characters of the human species are recombined. Like the suppression of crossing over in male *Drosophila* (that dogged problem Darlington had tried to solve in the early 1930s), it serves as a mechanism to reduce the number of meaningful chiasmata, thereby exposing less variation to selection, and maintaining a certain level of homogeneity. At another level, genetically determined differences between *individuals* influence the possibility of union between members of the opposite sex, shaping the outcomes of posterity.

As in the loss of fertility resulting from the breakdown of incompatibility, or the legitimate and illegitimate union brought about by distyly in *Primula sinensis*, infertile marriages (and divorce) in humans were consequences of genetic incongruity between pairs.

The important point was that, just like the genetic system's exploitation of uncertainty for the determination of adaptiveness of the species, of which individuals are unaware, the differentiation of society is based on genetics, of which the people concerned are likewise unconscious.

Man was immensely adaptable, Darlington conceded, but not through the plasticity of the *individual*. Rather, it was the *variability of the species*, disguised in individual plasticity, which made man seem so malleable. Individual adaptability was one of the great illusions of common-sense observation; it was responsible, Darlington claimed, for some of the chief errors of political and economic administration. For where character is determined, education, deterrence, goodwill, and reformation can only fail. Should irremediable criminals be segregated? Painlessly liquidated? Or could their births have been prevented

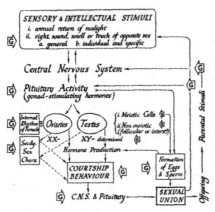

"The Diagram of Love," illustrating the genetic control of the process of consummation of love in humans. Darlington believed that humankind was at the mercy of its determining genes to a far greater extent than it was conscious of. From Darlington, *The Facts of Life* (London: Allen and Unwin, 1953), p. 327.

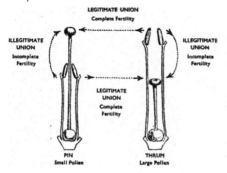

Heterostyly in plants. Darlington believed that the genetic processes in control of sexual union in plants had analogues (such as the previous figure) in humans. From Darlington and K. Mather, *The Elements of Genetics* (London: Allen and Unwin, 1949), p. 250, after Darwin, *The Different Forms of Flowers*, London, 1877.

in the first place? As in "Race, Class, and Mating," Darlington kept his eugenics implicit, but the suggestion that "there are certain individuals we can well do without," and the promise that "we could have known in advance," seemed to point where he was leading.[30] Could genetics be of any use to agriculture? Again, Darlington only hinted:

> In England . . . it is not the lack of research which limits food production but the genetic unfitness of a large part of the tenant farmers, the legally secured occupiers who are organized to keep better men off the land.[31]

This was no longer the same Darlington who years earlier, upon return to England from America in 1933, had been disgusted by the very use by his fellow upper-class countrymen of the word "they" to denote members of lower classes. The important means for increasing food production, he now believed, was to breed better farmers, and to put them in possession of the land. Could such breeding be maximized? Darlington did not put forward a detailed plan, but he did provide copious statistics from the practice of bull breeding, implying that, as in bulls, fertile unions could be engineered. After all, he argued, history had shown that marriage is an inadequate method of breeding.

"The state or mankind," Darlington explained,

> cannot accept a responsibility which goes back as far as the fertilized egg without one day claiming the right to go further and control the quality or proportion of gametes which go to make the fertilized egg. Fertilization is . . . an important event. But it has antecedents. It is not so important that all responsibility should begin after it, and none be allowed before it.[32]

How to exercise this control was the deepest problem confronting mankind. Darlington had some ideas of his own, but at this stage his primary goal was to make others acquiesce in the basic judgment that such control was not only desirable but essential to face the challenges of the future. Genetics could no longer be ignored with impunity. Despite Eddington's and Fisher's arguments to the contrary, free will *could not* be inferred from the indeterminacy of quantum systems.[33] Just like equality, immortality, and progress, it was an illusion of dangerous proportions. The Lord's Prayer, the Declaration of Independence, and the Communist Manifesto were all *obstacles* to man's understanding of himself. Until people (or at least the small percentage of those who could understand, and lead) were taught to hold their opinions on explicit grounds, this would not change. "It is easier and pleasanter to believe the incredible," Darlington warned, "than to pretend to understand the incomprehensible."[34] But the truth had to be spoken:

> As the nucleus builds up or creates a protoplasm and a cell suitable to itself, and the cells build up or create an organism, an individual, suitable to themselves, so finally the individuals build up and create an environment, including a society, suitable to themselves. The environment and the society are three removes of determination from the nucleus. But in the end the structure of society rests on the stuff in the chromosomes and on the changes it undergoes. And it is as processes of determination in their whole sequence that we must study them, seeking to understand the determination and to exploit the uncertainty to which it is subject.[35]

Unlike what the church, or Marx, or liberalism taught, free will meant the freedom to be yourself, the freedom to recognize necessity. No other freedom exists. If man supposed that goodwill, assisted by a little prayer, or organization of workers, or reform of education will remedy the forms or faults of character that determine whether a life is good or a marriage is happy, he would miss the chance for improving himself; he would forsake the gift uniquely given to him by natural selection: the conscious ability to exploit uncertainty for his own good. Paradoxically, by refusing to believe in free will, he would be able to make the "right choice." Darlington concluded:

> What we are now in the way to attempt is to make scientific knowledge an instrument of civilization it has so far failed to be. It

has failed because the picture of the world it has given has been disconnected: the disconnected parts have been mechanical and hence inhuman and even antihuman. Now we have a connected picture. . . . The connectedness and the humanity of science we owe to the new life which genetic principles have given to the whole of our knowledge.[36]

Now at Oxford, in a new setting, and a new life, Darlington hoped to begin to spread his gospel.

~ 13

The Breakdown of Classical Genetics

A FEW COLLEAGUES remembered that news of Darlington's appointment was greeted in the Botany Department at Oxford with a "stunned silence"[1]; his sarcastic, cutting manner and outspoken views were notorious. Under Darlington's predecessors, ecologists Arthur Tansley and Thomas Osborne, a biology devoid of genetics had become firmly entrenched at the Botany Department. Many traditional biologists at the time believed that if you wanted to know how living things worked, you did biochemistry and physiology; if you wanted to know how they lived and survived, you did ecology; and if you wanted to know how they grew, you did anatomy and morphology. Apart from that, you needed to know the range of organisms and their relationships (taxonomy and types). None of these had much to do with genetics. As for chromosomes, they were regarded as interesting organelles important in maintaining hereditary accuracy, and useful for taxonomic purposes, but little more. Even the Zoology Department had only one geneticist, eccentric lepidopterist and Fisher collaborator E. B. "Henry" Ford, since 1939 University Reader in Genetics.[2] The "one gene/one enzyme" hypothesis that would win a Nobel Prize for Darlington's friend George Beadle and Edward Tatum in 1956, and that really established the link between genes and cell physiology, was hardly yet acknowledged outside the specialized field of biochemical genetics. Everyone knew that the celebrated but infamous geneticist Darlington would want to change all that.[3]

Hugh MacDonald Sinclair of the Laboratory of Human Nutrition, one of the Electors responsible for bringing Darlington to Oxford, was more right than he knew when he wrote to him: "Perhaps you will allow me to mention in confidence that the view was expressed that you should allow nothing but cytogenetics in the Department."[4] For Darlington treated his cause like he did his driving: "Cyril assumed others would get out of the way."[5] Uncompromisingly disparaging of the attitude that genetics was "just another subject" or "irrelevant," he was blisteringly dismissive of colleagues who showed what he regarded as ignorance, and could be thoroughly offensive about other disciplines. "Ecology," he would say, paraphrasing Oscar Wilde, "is the pursuit of the incomprehensible by the incompetent."[6]

Upon arrival, Darlington made no secret of the fact that he thought William Owen James a fool (James was one of the foremost plant physiologists in the world at the time), and showed little sympathy to Jack Harley and Freddie Whitehead, "the woodmen." On the defensive, many old-school men wondered aloud how a cytogeneticist had been chosen to lead them. "How long has this man been interested in Botany I should like to know," colleague Bennet Clark asked Osborne shortly after Darlington's appointment.[7]

Within a few years of his appointment, James, Harley, Whitehead, and John Burnett had left the department, becoming what were called the "Darlington refugees." New men were changing the face of the department: Canio Vosa (who became Darlington's technician as Len La Cour had been at the John Innes), Leslie Crowe, Alan Bevan, Keith Lewis, Brian Cox, and electron microscopist Barry Juniper. Bringing about such transformations, and showing little consideration, Darlington succeeded in antagonizing much of his own staff. The situation was so bad that herbarium worker Frank White, an old-school taxonomist of shorter-than-average stature, took pleasure in the fact that Darlington, well over six feet high, and even taller in demeanor, had to bend to get through the herbarium door.[8] Overlooking the fact that his colleagues were objecting as much to his manner as to his genetics, Darlington wrote angrily to his colleague Guido Pontecorvo: "Genetics makes too much sense to be dispensed with and they know it!"[9] The resistance he encountered while trying to do so only affirmed his theory of the conflict between tradition and novelty, between the safety of the old and the disruption of the new.[10]

⁓ UNDAUNTED BY THE RESISTANCE WITHIN, Darlington continued to spread his larger message of a social genetics outside. The

Woodhull Lecture at the Royal Institution in London in March 1958 presented a further opportunity to expound his theories and advance his agenda.[11] Standing before a packed room, Darlington had a simple message to deliver. In order to safeguard the future of mankind, he argued, it was indispensable to study his past. Not as cultural anthropologists, and social historians had done, but with the unblinkered, impartial gaze of the scientist. Such candor had been impossible until now for *scientific* reasons. Since the Mendelian design of experiment required gene mutations of gross effect they were taken by Bateson, and later by Morgan, as the units of natural variation. Moreover, the pure lines of Mendelian experiments were taken as the raw materials of natural variation. The model systems that resulted, through the work of the mathematical population geneticists, were easy to handle in theory, but did not describe natural populations faithfully. It was in this period, Darlington claimed, that environmentalist explanations of human variation gained ground in the social sciences. Science itself had failed them. Mendelism's "character" was too crude and naïve to explain the subtleties of human variation, and because culture can sometimes be socially *transmitted*, it became commonplace to believe that it could never be racially or individually or genetically *created*.

At issue was a matter of method. Mendelian genetics had succeeded through what Darlington called the "analytical" approach: a static understanding of the genotype as the sum of its genes. But the approach broke down with natural populations. Here an "integral" approach became necessary: the difference between genotypes, after all, was not simply the sum of the differences between their genes, but dependent on interactions of a great assembly of independently varying units. Analytical breeding experiments revealed the reaction of a definitive heredity with a definitive environment (unit of variation: the gene). Fixed natural populations, like plants, exhibited the effects resulting from the environment's selection of heredity (unit of variation: the chromosome). Mobile populations, such as man, illustrated above all the integral results of the selection, in its turn, of the environment by heredity (unit of variation: the polymorphism). The opposed agents were not independent: they interacted in sequence, and sequence usually began with a variable heredity. Thus, connectedness was the dominant relation of "integral" genetics, pioneered by Galton; determination (with uncertainty), pioneered by Mendel, that of the "analytical" approach. Mendel's method had led to a more exact understanding of the ultimate determinants of heredity (Watson and Crick); at the same time, it made clear that in an organism like man, heredity can never be fully reduced to analytical terms. In order to understand its evolution and its relation to the environment—particularly significant in man—a return to Galton's integrative approach would prove necessary. "A century which relies only on analysis and seems to be afraid of synthesis," Darlington quoted Goethe in his diary, "is not on the right way; for only both together, like breathing in and breathing out, form the essence of science."[12]

The confluence of methods of the experimental breeder, of the cytologist, and of the naturalist, Darlington claimed, had established that the bulk of natural variation is due to the mutually adaptive combining of mutations and structural changes; that the mutations themselves are of a different order and size; that they are usually undetected elements of polygenic systems; and that genes expand or contract according to the size of the breeding group within which they are being exchanged. Polygenic systems, moreover, are adapted to the range of variation of the breeding group within which they subsist. Their evolution, therefore, is itself subject to changes in the size and character of the breeding group. Darlington's conceptualization of the genetic system, and of the method needed to study it, had convinced him that while classical Mendelian genetics had explained heredity, it had failed to explain variation. Genes and genotypes are related to breeding groups, and had to be considered together in examining the evolution of any natural species, *especially* man. Applying simple Mendelian analysis to properties such as intelligence, temperament, instinct, resistance or susceptibility to disease, and fertility was, in itself, bound to fail.[13]

Advances in human genetics had already demonstrated this insufficiency. Well known were R. A. Fisher's prediction and Robert Race's demonstration of the structural complexity of the genes responsible for the various blood groups, E. B. Ford's prediction of the genes' physiological complexity on the analogy of the polymorphisms of other animals, and John Fraser Roberts's demonstration of their complexities by statistical studies of their relations with susceptibility to disease. A particularly famous instance was the demonstration by Anthony Allison, Darlington's colleague at Oxford, of how malarial diseases had forced whole races of men to modify the composition of their hemoglobin as a means of avoiding extinction.[14] Taken together, all such studies had confirmed, as Darlington had been arguing, that it was necessary to begin the inquiry into the heredity of any organism much farther back in time than had previously been thought. To understand the trickling results from human genetics, and to be able to advance further, it would be necessary to examine the breeding history of man.

⌣ DARLINGTON WAS ALREADY CONVINCED by contemporary studies of twins and foster children, educational experience, mental deficiency and ability, criminal pedigrees, and animal breeding for temperament and intelligence, that heredity was more determinative than even Galton had imagined.[15] But before he could turn to history, he would need more experimental proof that natural selection was still working on important determinants of the evolution of man. It was plain that changes to the breeding system affect visible diversity of form. But how did they affect the invisible variables: intelligence and fertility, viability and resistance to disease?

⌣ IN A STUDY ON *Campanula persicifolia* Darlington had discovered that the species is normally outbreeding. The natural population, therefore, is highly heterozygous, and since interchange is a common structural change, and the character of meiosis favors their fertility, interchange heterozygotes are commonly found in nature. When such heterozygotes of the same interchange type are crossed, Darlington found, a proportion of structural homozygotes appear. The proportion is less than the Mendelian proportion, since one of the two reciprocal types of homozygote is totally eliminated. But when, on the other hand, such plants are selfed, all the progeny have rings at meiosis and no structural homozygotes appear; both types of homozygotes are eliminated.[16] This study had made clear to Darlington that, through selective elimination, the chromosome mechanism resists change in the breeding system; the genetic system behaves as a unit in maintaining an optimum hybridity. Could such an experiment be applied to humans?

Indeed it could, and was: man himself had been conducting it unknowingly. By collecting and studying hundreds of pedigrees and marriage registrars (he had advertised for pedigrees in the popular press), Darlington had found that a change from outbreeding to inbreeding leads, as in other organisms, to depression, especially affecting the invisible variables: first-cousin marriages lower the fertility of children in an outbreeding group.[17] Nevertheless, it also provides the means of rapid improvement: in small, inbreeding communities cousin marriages expose defects rapidly, and these are in turn eliminated more quickly through the reduced viability or fertility of those who carried them. Once the group became adapted to inbreeding, it could thrive. The results of the cousin marriage study showed that, just like in the simpler outbreeding *Campanula*, natural selection continued to operate on man.[18]

But if natural selection was still operating on man, man in turn was increasingly controlling the substrate on which it worked. Medical advances afforded artificial fitness, allowing men and women who did not hitherto have the chance to breed to do so. "Those who were saved as children," Darlington wrote very much in the vein of his *Heredity* partner Fisher, "return to the same hospital with their children to be saved."[19] What then, should man do? As usual and unlike Fisher, Darlington remained cryptic: the advantages and disadvantages of interference had to be carefully weighed.

> Meanwhile, let us note a larger principle: every branch of government in its own field controls the evolution of the people it administers. The punishment of crime affects the reproduction of the criminal class. Education affects the mating groups of those who pass through the system. Taxation and subsidies affect the relative numbers of children born in the different social classes and can be adjusted to vary the results.[20]

Once again, Darlington's interest lay in the inculcation of a principle. Just as in his cytological work, practical details and their ethical implications were of lesser interest. More than anything, Darlington was a theorist, a rationalist, a thinker, and *not*, as he had already discovered many years earlier during his arduous trip to Persia, a man of action. Sterilization and artificial insemination seemed to him sensible techniques with which to begin conscious control of evolution in man, but he would let others fight these battles.[21] To Darlington's mind, it was the *principle* that first had to be gotten across. "For mankind," he wrote, ". . . will not willingly admit that its destiny can be revealed by the breeding of flies or the counting of chiasmata."[22] To help him along would demand two things: "education of the public in regard to learning, and education of the learned in regard to the public."[23] It was crucial if the future was not to be forsaken for the present that laymen and leaders alike understand the selective effects their actions were likely to have. To that end, Darlington would need the help of the past.

∽ IN THE MEANTIME, the battles at Oxford continued. Old-school botanists were not the only ones resisting Darlington's unification program. Thanks to the untiring efforts of Osborne, who relinquished scientific research for the purpose, the Botany Department had found a new home on the edge of the University Parks in the

Science area in 1951, bringing the previously isolated botanists closer to the other scientific departments.[24] To Darlington's mind, however, the move had failed to spark the cooperation between forestry, zoology, botany, agriculture, biochemistry, anatomy, pathology, and clinical medicine made necessary by modern genetics. In each of these departments, genetics was being taught in such a disconnected manner that it could not possibly become the causal glue of the life sciences, as Darlington envisioned. Each department zealously, foolishly continued to guard its own turf and its insular culture.[25]

Darlington held firm, continuing his crusade for genetics in the university. A measure of relief came in 1961 with the arrival of John Pringle, the forceful and dynamic Linacre Professor of Zoology. Keen to bring in molecular biology, and to dissolve the conventional boundaries between the disparate life sciences, Pringle was a welcome ally, and a joint policy agreed between the two departments brought about the establishment of the Biological Prelim in 1963.[26] But Darlington wanted to go further. In 1956 he had proposed that a Chair of Genetics be created. Four years later he suggested a Centre of Biological Studies in which research and teaching in cytology, genetics, virology, immunology, microbiology, biophysics, and biochemistry would be carried out. As both initiatives met strong university and departmental opposition, the degree to which Darlington was hampered by his institutional reliance became abundantly clear. Increasingly frustrated, he spent less and less time on administrative committees, and more and more on what would be, at least in scope, if not in intellectual synthesis, his greatest project yet—*The Evolution of Man and Society*. Withdrawing, Darlington would need to pay a price in isolation for regaining his lost independence.[27]

Despite his withdrawal, and, perhaps, because of his dependence, Darlington did not forsake the opportunity to give testimony before Oxford's internal Franks Commission in 1964; indeed, he relished the occasion.[28] The commission had been set up to investigate the academic and economic activity of the university as part of the reforms in higher education instigated by the government. Ruthlessly attacking the university for its conservatism and neglect of the sciences, calling its admissions and appointment policies "crooked," its educational vision "reactionary," and the colleges "corrupt, self-perpetuating oligarchies," Darlington succeeded in creating a minor storm that made its way as far as the *New York Times*.[29] Former Vice Chancellor, and Rector of Lincoln College, Walter Oakenshott, despite (or perhaps because of) cheers from the students, was not amused, publicly attacking

Darlington, and calling his claims "wild charges" and his behavior "scandalous."[30] Darlington was not the only scientist in those days convinced that it was the universities that were responsible for the creation of C. P. Snow's *Two Cultures*, and the universities that bore the responsibility to remedy it. Others, too, believed that the disconnected science departments were as much to blame as the philistinism of the humanities and backwardness of the administration.[31] Others, to be sure, gave equally damning testimonies before the commission. Nevertheless, the derisive and mocking manner in which Darlington went about expressing this opinion were singularly characteristic, making him a particularly unpopular figure with the administration. No wonder. "I should say," he had told an interviewer in 1964, "that the strongest motivation in my character, apart from curiosity, was indignation."[32]

THE DRIVE TO COMPLETE his magnum opus continued. Darlington had argued that social differentiation and breeding differentiation are mutually indispensable, but his case had been largely built on inference from genetic principles, and experimental data from plants and animals alone. The backbone, or theory, was in place. Now the content, the body, needed to be filled in. This would be the task of history. For history was one grand, tumultuous human genetic experiment. If its lessons could be properly learned, it would also be an indispensable guide to the future.

If breeding systems in primitive man, as in plants, always favored maximum hybridity, recombination, and uniformity, how had differentiation and stratification come about in advanced societies? To answer this question, Darlington would have to return to the histories of nomadic populations in Arabia and Somaliland, consider the effects of the inventions of language, agriculture, steel, and navigation, and study the cooperative and commercial history of differentiated tribes. In what ways were such stratifications maintained, or manipulated? For this he would have to learn the sanctions of the great religions, show how Hinduism, Judaism, Christianity, and Islam diverged in their effects on the breeding system, and consider the fine points of Roman, Ecclesiastical, and Indian Law. How had outbreeding followed by rapid inbreeding brought about the rise and fall of empires and nations? Here he would need to scrutinize closely the expulsions, migrations, and conquests of peoples, the edicts of kings, the indiscretion of queens, the policies of governments, and the inventions of "Great Men"—from the time of the first civilizations in the Fertile Crescent through the Chinese dynasties, African kingdoms, Western and Eastern empires, European nations, until the present. In order to convince the world that it was now possible to take *conscious* control of man's evolution, it would be necessary to demonstrate the epic results of humankind's hitherto *semiconscious* and *unconscious* control of it. To that end Darlington would have to do nothing less than rewrite the entire history of *Homo sapiens*. "We have been told the moral solutions of genetic problems for 10,000 years," he wrote in his diary in 1965. "Now we have to discover the genetic solutions of moral problems."[33]

In this quest Darlington spared little effort. At Magdalen College, where he became a Fellow upon assuming the Sherardian Chair, Darlington spent many hours at high table and in the Senior Common Room talking to G. R. Driver and O. R. Gurney, and to Colin Hardie and K. B. McFarlane, learning from them many Semitic, Hellenic, and Mediaeval histories. With the Reverend Arthur Adams, who became a friend, he discussed the bible and religion. With Philip Tyler, and with Gwendolen's son, Paul Harvey—both historians—he conferred on wide-ranging historical issues. He frequented the Ashmolean and Pitt Rivers museums and corresponded widely with archaeologists, linguists, human geneticists, animal behaviorists, and historical specialists of every ilk, British and foreign. In 1963 he traveled to Africa, where in Kenya he met Louis Leakey, and where in Johannesburg he was shown around by Raymond Dart. Two years later he was in the Canaries, and a year later still, in America, visiting Georgia, New Mexico, and Washington. In 1968 he was traveling again, this time to New Zealand and Australia, for a more extended stay of three months. As always during his travels—and in the spirit of his hero Galton—Darlington took pains to study the history, religions, professional practice, breeding habits, and migrations of the inhabitants, and to note them down carefully. Above all, Darlington read voraciously. In all these ways he was preparing himself to write the book that would be his greatest monument. Just as with his studies at St. Paul's, and his scientific work on chromosomes, Darlington was going about this task in his own unorthodox, self-taught, and idiosyncratic way.[34]

⌒ DARLINGTON CAME TO SEE THAT HISTORY COULD be thought of in different ways. The Jews and Greeks took a teleological view of it, believing that everything happens by design and purpose, and converges to an appointed end. The Chinese, by contrast, held strongly that nothing has any demonstrable cause or reason. A third view, that of the Puritans, of Marx, and of Spencer, imagined history as the evolution of man by progress, while others, such as Toynbee and Spengler, understood it to be a cyclical phenomenon in which civilizations rise and fall, only to rise and fall again. Darlington, however, came to hold a fifth view: history was a series of reciprocal interactions between individuals and their environments. Neither progress nor decay was inevitable, but no individual was unaffected by what went before him, or without effect on what comes after him. In history all events and processes were connected. Some of these connections were deterministic, and others "nondeterministic" due to the organized uncertainty of genetic recombination. The causal connections could thus be entirely ascertained in retrospect, but not entirely predicted prospectively.[35] What remained for Darlington to do was to *prove* that this was so.

A massive book of over seven hundred densely typed pages, filled with twenty-two maps and diagrams, forty-four charts and tables, and eighteen genealogies, *The Evolution of Man and Society* was a major publication, and was translated into most European languages and Japanese.[36] His publisher, Allen and Unwin, considered the book comparable in importance and scope to Toynbee's *A Study of History*, Bertrand Russell's *History of Western Philosophy*, and even Darwin's *Origins*.[37] Here, finally, Darlington produced what he took to be the historical evidence for his genetical claims. Here he explained everything from why early Christianity embraced monasticism and nunneries (low fertility due to overly aggressive outbreeding); why the Jews collected taxes for mediaeval Europe, to how it is that the Irish genetic propensity for story-telling made its way into Norse mythology (through the Irish settlement of Iceland). Here he ascribed the fall of Byzantium in part to the large role of eunuchs in its governance: by coupling social dominance with biological sterility, they ensured that the qualities socially selected for would be biologically selected against—surely an unstable system; the corresponding system in the Western world depended on the *nominal* celibacy of powerful clerics who in fact often reproduced themselves luxuriantly: they only relinquished the possibility of having legitimate offspring. Here too Darlington showed how inbreeding, stabilized by long-term selection, produced groups with specialized abilities, and how these groups affected the fate of nations. Spain's expulsion of the Jews in 1492, for instance, quickly became Holland's priceless gain; the French expulsion of the Huguenots in the seventeenth century likewise meant the gain of a highly skilled and accomplished population by Britain and South

Africa. Cornish miners, probably originating in Anatolia, became successful miners the world over, not because of a fortuitous culture, but due to those genes selected in them that furnished the capacity for such a skill. Since genetical admixture was always the main source of cultural and institutional development, Darlington concluded that where it is prevented, as in the success of the modern anticolonial movement, regression is likely.[38]

Following Darwin, Darlington treated man, in his heredity and variation, in his physical, emotional, and mental properties, his individual, social, and racial character, in his diseases and his speech, in his behavior and his beliefs, as a proper object of experimental study. Following Galton, he applied man's differential intelligence to the problems of history and of primitive and advanced society. After Alexander Morris Carr-Saunders, he showed how human, like animal, societies regulate their reproduction by instinctive moral controls that reduce competition and avoid conflict, and how such innate morality may be distorted or suppressed by religion. Following Fustel de Coulanges, Darlington demonstrated how stratified societies derived from the cooperation of different races. Following Nikolai Vavilov, he showed how in the domestication of plants lay the secret of man's history, and after Alexander von Humboldt and George Perkins Marsh, how indeed that secret lay as well in the *destruction* of man's habitat by agriculture, and later by invention. Like H. G. Wells and Carleton Coon, Darlington recognized that the history of man demands many methods of inquiry, and following the historians Lord Acton and Henri Pirenne, and the archaeologist Flinders Petrie, he placed genetic assumptions at the center of his claims. Darlington acknowledged his debt to all these men, as well as to the anthropologists William Halse Rivers, Baron Fitzroy, Richard Somerset Raglan, and Vere Gordon Childe, and to the historian and philosopher of morals William Edward Hartpole Lecky.[39] Coming after them, and appreciating their labors through the gaze of modern genetics, and with the help of modern dating techniques in archaeology, Darlington felt he could now achieve a greater connectedness than anyone before him. Language and religion, economics and technology were all inextricably entwined as causes and consequences with race and class, mating and fertility. Scientific judgment and integration would banish bias and disarray.

"Man has to go forward," Darlington wrote in his diary in 1953. "He therefore flatters himself that going forward is good. It is no fallacy to suppose that going forward is good, but merely that progress is good. Our good fortune, or ill fortune, is that progress is inevitable. And also irreversible."[40] Now, sixteen years later and in the year America put a man on the moon, the history of humanity seemed to him to vindicate this view. What it also taught, however, is that the appreciation of the great connectedness in human affairs could finally allow for prudent and humane planning of man's evolution; it could be used to direct and safeguard his destiny. Indeed, never before had such a monumental history of the past been written in order to shed light on man's path into the future. If history had shown that man's success lay in his variability, and that his failures stemmed from shortsighted exploitations of the environment; if introduced genetic differences have always sown the seeds of social differentiation and the cooperation of dissimilar individuals out of which the richness of human culture has arisen; if Great Men always resulted from outbreeding, and could change the course of history; if stratified societies always competed favorably with unstratified ones; if civilization always evolved most constructively in the center of its distribution, where movement brings the greatest chance of outbreeding and competition the greatest scope for selection; if culture was never just transmitted through ideas, but, through migration and marriage, in the genes; if the heredity of the group was the environment of the individual, and if the environment selected man as much as man selected the environment, the world's leaders would need to drastically change their thinking, and ways, in order to plan for a better future.

Taking all these together, Darlington had one central message. Man's diversity and diverse environments—intimately connected to one another—had to be maintained at all costs. The two great uncertainties affecting his evolution—the unique recombination creating the Great Man and the unforeseeable future of infectious diseases—made this imperative even stronger. The fact that misfits, criminals, and delinquents were the necessary price to be paid for hybridity did not change the basic truth. "Man's future prospects," Darlington concluded, speaking of the Bushmen, the aborigines, and other primitive tribes,

> are proportionate to the amount of genetic diversity he maintains among the inter-fertile members of his own species. In this respect, more than in any other, the loss of any primitive and apparently unsuccessful tribe affects the future of mankind as a whole. In this respect, mankind is one; and for us men are undoubtedly the most precious of animals. In order to preserve him it is not enough that we refrain from killing him. We have also to preserve the diverse habitats which diverse peoples need for their survival. To be sure, the restricted and specialized habitats of

civilization give the greatest opportunities for what we (in an ignorance we have come to share with Galton) are pleased to call intelligence. But we have now learnt that intelligence is of many kinds. It has to be measured not on one scale but on many. And its diversity, if lost, cannot easily be recovered. We have therefore to preserve these diverse habitats, along with their diverse inhabitants, from damage which civilization has so far so wantonly wrought upon them.[41]

∽ AS A YOUNG MAN IN 1926, Darlington had rejected Julian Huxley's eugenics. Now, more than forty years later, having detailed a complete philosophy of biological determinism, and rewritten the history of mankind from a biological point of view as against the Marxist, the egalitarian, and the progressive, Darlington had, in effect, turned the historic eugenic argument on its head. From a rigid hereditarianism, he had arrived at conclusions that could be read as manifestos of environmental groups. For the environment, Darlington now believed, even more than man, and more than ever before, needed attention. Man was not the landlord, but the tenant of the earth, and one with a limited lease.[42] If he failed to protect his environment, by destroying the variability it helped preserve in mankind, he would ultimately be endangering himself. Ministers of education, agriculture, immigration, finance, and law; farmers and teachers, heads of state, developers, scientists, and university chancellors would all have to learn this lesson well if mankind were to save itself. For his part, Darlington, or history, had laid the facts bare.

～ 14

On the Uncertainty of Determination

B<small>Y THE EARLY</small> 1940s, Darlington had satisfied himself intellectually that biology was applicable to man, and could help him understand the past and plan for the future. In order to ensconce a new social genetics Darlington had needed to proceed in parallel down a number of tracks. To gain an independent voice, he had founded *Heredity*; to gain a wider audience, he had moved to Oxford. There he had waged a full-scale battle to integrate genetics into a university culture beholden to tradition and averse to change. But all this time a greater conflict still had been raging outside the cloistered walls of Magdalen College and the old stone buildings of South Park Road. As he turned to a wider audience, Darlington had been waging a ruthless social and political battle precisely when much of what he was saying was anathema.

～ I<small>F</small> M<small>ARX MADE</small> "Class," and Hitler "Race," taboo, in Darlington's view the two men had been unconscious collaborators. To him, after all, race and class were the same thing. Despite their contradictory political philosophies, Darlington wrote in his diary, Nazi Germany and Lysenkoist Russia were therefore both practicing "biology witchcraft."[1] Darlington had been more vociferous in his criticism of Lysenko than he had of the Nazis, most assuredly because of his hereditarianism, but also because he always chose less obvious battles. Still, he had drafted a condemnatory Statement on Nazi Race Policy for the

Conference of the Allied Ministers of Education in 1945, in which the notion of a genetically superior nation was dismissed on scientific grounds.[2] Between these two extremes Darlington found to his dismay that liberalism too had lost its way. "No Scientific Basis For Race Found by World Panel of Experts," read the *New York Times* caption following the release of UNESCO's 1950 Statement on Race, to Darlington a prime example of such misguidedness. The fruit of a U.N. Economic and Social Council resolution asking UNESCO "to consider the desirability of initiating and recommending the general adoption of a programme of disseminating scientific facts designed to remove what is generally known as racial prejudice," the Statement denied any relation between culture and genes, emphasized the genetic dynamicism rendering the concept of "race" scientifically meaningless, rejected the notion that races differed in mental capacity, and emphatically put out of court a hierarchy among groups of peoples, or any danger resulting from miscegenation. Contributors, among others, were Julian Huxley, J. B. S. Haldane, Leslie Dunn, and Theodosius Dobzhansky, all world-leading geneticists.[3]

To Darlington, this was a mistake. Different races never had been, and never could arise, in similar environments. Endowed with different gene complements, their environment, *by definition*, would always be different. Groups did differ in their innate capacities. Diseases, temperaments, and proclivities did differ in proportion from one group to the next. While intelligence could be modified by learning, heredity was predominant, and did vary among races. Culture, therefore, was very much influenced by genes. Race crossing did produce progeny different in innate capacity from either parent, and could be either advantageous or disadvantageous, depending on the environment. There might therefore be a biological justification for prohibiting intermarriage between races *in certain cases*.[4] To the anthropologist Alfred Metraux, head of the Division for the Study of Race Problems at UNESCO, Darlington wrote:

> There is a danger that any statement about Race issued by people who disagree with the Nazi views on Race . . . will be designed as a reply to those views. Since the Nazi views were emotional in expression and political in purpose, any discussion of them by scientists should be explicit, and explicitly separate from the expression of scientific opinions. Otherwise their opinions will be confused by the emotional and political issues. This confusion is

found throughout the UNESCO Statement. . . . By trying to prove that races do not differ in these respects we do no service to mankind. We conceal the greatest problem which confronts mankind (and particularly in respect for the organisation of UNESCO) namely how to use the diverse, the ineradicably diverse, gifts, talents, capacities of each race for the benefit of all races. For if we were all innately the same, how should it profit us to work together? And what an empty world it would be.[5]

The anthropologist Ashley Montagu was unhappy with Darlington's article. Such determinist conclusions were an illustration of "the dangers of uncritical application of the findings of geneticists on plants and lower animals to the explanation of the difference between human groups."[6] Metraux, for his part, was furious. Any opinion contrary to the UNESCO Statement, in his mind, was racist. Darlington, therefore, was nothing but "an extreme racist."[7]

Such talk alarmed Darlington's old friend Hermann J. Muller, who thought the egalitarian statements of UNESCO were not only fallacious, but also dangerous. Should something, behind the scenes perhaps, be done? Could Darlington organize a British group to try and change the statement? Darlington was pessimistic. In his view no such group could be formed. "Half the British geneticists are Fellow Travellers rather than geneticists!" he wrote back.[8] Determined, Muller—himself once much more than a mere "Fellow Traveller"—circulated a critique of the Statement among geneticists, but failed to bring about any big changes; the postwar intellectual climate was marked by a strong cultural determinism, and violently opposed to any unorthodox talk of race.[9] "'Race,' they said," Darlington recalled years later, "who has ever heard of such a thing! 'Species' we believe in. 'Ethnic groups' . . . we will allow. 'Cultures' we admit and admire. But races, never! And so far as Classes and Heredity are concerned, isn't 'Background' good enough? Why commit oneself to an unpopular opinion? . . . By 1945 . . . the result was everywhere the same: a unanimity such as no inquisition had ever established."[10]

∼ IF LITTLE EVIDENCE existed for Darlington's strongly determinist claims, his sociological allegations nevertheless were very close to the truth.[11] In the 1930s the Nazi Party in Germany began its vicious perversion of Mendelian genetics and Darwinian evolution to degrees of evil that had never been experienced before. The rest of the scientific community, rather than depoliticizing genetics and evolutionary theory, as it had attempted to do in the case of Lysenko, responded by bending Mendelism and Darwinism in the service of antiracism. Before the 1930s leading geneticists had openly expressed racist views; Julian Huxley and J. B. S. Haldane are examples. Since that time, however, British and American liberals were forced, in no small part due to Nazism, to reconcile their overall worldview with their racial prejudices, and the two old friends gradually became symbols of antiracism.[12] They were joined in Britain by Lancelot Hogben, and in the United States, where racial rather than class-based problems were the issue, by an overwhelming majority of researchers. Lionel Penrose, chair of the Galton Laboratory of Eugenics and the leading human geneticist in the world, proclaimed in 1945 that the only "racial" issues with which his discipline ought to be concerned were those relating to the human race as a whole.[13]

Darlington's claims for the genetic basis of classes and races were not, and could not have been, supported by sound genetic evidence. There simply existed so few known, or approximately known, genetic markers of variability (Dobzhansky had listed a grand total of twenty-two, including three based on blood grouping, in his presentation at a 1950 Cold Spring Harbor conference on the subject) to render any discussion of genetic differences between races meaningful.[14] And yet, despite the fact that evidence to the contrary, namely that there existed no genetic differences between races, was likewise still lacking, a powerful consensus against discussion of genetic determinants of racial differences began to emerge. Guided as much, if not more, by their liberal, antiracist sensibilities than by scientific considerations, many geneticists now argued that prudence demanded that such inquiries be abandoned and condoned. After all, the Nazis had made abundantly and tragically clear to what ends such ideas could be directed. To Darlington's rigid mind, this position spelled nothing but outright cowardice. Although genetical practice and theory had in the 1930s grown more sophisticated, producing sensitivity, hitherto lacking, to the complexities accruing from its application to man, in Darlington's estimation the impact of Nazi racialism, and the resulting popular sentiment, upon geneticists' self-expression could not be overstated. In the debate between either attempting to study, identify, and characterize any existing genetic differences among races, or denying *a priori* that such evidence existed in the service of antiracism, choosing the latter seemed to Darlington positively dangerous.[15]

◡ ON THE FACE OF IT, Darlington's first book on man, *The Facts of Life*, should have been read and championed by every geneticist the world over. Darlington was a talented writer, and the book was an eloquent, powerful panegyric to a science deeply in need of one. With abundant wit, and attention-catching turns of phrase, Darlington dismantled Lamarckian notions of heredity, clearly explained modern genetics and its history, and compellingly argued how important it was for laymen and leaders, historians, philosophers, and social scientists to take it into account. The text was accessible, forceful, and clear. There were few world-class geneticists who could produce such a book.

A reaction to a reaction, *The Facts of Life* was a reply to the UNESCO Statement on Race, and in the highly flammable postwar political atmosphere, the backlash from within Darlington's own camp followed almost instantly. Every one of the reviewers, even those more sympathetically inclined, denounced Darlington's determinism as *scientifically* unfounded, his argument and presentation of the facts as one-sided. Dobzhansky at Columbia and Penrose at University College London accused Darlington of making no distinction between "authentic biology and gratuitous fancy." His portrayal of the evolution of social class in Europe over a period of fifteen hundred years as though it were known to be due to a physical polymorphism seemed to them ridiculous. His reasoning for the predominance of genetics over environment—man chooses his environment, the environment chosen is determined by the genotype of one who chooses, *ergo*, the environmental effects are not environmental but genetic—was the height of casuistry. His claim that it is not individual but species plasticity that allows for adaptation was with no scientific foundation; in fact, evolutionary principles stipulated that adaptation to rapidly shifting environments required genetically determined *individual* plasticity. Darlington's dismissal of evidence of identical genotypes in man reacting differently with varying environments was irresponsible. His assumption, wherever there was any evidence of a genetic component in a human difference, that the difference was wholly genetic, amounted to scientific recklessness. While the opposite inference may have been the fault of many social scientists, *he* should know better.[16]

Even Darlington's friend Muller, to whom the book was dedicated, could not endorse it wholeheartedly. Hearing of the dedication, Muller wrote to Darlington:

It would be rash to predict in advance my reaction to the book's ideas, although I suspect that, considering how heterodox we both are, we agree to a surprising extent. I imagine that my chief difference would concern itself with matters of *race* differences, as expressed in the behaviour of individuals and peoples, since despite my stand against the exaggerated UNESCO Statement on the subject I believe that differences of historical background have played much the largest role in the observed behaviour differences between peoples. On the other hand, as you know, I think that the importance of genetic differences between *individuals* is usually grossly underestimated (emphasis added).[17]

After reading the book, Muller wrote to add:

I think our area of disagreement would lessen markedly if you were to spend a year in Hawaii as I am doing, and see at first hand how much different races come to appear alike when they make an effort to ignore their differences, and how salutary that appears to be.[18]

Muller was one of the few close geneticist friends Darlington retained, and yet even he was saying (politely) something akin to what Hogben had said. Darlington, isolated behind a microscope at a horticultural institute, and now at the ivory tower of Oxford, was not in touch with the world. Like the models of the population geneticists he himself had criticized, Darlington's simple and clean genetic deductions were far removed from an exceedingly messy and complex reality. His strong group determinism, and his dismissal of economic and social factors, seemed to Muller misguided. Was it not more scientifically tenable to emphasize individual variability than to speak of racial groups? Was it not true that men couldn't always choose their environment, and that some environments were a lot more stifling than others? In any event, could he make such authoritative statements when so much remained unclear?

Postwar genetics was being affected by two intimately connected processes: the metamorphosis of eugenics into the rigorous and rapidly growing science of human genetics, and a parallel heightened sense of caution in its dealings with issues of race. The eugenics movements, led in the United States by Frederick Osborn and in Britain by C. P. Blacker, had been cleaning up their acts since the war, emphasizing Galton's nineteenth-century liberalism and distancing themselves from Nazi race-based policies. Prominent proponents of eugenics such as Muller and Huxley, while remaining elitists, had become champions of a eutelegenesis, or germinal choice, ostensibly purged of class and race biases.[19] Darlington, neither a part of the human genetics community[20] nor cautious in his statements on race and class like the postwar eugenists, was knowingly driving straight into a brick wall. Convinced that political dictates were playing a bigger role than scientific commitments in attitudes both toward race and individual determination among scientists, he saw no reason to back down. Had not the same thing happened in Nazi Germany and in Soviet Russia? The longer falsehood was appeased, Darlington thought—as Munich and the reaction to Lysenko had both shown—the worse the encounter with truth would be.

MOST OF DARLINGTON'S OLD-TIME COLLEAGUES, whether staunch anti-determinist like Dobzhansky and Penrose, or increasingly cautious, though firm, determinists like Muller, knew better than to call him a racist. They knew that his was a characteristically prickly overreaction to postwar hypersensitivity; that although deplored by liberal progressives as anachronistic, or worse, Darlington's rather nineteenth-century self-assurance, like that of Galton before him, professed a genuine belief in non-sentimental rationalism. Moreover, Darlington's was an unusual endorsement of the virtues of both inbreeding (race formation) and outbreeding (race submersion): the ability of races to keep apart maintained civilization, but their ability to mix allowed evolution and promised survival. Most colleagues were aware that Darlington believed the idea of a hierarchy among peoples nonsense. To John Baker he wrote:

> Has equality among men any scientific meaning? The concept is moral. An ape may be equal to a man or a chair equal to a table morally. But not scientifically. All races and all individuals among men (even one-egg twins) differ genetically and are unequal genetically on every scale of comparison that may be taken. We may equate them by an integration of scales but this involves a moral judgment.[21]

To himself Darlington noted: "The distinction between races and classes is a real one. But any order of merit is merely relative to past conditions: it may always be reversed in the future. . . . Do we want an aristocracy? No: a diversified, stratified, balanced community."[22] This was hardly mainstream racism.

But Darlington often unforgivably slipped into language that was difficult to interpret as anything other than outrageously racist and chauvinist. Argentineans, he told a London audience, although good cattle breeders, are "an inferior race." "No," he answered a man in the crowd who asked whether it is essential that all people must be taught science, "I would not teach the lower orders anything very much!" Lesbians, he wrote in *The Facts of Life*, should be barred from the teaching profession because they championed a feminism that disregards the manifest differences between women and men, which could only prove dangerous to impressionable minds. The list went on and on.[23]

Such egregious pronouncements quickly led to a justified challenge to Darlington's professed impartiality. In the absence of clear-cut data, could his confident proclamations on racial and class-based genetic differences really imply no value judgment? Penrose and Dobzhansky rightly didn't think so. To them the determination of racial and class character was a fallacy already "hoary with age" when Count Arthur de Gobineau tried to give it an aura of scientific respectability in the late nineteenth century; presenting it now as a revelation arising from modern genetics was more than pure farce:

> Biological Racism reached its most extreme development so far in the racial myths of the Nazis. The revulsion from the horrors of Hitlerism made racism temporarily abhorrent to most people. But it would be too optimistic to hope that racism died out with the Third Reich. It is deeply rooted in the insecurity feelings of too many individuals and groups. The post-war political reaction creates a favourable climate for a revival of racism. The book under review aims to show to social scientists how important it is for them to take 'the facts of life' into account . . . it can only show how naïve a biologist can be when he starts to deal with social problems. The author does not favour wars between classes and races, he favours co-operation. But what kind of co-operation he has in mind is clear from the example which he gives: that of the Negroes and whites in the southern United States. The book may, then, accomplish a purpose the author does not intend: to give the

renascent racism a specious scientific basis, elaborated with a vigour and brilliance worthy of a better end.[24]

⌐ THE COMBINATION OF the professionalization of human genetics, the sensitized political atmosphere, and a martyr's reactiveness did not bode well for Darlington's views. If he was convinced that man needed to exploit the determination of uncertainty inherent in his makeup, most everyone else remained very much uncertain about the degree of that determination. Many accused Darlington of merely trying to provoke his political adversaries. "It is true that, especially since Hitler's rise to power, there has been a prejudice against recognizing genetic differences among men," Dobzhansky and Penrose wrote, but wondered, "Will this be diminished by the author's attempt to swing the pendulum even farther to the other side of reason?" Alex Comfort of the *Guardian* concurred. Darlington had a case, but his "overstatements transcend the bounds of proper provocation and breed irritation or a wish to apply depth psychology to the author."[25]

Everyone knew that Darlington thrived on controversy, but most geneticists felt that in the present climate, he was doing more damage to his cause than good. Many of them conceded, as Curt Stern put it, that "Darlington is nearer the truth than many whose caution makes them shy away from a recognition of the fatefully great extent of genetic determination of human variability," but the fact remained that, in postwar Britain in a field led by Penrose and his disciples at the leftist Galton Lab, he was quickly labeled a reactionary, the more outrageous of his pronouncements magnified, the finer points lost. Accusations of racism were inevitable. Anthony Allison, whose discovery of the heterozygote advantage of sickle-cell Darlington so liked to quote, called parts of *The Facts of Life* "a surprisingly short step to . . . *Mein Kampf.*" Before long, precisely as Penrose and Dobzhansky had feared, Darlington's views began being used as scientific justifications for a radical right-wing agenda in search of respectability, a trend culminating in fascist Oswald Mosley's reference to his work on cousin marriages to bolster a campaign in the early 1960s to pass segregation and anti-miscegenation laws in Parliament.[26]

⌐ FOR DARLINGTON, what was at stake was the question posed by Thomas Henry Huxley in his famous 1863 book, *Man's Place in*

Nature.[27] The failure to recognize, as Huxley had, that man was still a subject, however unique, of the law of natural selection was always either due to misguided idealism, or worse, political fashion. When Huxley's grandson, Julian, sent him a letter reversing his grandfather's argument, Darlington was disgusted. "I think you might find it profitable to think of cultural mechanisms as constituting a new 'level' just as, clearly, genetic mechanisms constitute a new level as against the underlying chemical level," Huxley wrote, adding:

> Cultural evolution does have its own trends and results which cannot be (at the present stage of our knowledge at least) wholly deduced from the underlying genetic mechanisms any more than biological evolution can be deduced from the underlying physico-chemical mechanisms below the gene level. I once tried to put it epigramatically by saying that the new level of life was based on the self-reproduction of material units—genetic—that of human societies and cultures on the self-reproducing of mental units such as ideas.[28]

Huxley's psychosocial explanation of evolution, in which man singularly extricated himself from conventional pressures of natural selection and assumed his leading role in evolution through the development of his mind smacked to Darlington of Lamarckism: psychological selection depends on different prospects of survival of different ideas—how do people and populations choose ideas?—to meet their evolutionary requirements. But to Darlington it was plain that man doesn't select ideas, rather natural selection selects man plus his ideas. The grandson of Darwin's bulldog was selling out scientifically in order to bolster a case for purpose and direction in human evolution. This made Darlington seethe in anger. "Since 1937," he wrote in his diary years later, "Huxley's method was to take man out of the discussion. My method was to put him right into the middle of it!"[29]

As Darlington read former Haldane student and Nobel Laureate Sir Peter Medawar's anti-eugenic arguments for the primacy of cultural, *nonbiological* evolution in human affairs, his consternation only grew.[30] He was convinced that those entrusted with the responsibility of fulfilling T. H. Huxley legacy had caved in: Huxley was deriving his biology from an atheistic religion, Dobzhansky from a theistic one, and Medawar, worse still, from contemporary politics. One should only arrive at one's religion, Darlington held, from biology. Making believe that man had transcended his was dangerous self-delusion.[31] Medawar, for his part, assaulted Darlington's gross "geneticism," his simplification of incredibly difficult problems, and his claims of scientific certainty where none existed. If anybody, it was Darlington who was extrapolating. Less sympathetic than Huxley, Medawar extricated Darlington from the community, speaking of the "self-taught quality" of his writing, and disputing his right to speak for science at all.[32]

The same rigid anti-authoritarianism and prickliness that had led Darlington to his peculiar, anti-consensus reaction to Lysenko in the mid–1940s was now pitting him against a new kind of consensus, this time concerning biological determinism and race. But whereas hindsight tends to vindicate Darlington's unrelenting stance against Soviet subordination of science to politics, developments in human genetics and evolutionary theory soon began to discredit his claims on man. Though Dobzhansky, Penrose, Medawar, Huxley, and like-minded geneticists consciously adopted the strategy of using modern evolutionary biology and genetics as a means of combating racism and shattering stereotypes, they were right to attack Darlington's unfounded pronouncements. Ironically, however, while the liberal's unsubstantiated counterclaims of the 1950s slowly gained supporting evidence in the years that followed, Darlington's smug righteousness only grew. Increasingly, the gap between his professed partiality and actual biases became abundantly patent.

〜 GRADUALLY MORE ISOLATED, Darlington turned to more receptive quarters. As the 1960s ushered in a conservative reaction to the social legacy of the 1940s—the welfare state and full employment—and to the large influx of immigration from the West Indies, and later from Africa and Asia, the voice of a "New Right" made itself heard in Britain.

In search of respectable scientific authority for its anti-egalitarian claims, the movement was glad to present a forum for biologists. Despite a contempt for his old rival Ruggles Gates, and despite his differences with the anti-planning John Baker of the SFS, Darlington joined their ranks, becoming involved with the journal *Mankind Quarterly*, and, together with a wider international circle that included E. Raymond Hill and Charles Tansil, the Italian Corrado Gini, the German Eugen Fischer, and Americans James Gregor, Henry Garrett, and Donald Swann, with the International Association for the Advancement of Ethnology and Eugenics (IAAEE).[33] A nonprofit educational

PUNCH, February 13 1957

LAND OF OPPORTUNITY

An alarmist Punch cartoon, 1957, illustrating the prevalent fear
among the "New Right" that England was losing her own white
population to migration, while immigration of African and Asian
peoples was on the increase. Notice the sign pointing the new
immigrants to the Welfare State. Found in the Darlington Papers,
Bodleian Library. Punch Limited copyright.

organization, the IAAEE sought to promote the interdisciplinary study
of race and race relations. Gregor, its secretary, wrote to Darlington to
explain:

> The Association was originally formed by academicians who
> chaffed under the express and subtle restrictions imposed by pre-
> vailing sentiment. The Association seeks to restore true freedom

of expression in an area where that freedom is seriously curtailed. You are, I know, familiar with some of the agencies which have affected that curtailment (I refer to UNESCO 'Statement on Race' which seeks to establish orthodoxy where freedom of expression and inquiry should prevail.)[34]

Glad to have found a forum of like-minded iconoclasts, Darlington became from afar a member of the Executive Committee, and, a year later, Titular Head of the Association. Adopting the fight for freedom of expression, he joined the International Committee on Science and Freedom, chaired by Michael Polanyi, that same year.

But Darlington remained uncharacteristically cautious of his association with the IAAEE, a sure measure of its pseudoscientific nature. "How much genuine scientific and academic support or driving force do they have?" he wrote to Gini, "Or is their support and driving force largely political?" Darlington would have nothing to do with politically motivated racists, like Robert Gayre, editor of *Mankind Quarterly*, and a member of the association. Gini replied with a calming letter, and Darlington's apprehension quelled, though he never published in Gayre's journal. Friends and family thought that Darlington, who was also in touch with a number of race-studying educational psychologists from Georgia since his trip there in 1966, was being naïve. After all, Gini had been Mussolini's demographer and was a known fascist; Fischer was an old Nazi. The liberal-minded mainstream, for its part, was not fooled by the association's respectable academic title and pretence. Rejecting an invitation to join, one British scientist wrote: "Your aim is immoral, the means to your ends are positively sinful."[35]

⌒ AS EVIDENCE TENDED TO CLARIFY the somewhat scientifically vacuous debates of the 1950s, and to refute Darlington's naïve brand of strict biological determinism and racialism, political developments increasingly reintroduced these debates into center stage. This was deeply ironic. What could be dismissed in the 1940s and 1950s, with little scientific evidence to the contrary, as an outcast view now returned in the midst of the Civil Rights Movement and immigration debates of the mid-1960s and 1970s, despite mounting proof against simplistic linkages between race and complex socially conditioned traits, to ignite a full-blown cultural war between Right and Left.

Racial tensions in Britain had already escalated to such a point that

Parliament felt obliged to pass a Race Relations Act in 1965. Enoch Powell's notorious "Rivers of Blood" speech at Birmingham in 1968—in which he argued that Britain must be mad to allow 50,000 dependants of immigrants into the country each year—although leading to his unceremonious dismissal from the Shadow Cabinet, had had an enormous popular impact and forced the government to be highly restrictive in its attitude to immigration from then onward. While overtly Fascist organizations such as the National Front had doubtless made political capital of Powell's impact, racial populism, with the rise of immigration numbers from 2,000 per annum in 1953 to 136,000 in 1961, had moved from the gutter to the center of politics in inner-city Britain.[36]

Increasingly, this cultural and political debate was being fought on the battlefield of science.[37] A growing number of research programs and popular publications began to reassert the use of science as a means of distributing social roles, rewards, and punishments, or at least of justifying the way they were distributed.[38] If Darlington had become involved with a rather marginalized, outré crowd at the IAAEE, under such conditions the publication of *The Evolution of Man and Society* nonetheless made a splash right in mainstream intellectual life. Reviewed widely by leading biologists, historians, sociologists, and economists, it too became part of the debates then raging on human nature and its political offshoots. As conservative social thinker Max Beloff put it in a review of the book: "This view of history and the nature of the social theory that results from it are neither of them likely to receive a warm welcome in the present intellectual climate. They challenge not so much the facts as seen by historians . . . but the framework of assumptions of most of the liberal intelligentsia of the West. Dr. Darlington's view of history stresses what they would rather forget."[39]

Accordingly, *The Evolution of Man and Society* demarcated a divide between hereditarians and environmentalists, more generally translatable in political terms to conservative versus "liberal" or "radical."[40] The eminent, yet-to-be-discredited hereditarian educational psychologist Sir Cyril Burt thought the book "ought to be read by every educated man and woman who has at heart the welfare of his country and mankind;" hard-nosed biologist Garrett Hardin lamented that it didn't contain *enough* biology to bolster genetic claims that were surely true, however unpopular; William Hamilton, a leading theoretical biologist and father of the concept of kin selection, gave Darlington his reward for writing the "outstanding non-fiction book of the 1960s;"

and race-theorist Carleton Coon added a sterling review in the *Boston Globe*. On the left, eminent physicist and historian John Ziman called the book "Scandalous!" in reply to his own question: "Do some theories stink?" locating the real danger not in ideology in science, but in scientism in ideology; Marxist historian Robert Young gravely called the book a "take-over bid," and more jokingly summoned in a psychiatrist to diagnose the author's fancies as nothing but the product of the senile delusions to be expected in a man exhausted by long years of devotion to microscopy; Medawar, too, was not impressed.[41]

Despite its polarizing effect, however, what seemed most salient among level-minded critics, just as it had been with *The Facts of Life*, was an overwhelming consensus that Darlington's theory, however provocatively stated and authoritatively presented, was not strictly scientific. Some areas of human genetics had most assuredly expanded greatly in the previous few decades. Much was known about human genetic diversity: more than a hundred genetic markers in human blood alone were on record by the mid-fifties, and they could be shown to be distributed unevenly in different human populations. A reasonable understanding of the chemical basis of gene action in humans had also been achieved: many genetically controlled, disease-producing enzyme defects were known. Certain chromosome defects had been established as causes for mongolism, abnormalities of sexual maturation, and other disorders of development. That there were inherited differences in susceptibility to disease, and that natural selection *can* operate on human populations, as it does in lower animals, was not debated. But Darlington's insistence that cultural evolution is dependent on genetics, that races differ genetically in cognitive abilities, and that selection is Darwinian rather than socioeconomic, had gone further than the evidence would allow. Any honest, well-informed researcher would have to admit, regardless of his political leanings, that science just did not have answers for such claims. Had Darlington been able to support his theory? Most everyone rightly agreed he had not. Sir Karl Popper, for many the arbiter of what could and could not be counted as scientific, agreed:

What is your evidence for a genetic basis of professional skills? And what can be your evidence? That such skills, and ways of life, are inherited (rather than transmitted) is a possibility. But how can we test it? The idea that the ability to rule, to administer, is inherited is obviously one which could be easily misused. . . . You rightly stress hybridisation. But this opens all the doors, and any test of your theory becomes impossible. I admit that your theory may be right, and I admit that it may become testable one day, but at present it seems to me that though it is much more interesting and much more plausible than Toynbee's, it is almost as little defensible on 'scientific' grounds. I feel very sad that I have to say this, just because your theory is both extremely interesting and not implausible. But it seems to me that you rely on a kind of argument that I regard as invalid: that your theory of history allows us to put the jigsaw puzzle together. It is a dangerous argument: Freud relied on it, and Toynbee. It is particularly dangerous if the pieces of the puzzle are somewhat elastic and plastic.[42]

Beloff concurred. The book was important, but, in the absence of any real scientific proof for its claims, "perhaps a dangerous one." In the hands of some, it might tend to corrupt.[43]

Darlington was not in the least surprised. After all, he had heard cries of "danger" in reference to his work before. Then, like now, he thought, such cries were nothing but the creaking sounds of the great sociopolitical machine as it encountered new, disruptive knowledge. The irony was that Darlington's once "dangerous" cytological work of the 1930s was now cited by critics arguing that a man *secure* in his reputation in one field should be wary of extrapolating uncritically to another.

⁓ DARLINGTON'S ADVOCACY OF RACIAL, CLASS, and individual determinacy, however—and here lays the key to understanding his reaction to the particular social, political, and intellectual context of his day—was far from straightforward. Claiming more than could be legitimately extrapolated from the evidence was, in fact, a *deliberate strategy*. "Many Critics," Darlington wrote,

tell me I cannot *prove* my contentions to be true. The more naïve charge me (rightly) with not distinguishing between "fact" and "hypothesis"—which is why they think me so bad for the young. To all these I must say: I have never proved anything. I do not count on doing so.

Laying bare his tactic, he added:

What I do count on is to assemble such evidences and arguments

as will make those who disagree with me feel more and more
uncomfortable.[44]

Just as with his scientific work, deduction from first genetic principles
directed Darlington's hypotheses on man and culture. If this meant
claiming more than could be actually perceived at present, posterity
would act as final vindicator. Such had been the case with cytogenetics;
such would also be the case with the evolution of man and society. What
demarcated the two enterprises was a matter of degree rather than of
kind. When it came to man, the conflict of science and society stirred
even greater waves of resistance. If, as Darlington believed, "science
advances as though by the pulling out of a drawer which gives on one
side only to jam on the other," then, *especially* with man, "there is
nothing to be gained except by pulling on the other side."[45]

~ IF DARLINGTON WAS consciously overstating his case at the
expense of propounding a wrong-headed gross geneticism, the principle of resistance to new, disruptive knowledge nevertheless still held
true. Popper understood this. "The theory of discovery in *Conflict of
Science and Society*," he wrote to Darlington, "seems to be almost identical with the one I have been proposing for a long time. I mean the
role played by destruction (refutation) of old knowledge." But Popper
nevertheless could not agree with Darlington's method:

> I cannot agree with your treatment of equality. It only speaks of
> the natural inequality of men, but not of the principle of equality
> before the law. Men are born unequal, but we ought to treat them
> as far as the law and public institutions are concerned as equal.
> This leads to a lot of difficulties, obviously, but it is a general
> regulative principle. (It does not for example imply equal votes: we
> might, say, give one additional vote for knowledge of an additional
> living foreign language or—though I should not recommend
> it—for every Olympic medal). Similarly we are not born free. But
> we should be as free as is compatible with other people's freedom.
> The real trouble about equality is this: who should be the judge?
> Your own analysis shows that we cannot trust officials; not even
> geneticists.[46]

Popper, Darlington felt, just like everyone else, had failed to see the point. Just as with his attitude to the cytoplasm, which he thought profitable to study only after the role of the nucleus was sufficiently understood, Darlington thought the discussion of political and moral questions premature before the biology of behavior and mind was clarified. "With regard to equality," Darlington wrote to Popper,

> I am afraid I was speaking only of innate character. The profound problems of legal and political equality are for me derivative and unprofitable to discuss until the biological argument has been thrashed out. Again the question as to who should judge merit on what scale can arise only, I feel, in relation to particular social activities and is therefore similarly a political and derivative problem. With regard to the application of genetic principles to the study of society, you say the results could be misused. But have you ever known an idea about human beings that could not be misused?[47]

"Injustice is stronger than you," Darlington liked to quote Alexandr Solzhenitsyn, "It always was and it always will be. But let it not be committed through you."[48] Other men concurred, but held a widely different view of what precisely injustice constituted. Unlike Darlington, they were much less certain of the fact of determination.

~ 15

One Final Hurrah

On January 9, 1970 tragedy struck the Darlington family. Andrew Jeremy, Darlington's second son, had chosen to follow in his father's footsteps. Showing early aptitude and a clever scientific mind, Andy had followed the Natural Sciences Tripos at Queens' College Cambridge, and gone on to take a Ph.D. on the genetics of *Aspergillus*. Bright and promising, he was invited by Walter Bodmer and Nobel laureate Joshua Lederberg to become a postdoctoral fellow in their Stanford University lab in Palo Alto, California in 1965. Andy seemed poised for a stellar scientific career. But the madness inherited through both sides of the family (Nellie's father and Margaret's uncle) afflicted him, too, and Andy began to suffer prolonged bouts of depression, beginning as early as seventeen. After four relatively happy years in California, and just shortly after assuming a research fellowship at the Imperial Cancer Research Fund at Lincoln's Inn Fields back home, he put an end to his life.

Darlington was shattered, too upset even to attend his favorite son's funeral. Rachel, Darlington's youngest daughter, remembers coming to visit him and Gwendolen after the funeral, and for the first and last time, seeing her father weep. Andy had, upon returning to England, made clear his abhorrence of his father's reactionary views and domineering manner. Darlington was especially hurt when his son gave a cursory glance at *Evolution of Man and Society* and put it down demonstrably. As profound feelings of remorse and guilt flooded him, Darlington became inconsolable. Those who knew him well said that from that day on he was never the same man.[1]

⌒ LIKE CELL DIVISION AFTER DISJUNCTION, life continued. Darlington had spent the better part of the last twenty years integrating his scientific knowledge with the study of man. Now it was time to rest. On June 10, 1971, the sixty-seven-year-old Emeritus Fellow, looking distinguished and still strikingly tall, stood up to give his farewell speech at Magdalen College. "Some of my friends probably think that I am able, or at least willing to talk at any time, at any length on any subject—and without any notes," he began, smiling.

> But there are two sides to this question. Some of you may recollect a shameful episode in this college. It was at the Restoration Dinner 18 years ago. I was called upon to reply to the toast of the New Fellows. Alas, I rose to my feet, confused with the wit of my colleagues (and the wine of the college). I heard admonitory words from my neighbour. It was the voice of Henry Whitehead. Desperate at the tedious possibilities before him, he was saying "cut it short, Cyril, cut it short." From this fraticitial blow (for, after all, Henry was a fellow professor and that was why I was sitting next to him) I failed to recover. I made the shortest utterance ever heard on such an occasion since the abdication of James—I mean James II. My collapse was not due to lack of experience, or of speaking admonition, not at all. For my first appearance here was not 18 years but 45 years ago. Forgive me recalling what happened. The occasion was a meeting in Oxford of the BA presided over by an illustrious member of this college. I did not attend his presidential address—perhaps on republican principle, though I can't be sure. But I gave my little talk, on the use of the chromosomes in the interpretation of history. As I timidly mounted the rostrum, sure enough the dreadful words fell on my ears. This time it was the chairman. "Speak up, Mr. D.," he said, "speak up." So you see my dilemma. All my life I have been trying to steer a safe course between Speaking up and Cutting it short. Now, however, the way is clear. I must cut it short. I must take the classical advice—I must cultivate my garden. I must withdraw to the post of assistant gardener. To whom? To my wife.[2]

⁓ WITH AGE, a certain mellowing was taking place. In the wake of Andy's suicide Darlington was spending more time with his growing family. He delighted in showing his children and grandchildren his new garden in South Hinksey, rattling off the Latin names of his plants and supplying cuttings to fill their own gardens. He seemed to take particular pleasure in showing off the glorious Pelargoniums with their fancy leaves.

In reality, Darlington was leaving the university a broken and disappointed man. Crushed by the suicide of his son, he was also privately despondent over an Oxford legacy that was far from secure. During his tenure, botany had faltered at the expense of other, quickly growing, and more attractive biological sciences, and the grandiose plans for unification had withered.[3] There were achievements to be sure: the growing power of scientists in university government; the establishment of new colleges for inclusion of Science Fellows; the Biological Prelim; the new course in human sciences, begun in 1971; the establishment of a lectureship in the history of science in 1953, followed by a chair in 1973; and, finally, the creation in 1969 of a Department of Genetics, were all developments which Darlington could look back on with satisfaction. Due to his abrasiveness, impatience, and hatred of committees, however, they were all, with the exception of the Prelim, achievements in which he had not played a direct and decisive role. While some friends and devoted students had been gained, many enemies were made too, and, retiring to his garden, Darlington seemed in retreat.

Darlington was getting older. Since the late 1950s and early 1960s, molecular genetics had completely taken over his science.[4] James Watson thought him clever, but already completely stuck in 1930s biology when Watson arrived in Cambridge in 1952 to embark upon his famous collaboration with Francis Crick.[5] Chromosomes were out, genes and their constituents in, and Darlington was a chromosome man. Indeed, plans for the publication of a last scientific book, *A Diagram of Evolution*, were scrapped when reviewers recommended that it would not do much for Darlington's reputation.[6] At the same time a quantitative psychology, foreign to him, was dominating the debate over race and intelligence. Although not a principal actor in the race-intelligence controversy, Darlington corresponded with, and appeared several times on joint panels with, Hans Eysenck and Arthur Jensen, leading proponents of the view that I.Q. varied with race and was strongly heritable. He also befriended Cyril Burt. Though he believed attempts to quantify the relative importance of heredity and environment were misguided, outwardly Darlington was glad to marshal these men's studies in public talks and articles as evidence for the claims he had been making all along.[7] Privately, however, Darlington knew that the professional study of integrative human heredity was passing him by. "Although the maths is beyond me (on I.Q.)," he admitted to his trusted diary, "I find the genetic assumptions made entirely sound."[8]

Advances in molecular biology, psychometrics, and human genetics had progressively pushed Darlington aside from the center of a debate he had fought with archaic tools. Evolutionary biology too was becoming a professional discipline with strict scientific criteria, and a much clearer divide between professional and popular argumentation.[9] Most of Darlington's old-time colleagues—Fisher, Haldane, Muller, Huxley, and Dobzhansky—all intensely political men, had by 1975 passed from the world. A new breed of geneticist and evolutionary biologist— represented by Richard Dawkins, Geoffrey Parker, George C. Williams, Nicholas Davis, and William Hamilton—was taking over. Politics and the relations between science and society, if still very much in the background, were increasingly extirpated from their language.[10] The shift in world political sensibilities, the change in the standard for scientific expression and practice, and the outspoken nature of Darlington's claims together rendered him a man out of time. Both scientifically, and in doctrine, he was perceived to be a dinosaur, a crank. Increasingly isolated and irrelevant, the assistant gardener, chromosome man, and rewriter of history was gradually being forgotten.

⌒ But despite his son's suicide, despite his retirement, and despite the fact that new forms of knowledge were passing him by, Darlington had yet to say his final word. If he still chose to use his intellectual weight at all, it was not for its actual scientific contribution, but in the fight against the perceived academic and social witch-hunt against the hereditarian enterprise. Public censure, peer admonition, funding discrimination, and *ad hominem* assaults on hereditarians had, in a politically adverse climate, become a matter of course. In retaliation, Darlington, together with fifty prominent scientists, signed the "Resolution on Scientific Freedom Regarding Human Behavior and Heredity" in 1972, in which the legitimacy of applying rigorous scientific, especially genetic, methods to the study of man's behavior, was staunchly, somewhat angrily, defended. He voiced spirited attacks on the treatment of hereditarians by a hostile academic environment. In parallel, he critically reviewed genetical studies, like that of Oxford's new Genetics Chair, Walter Bodmer, which he claimed shied away from determinist conclusions. Of Bodmer and Luigi Cavalli Sforza—leading, and much respected, human geneticists—he wrote: "Are we to derive our science from our morals, or our morals from our science? The first is the established view of authority. The second is the unestablished view of scientists. Our authors are proposing that, to avoid disturbing authority, the measurement of intelligence and its public discussion should be forbidden. This has happened in the Soviet Union. It might happen in the U.S. But it will not, I hope, happen here." In turn, Darlington championed studies, like Oxford colleague and Social Freedom of Science champion John Baker's 1974 book, *Race*. This was a vast, decidedly typological work of the old anthropological-school type that viewed races as fixed, biological categories strongly determinative of ability and behavior. Darlington's praise for the book was voiced in the face of almost unanimous condemnation.[11]

Although increasingly isolated and forgotten, Darlington reserved one final hurrah: *The Little Universe of Man*, the third leg of his trilogy on man, published in 1978. Despite his former prescriptive caution, and despite what he had written to Popper, Darlington was now, in the twilight of his life, ready to give the world some uninhibited advice. Here was a complete conservative manifesto: devolvement of the welfare state; endorsement of the return to guilds, trade unions, Friendly Societies, and local voluntarism; withdrawal from growth and the return to competitive management and labor; segregation between bright and dull in education, and the creation of tailor-made, targeted curricula directed by economic occupational imperatives; restrictive and calculated immigration laws; and the endorsement of European involvement in the Third World. Far from dealing with known scientific facts, Darlington marshaled a by-now wholly discredited, rigid determinism to advocate radical changes in national, and world, politics.[12]

To Darlington the perceived fact of determination made certain social and political policies necessary, however painful, and repugnant to Western liberal sensibilities. This meant intelligently applying sterilization and abortion to reduce mental deficiency and its attendant ills of pauperism, crime, and prostitution; recognizing that literate education beyond the age of twelve was wasted on the bottom fraction of education in stratified societies, and needed restructuring; comprehending that forced integration in education, intended to remove conflict, in fact exacerbates it, and must be reexamined; understanding that the

world was indebted to two European achievements—the control of
nature, and the understanding of nature and man—and that the solution of its terrible population and resource problems must therefore
ignore the moral indignation of colonialization; and facing the awful
truth that there is not going to be enough room for everyone, and that
some people are more useful to society than others. None of these
problems, Darlington claimed, could be solved by pusillanimous liberalism. None could be approached before deeply ingrained taboos were
obliterated.

> The European taboo on the discussion of man as an animal lasted
> until it was eventually broken by Darwin and Huxley. But in our
> dynamic world its place on the forbidden list was already being
> taken by another taboo, that on the discussion of sexual behaviour
> and this lasted, as we can still remember, more than another fifty
> years. In due course in the 1930s there followed a third taboo
> which now dominates the study of human problems and is likely to
> continue until another generation rejects and ridicules its parents'
> prejudices. This third line of defence against the understanding of
> man is the taboo on the study of hereditary differences. The one
> belief or emotion that unites the jarring nations today appears to
> be the need not to notice, and certainly not to discuss, the exist-
> ence of differences between them in terms of their permanent
> underlying causes. Innate heredity or genetic differences must not
> be admitted between individuals or groups, between classes or
> races, or even between the sexes. But above all what must not be
> discussed, what must be rejected, are differences in the founda-
> tions of human behaviour, the study of brains, of instincts, and of
> intelligence. These foundations are complex and happily con-
> cealed from the public view. They must remain concealed. In a
> world already overcrowded and over-troubled they might cause
> more trouble. Concealment, of course, means deception, pre-
> tence, and confusion. Successively with evolution, sex, and human
> affairs it has meant that scientists are pushed into deceiving, first
> themselves, then their pupils, and then the public. Thus fear of the
> truth, on the part of the public or of the establishment, which used
> to protect the central mystery of religion, has now shifted to fields
> of inquiry which are, next to religion, the most difficult. For in
> heredity and intelligence we have two subjects which inherently
> we can never entirely understand. Established opinion is objecting
> only to our making a start.[13]

Why, then, did Darlington's infatuation with man and his future not lead him to become a eugenics activist, as Muller, Fisher, and Huxley had? Darlington himself recognized early on that his was to be an intellectual life, not a life of action. But beyond his rather weedy energy level a more disturbing truth lurks. Darlington despised duplicity, cowardice, and self-protective caution. He hated fanaticism and blind, irrational faith. More fundamentally still, Darlington suffered from the same malady he once ascribed, in anger, to Haldane. "Any people but my own I dislike except as objects of anthropological study," he wrote in his diary.

> Of my own people I detest the urban proletariat and despise their petty bourgeois betters. I detest the old aristocracy, and as for the new plutocracy I find their money stinks. The specialist craftsman bores me to death. The intellectual seems to me a humbug even when he is not a careerist. Altogether I cannot accept by irrational instinct to love the human race, and those who do pretend to love any but a few choice friends I think are deceiving themselves.[14]

However fascinated by man intellectually, Darlington was not, as Muller and Huxley surely were, a great lover of mankind. It is perhaps fitting, therefore, though not without a touch of irony, that in his final *cri de coeur* Darlington acknowledged an old, lost friend, for whom he always possessed kindred feelings. "Although I parted company with him over communism," he wrote in the introduction, "it was in the long youthful discussions with the late J. B. S. Haldane, twenty years earlier, that my three books on Man had their roots."[15]

⌒ *The Little Universe of Man*, though widely reviewed, made a considerably lesser impact than *Evolution of Man and Society*. Darlington's dogmatism and his disregard for evidence contradictory to his claims were painfully obvious. The "pulling out of the drawer" finally had transgressed all acceptable political boundaries. So glaring was the gap between Darlington's professed partiality and his actual bias that the publishers, George Allen and Unwin, were forced to ask him, upon consultation with their solicitors, to make seventeen changes in the book manuscript in order to avert risk of prosecution under the Race Relations Act. Darlington's contentions that "the Negro has ruined the American cities," his implication that in nonwhites "behaviour is guided by pre-logical processes," that "the profession of witch-doctor dominates African life," and his sad admission that some of his ideas might even lead to violence all had to be dropped.[16] But his use of Cyril Burt's data without acknowledgment of the recent strongly evidenced allegations of fraud against him,[17] his uncritical marshalling of the widely debated cases of XYY criminals, and his offensive determination of Negro lack of intelligence from the point of view of Western culture nonetheless remained in the edition finally published. This was six years after Richard Lewontin had disposed of most serious scientific racism by showing emphatically that genetic differences between individuals within a "race" were greater than those between individuals from different "races."[18]

A small number of hereditarian biologists hailed Darlington's cause. Despite acknowledging that what he was doing was "far more than speculation, it is even more than conjecture, it may be called daring deduction," Konrad Lorenz agreed with most of the book. Ernst Mayr similarly expressed gratitude, saying: "I am delighted you have said all these things which are so true but which are simply suppressed in the 'egalitarian' mass media. Of course you are biased, but so is everybody else and, I am frank to say, your bias, or I should say government policies based on your bias, should promise a far better future for mankind than the ruling bias." But both men tellingly expressed these views in private letters. Published reviews tended to be highly critical, or worse. This time, Darlington had overstepped the boundary. Some colleagues and peers responded with a smile or a shrug. Less-forgiving critics were abusive. In the estimation of one anonymous reviewer for the publishers, *Little Universe of Man* was a disgusting display of the "worst sort of racial chauvinism," filthy dirt "unfit for human consumption." Such sentiment was buttressed by a not insignificant number of congratulatory letters to Darlington signed "from one racist to another." Even better (or worse) was Darlington's own endorsement. Asked by Anthony Storr in *The Sunday Times* whether he was a racist, Darlington replied: "Well, I'm regarded as one by everyone except the Jews, who *are* racist, and who utterly agree with my views."[19]

❧ THE TRANSLATION OF IDEAS from plant and animal genetics to man had finally come full circle. That it flew in the face of political and social trends, that it disregarded both current attainments of knowledge and standards of scientific procedure and responsibility, and offended individual and public sensibilities did not matter. To the aging andpolarized Darlington, all such obstacles were merely the stage props of the great drama of the "conflict of science and society." *He* would consciously carry on overacting his perceived main role, and the strategy, until the last, would remain the same. "After my last book someone wrote 'Darlington stinks' which isn't very nice you know," he said in a bittersweet mixture of disappointment and pride. "But I've got to do it. We need controversy to get out of the mess we're in."

Darlington doubtless believed in the 1950s that his geneticism was right on target. Though little, or no, direct supporting evidence existed at the time, lack of evidence to refute claims of racial and individual determinism in man on the one hand, and postwar hypersensitivity tending toward *a priori* pronouncements in the service of anti-racialism on the other, might well have fostered such a belief in a mind such as that of Darlington. But as human genetics grew and developed the kind of gross geneticism and racialism Darlington had advocated was resolutely refuted on scientific grounds. Mounting evidence of the interaction between genetic and environmental factors and their complex influence on behavioral and mental traits, and a growing awareness of the extent to which previous, simplified designations of traits such as intelligence were in fact socially constructed, had rendered Darlington's confident deductions void.

Polarized and somewhat blinded by his theory of the interaction of science and society, Darlington was unable to see how science was quietly passing him by. As the often gross discrepancies between classical phenotypic and modern genotypic categorizations of humans became apparent, as the concept of "population" replaced the old notions of "type" and "race," Darlington seemed too busy reacting against liberal authoritarianism to notice. Convinced that it was he who was proceeding with the unblinkered gaze of the impartial scientist, Darlington miserably failed to recognize the extent of his biases. His hero, Galton, had suffered a century earlier from the same shortcoming, though Galton may perhaps be excused on account of the general concordance between his views and the attainments of the science of his day. Darlington, though he rightly fought a juvenile and unscientific demagoguery from the Left in the decades following the war, cannot, in the end, be excused as lightly. By the 1970s it was patently evident that, an intrepid and brilliant cytologist in the thirties and forties, Darlington had, on man, become late in his life little more than an ideologue masquerading as a scientist.[20]

Conclusion: Paradoxes

Biographers always want to explain their characters, especially the irrational elements. These are chiefly of genetic determination, but since heredity is a mystery to them they always attribute oddity to environment just as an illiterate person would, at least one who did not know the family well.

~Cyril Dean Darlington

T HAT SO MUCH DETERMINACY WAS organized to express the results of uncertainty; that the study of units less than the organism gave birth to a theory of units more than the organism; that individuality and continuity have been sustained together throughout evolution; that selection works, not only on individuals, but on processes of fertilization and meiosis which occur before individuals or individuality exists—such were all paradoxes arising from the chromosomes. Cyril Dean Darlington, the man responsible for recognizing and crystallizing them, died quietly in a small room in the John Radcliffe hospital in Oxford on March 26, 1981, himself a man of paradoxes. He was a cytologist who was profoundly a theorist, a rationalist, and *not* an empiricist. He was a prophet of genetics who rejected the classical gene, a revolutionary left behind in the revolution of the new molecular age. He was an atheist who quoted the Bible; an antitotalitarian by conviction, and autocrat by temperament; a eugenicist who turned the eugenic argument on its head by arguing for the maintenance of the environment. He was a man obsessed with breeding and heredity who married a woman from manic-depressive stock, and who suffered the suicide of two of his five children; Debby killed herself in 1986, after his death. Indeed, Darlington was a fanatical determinist whose own life began accidentally, and continued serendipitously from bleak prospects to world fame and finally from rebuke to isolation to irrelevance.

A man in many ways both behind and ahead of his times, Darlington not only conceived, but also *lived*, the profound paradoxes of the human condition.

In some sense the themes—one relating to content, another to philosophy, and a third to method—unifying Darlington's life and work are each tinged by paradox. The first of these, tying together his work on chromosomes, genetic systems, evolution, and man, is Darlington's resolute critique of classical genetics. Genetics had originated through breeding experiments with whole organisms, and, until Darlington's *Recent Advances in Cytology*, suffered from a strong organismal bias. Its integration with evolutionary theory, on the other hand, put the gene, as an abstract mathematical unit, in the center of biological thought. The mechanical question as to how the chromosomes did their job was, if not a matter of indifference, certainly subordinate in importance. Darlington's insight that the mechanisms of heredity themselves underwent selection and evolution, helped him describe a much more dynamic and complex genetic system affecting the origin and propagation of species, a system that promises to challenge geneticists for generations to come.

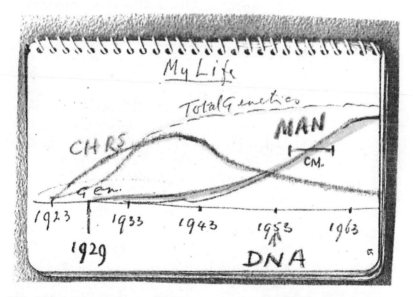

"My Life," a sketch from a Darlington notebook. CHRS—chromosomes, Gen.—genetics, CM—cousin marriages. Notice the change of focus in 1953, after the Watson-Crick discovery of the structure of DNA. Darlington Papers, Bodleian Library. Reprinted with permission.

Over thirty years after the publication of his great cytological tome, Darlington stepped up to the podium at the First Oxford Chromosome Conference in 1964 to give the keynote address. With the advent of molecular biology in the 1950s, much of his work had lost its relevance; the emphasis had in a flash of excitement following Watson and Crick quickly shifted from chromosomes to their constituent parts. As more became known of the biochemical function of enzymes in the nucleus, of the control of gene expression, and as the electron microscope allowed for higher resolutions, Darlington's mechanical and structural explanations of chromosome behavior had become both less convincing and less germane. His seminal scientific contributions already long behind him, he was finding it excruciatingly difficult to keep up with the new developments in biology. Darlington, it was clear, was quickly —almost too quickly—being overtaken by younger, eager molecular men. Still, "the man who invented the chromosome" was not convinced. Facing an audience of cytologists now considered to be, and in truth very much at, the periphery of biology, he said:

> For me it means that the sequence, molecule-gene-chromosome-organism-community, is a physical hierarchy corresponding to a living hierarchy. In evolving this adaptive hierarchy—as in the sexual cycle—chromosomes always come first; organisms come second. How different from this view of ours are the views of our colleagues in kindred sciences! They have heard of chromosomes, but to describe them they use a different language. The *anatomist* thinks of the chromosomes as very small rod-shaped bodies possibly present in all cells, visible with difficulty during the process of cell division and believed to contain the hereditary factors. The chemist is more definite, and sees the chromosomes as representing a chemical structure and a genetic code which he has himself discovered. The structure and the code together place the chromosomes in his total scheme of nature: from them the chromosome, as well as all the rest, can be deduced. The *experimental breeder* approaches the question from the other side. He can tell us what the chromosomes are like by looking at the whole organism. For him the chromosome contains a group of genes identified by, and deduced from, his own experiments, and enabling him to predict what will happen in whole plants and animals. The *mathematical geneticist* formalises these notions; he sees the chromosome and its constituent genes as a mechanical model obeying

rules of recombination and mutation, interaction and selection. These rules he knows how to express in mathematical terms and from them he can ascertain the laws of evolution, laws derived from, and applied to, whole organisms. Lastly, the *naturalist* brings us back to the view of the anatomist; he sees the chromosomes as properties belonging to whole organisms which he knows. . . .

Thus all our colleagues who are busy either with the structure of molecules or the appearance of organisms, regard the chromosomes as doing what various theories, such as the chemical theory of the chromosome and the chromosome theory of heredity, require it to do. They find that it is doing their job, smoothly and well—so smoothly and well that they can take it for granted; they can deduce its properties; they do not need to observe them.

We must applaud the success achieved by our colleagues on the basis of these assumptions. But they see the chromosomes through the mind's eye. We, who believe we see actual chromosomes through the microscope, must explain what we have seen, and point out that it is not always what our friends expect. For us, neither the chemical code, nor the linkage map of the chromosome, nor the genes embodied in it, are enough.[1]

Had a molecular biologist chanced to be present in the Oxford hall, Darlington's call would have elicited from him a smile. Depending on his character, it would have either been smug or lined with a delicate trace of commiseration. There is nothing worse professionally than the dishonorable shame borne of irrelevance. Darlington, after all, had been a great scientist in his day. A Crick or a Benzer or a Brenner looking in might well have felt a tinge of sadness for the man.

Two generations later, Darlington's words would surely today gain a more careful ear from a thoughtful geneticist or evolutionist, even a molecular biologist. As the limits of the strictly gene-based approach are rapidly being realized, Darlington's dynamic chromosomal perspective suddenly sounds much less anachronistic than it once did. The classical view of genes as localized sections of DNA that code for phenotypic traits is being replaced with a more complicated notion, encompassing the discoveries of overlapping genes, "genes in pieces," alternative splicing, developmental genes, complicated gene-gene, gene-cell, and gene-cell-environment interactions.

Obviously, Darlington could not have foreseen all such developments. However, considering the great promise of reductionism that rendered his own science obsolete during the latter part of his life, it is remarkable, and in some deep sense paradoxical, that the echoes of late–nineteenth-century theories based on a unifying, nonreductionist approach to the problems of heredity, development, and evolution are once again reverberating. From the bowels of reductionism are emerging the first threads of a renewed holism. Chromosomal phenomena such as inversions are more and more thought of as co-adopted systems; whole parts of the chromosome, such as the telomere and centromere (a name Darlington coined in 1936), are being considered both phylogenetically and ontogenetically; evidence is mounting that certain mutations are somehow directed; the "gene" is giving way to the "genome." The combination of the material basis with the evolutionary framework is indeed the only way to make sense of biology as a whole, even as we continue today to stumble forward in the dark in quest of solutions to unanswered problems. What this means is that in order to understand the development and evolution of organisms, we must think of the genotype as a coherent system with a long history, not just a linear aggregation of functional genes. We can place Darlington firmly on a short but distinguished list of biologists who from the outset pointed to the limitations of the power of the gene concept alone to explain biological phenomena.[2] Among them, he was most attuned to the role of history, or evolution, in the development of nature. It is a measure of the depth of his intuition that Darlington went to his grave believing that the full implications for evolution of his chromosome-based genetic systems—connecting the law-governed physical sciences from below to the historical and contingent biological sciences from above—had not yet been grasped.[3]

Darlington's theorizing of integral versus analytical genetics in relation to the study of man and culture stemmed similarly from the same critique of the classical mold. When we speak today of a biological determinist, we have in mind someone who believes in the omniscient gene. Yet it was precisely Darlington's rejection of the single-gene concept, and embrace of the much wider, more integrated genetic systems approach to biology, that led to his extreme determinism. The great paradox of genetics, and therefore of evolution, was that so much determination, in the form of mitosis and meiosis, chromosome pairing, and crossing over, should be organized to express the results of uncertainty. To Darlington's determined mind this was equally true for the lilly, for the frog, and for man.

Darlington's rise to fame began with his immaterialist teacher's reactionary rejection of the chromosomes, and ended with his own, similarly perceived reaction against the gene at the expense of the chromosome. As a champion of the chromosome, Darlington may have started off on an opposite path to that of Bateson, but he ultimately ended up with a conspicuously similar biological worldview. Programmatically striving to prove more than they actually could, both men were attracted to the cytoplasm, both emphasized the "system," and both gradually adopted a worldview based on the genetics of society, in which "Great Men" springing from mutation or recombination figured prominently. Paradoxically, both remained great believers in genetics while championing a kind of antireductionism not too far from holism itself. Both men's brand of extreme determinism must be judged harshly by the passing of time. Still, Bateson's obstinacy with respect to the chromosome cannot be vindicated in hindsight; there are signs today that Darlington's obstinacy with respect to the gene might ultimately be.

The second unifying theme of Darlington's life deals with the relationship between fact and value, or the derivation of ethics from science. To most minds, the function of scientific theory is to account for experience; the function of myth—to provide a source and justification for values. There are a number of ways to understand the relation between the two. One might hold—and this I take, sadly, to be a majority view—that the very same mental constructs should serve both as myths and as scientific theories. It is this approach that underpins the equation of what is "natural" with what is "right," and which has led to everything from robber-baron social Darwinism, to the critique of evolutionary theory by socialists, to its dismissal by some gay activists due to the lack of evidence for fitness of gays. A second approach is that myths are unnecessary, and scientific theories are all that we need. Many strive to believe in such a progressive scientism, but often wonder whether, if scientific theories can say nothing about what is right, but only about what is actual, some other source of values may be necessary. Finally, there are those who believe that mankind needs both myths and scientific theories, but that it must take pains to carefully distinguish the two. This approach holds that while there is no place in science for teleology, or value-laden hypotheses, nevertheless to do science one must first be committed to some values—not least to the value of seeking the truth. This value, however, cannot be derived from science, but rather is a prior moral commitment needed to make science possible. Values, then, are not dependent on science, but rather science dependent on values.[4]

Darlington presents an antithesis to this third position, without adhering to either the first or the second. To his deterministic mind, all moral and ethical systems must derive from fact, science being the ultimate system for its description. "Science is concerned with what a man (or a thing) must do, ethics with what he thinks he should do," he had written in 1941. "Until the contrary is proved, therefore, we must suppose ethics to be derivable from science." Both in his divergent reaction to Lysenko and in his work on man, Darlington zealously followed this precept. In so doing, he takes his place in a long-standing debate about nature and nurture, heredity and evolution, and their relation to man and his social and moral order. T. H. Huxley thought the "survival of the fittest" unsuitable a guide to moral behavior. Man should study nature's ways, he argued, and follow the very opposite strategy in his own social and political machinations. *Mutual Aid*, a book written by a gentle Russian anarchist, Petr Kropotkin, was an impassioned response to Huxley's discounting of nature's pertinence to human affairs. Kropotkin enumerated countless instances of occurrences in nature of what can only be described in human terms as compassion, cooperation, harmony, synergism, and kindness. Nature, then, far from being a negative blueprint for civilization, could become the source and guide to its good morals. Many years later, Stephen Jay Gould continues the debate. To Gould's mind, nature is not intrinsically anything that can offer comfort or solace in human terms. In fact, it is nonmoral by definition. The answers to moral dilemmas, Gould claims, reside, like the kingdom of God, within us.[5]

Darlington was convinced that the time had come to banish this age-old dispute. To be sure, scientific fact was one thing, moral reckoning another, but to deny any connection or interdependence between the two was deeply counterproductive. When his intellectual proclivity for synthesis became wedded to a psychological obsession with the role of the Great Man in history[6] Darlington began to think of this denial in almost mythic terms. Ate, the Greek goddess of infatuation, mischief, delusion, and blind folly, had from time immemorial rendered her victims incapable of rational choice and blind to distinctions of morality and expedience. From Paris's obstinate abduction of Helen to "Caesar's spirit, raging for revenge with Ate by his side," folly has been the curse of mankind. The inability to transcend morally ingrained, or politically inculcated, injunctions that run counter to conclusions made necessary by our biological nature, seemed to Darlington's troubled mind to epitomize the modern strain of this incapacitating affliction. Translating genetic principles from plants and animals to human history and culture, Darlington believed that from the very small and minute, sense could be made of the larger picture. Chromosomes and chiasmata explained race, class, and mating, and would ultimately need to be used as a guide to planning the future of mankind.

Here, too, a paradox looms. For the more he insisted that the conclusions he derived from the science were necessary, the less were the conclusions therein derived. The stronger he clamored against the inability to recognize fact as fact, and act upon it with moral resolution, the greater the underlying bias driving his pronouncements became. As Darlington's fight for a worldview based on scientific truths grew more intrepid, the gap between his own worldview and the science upon which it was supposed to be based only widened. To compensate, Darlington ended up making claims for science that, put simply, were not known to be true. At his worst, he sometimes made claims that were patently false.

The paradox is interesting, but is it instructive? Is it a comment on the relationship between fact and value, scientific theory and myth, that he who insisted more than most that the two are neither identical, nor interchangeable, nor wholly distinct and unconnected, but rather necessarily derivable one from the other, ended up betraying his very own claim? Does this mean that value can never really be derived solely from fact because the act of judgment necessarily entails a step unaccounted for, and not contained within, any fact, be it what it may? Or does it simply mean that we should be tirelessly careful before we call a fact a fact? Darlington was wrong on race and wrong on determinism. Is that necessarily why he ultimately reached conclusions we rightly reject as outlandish and morally unacceptable?[7]

The question is perhaps more germane today than ever. Determinism, or the possibility of determinism, is constantly hovering above us. The nature of genetic research today makes this inevitable. Clearly, the explosiveness attached to the matter stems from the fear that certain empirical possibilities open up the door to social and political evils. The fear is grounded in both our ancient and recent history, and is justified on that basis alone. Deterministic philosophies and sociologies have fuelled social programs from slavery to colonialism to eugenics; to the opposition to civil rights, to psychology and education. It is a tragedy that a warranted sense of heightened caution when dealing with such issues, both in the scientific and public spheres, had to await the terrible genocide visited upon us by the Nazis.

Still, Darlington's story makes clear that immediately after the war political sensitivities accounted as much as, if not more than, scientific considerations for the strong reaction against his ideas. That UNESCO refuted at mid-century the category of "race" without succeeding in achieving a consensus on the very definition of the category itself is telling. If Darlington was overstating his case, there is little question that what was widely accepted in the two decades after the war as the orthodox position with regard to both determinism and race equally stood on dubious scientific ground. Hindsight has tended both to vindicate and to castigate such orthodoxy: While race is no longer believed to be a viable biological category, the claim that genes have nothing to do with behavior has had to be abandoned. It is now clearer than ever that the reason why human society is so difficult to understand is because it is determined both by cultural and genetic inheritance. What people believe, and how they believe, is influenced by culture, but the ability to be influenced by culture and to build large stratified societies depends on genetic makeup. The reason why humans have developed complex, industrialized societies and a written information culture, whereas chimpanzees have not, is ultimately that humans and chimpanzees differ genetically. This doesn't mean that society must become a branch of biology, any more than biology should become a branch of chemistry. But just as biologists are expected to understand chemistry, so, too, would it be a good idea for social scientists to understand some biology. It may just be that sociology has as much to gain from biology as biology has already gained from chemistry. In his own application of biology to human history and culture, Darlington took this way of thinking too far, but he was pointing in an important direction.

Notwithstanding Darlington's own shortcomings, then, the basic question he posed remains unanswered. Is it indeed possible to create an honest science of human nature without a Pandora's box of negative consequences? Is the society of mankind on this earth responsible and mature enough to face up to certain possibly uncomfortable outcomes of our own quest for self-knowledge? Regardless of the ultimate fate of determinism, the attempt to deal with this question is doubtless a necessary step toward a healthier, better informed, more consensual, and more conscientious future for us all.

Finally, we can consider the most determinative factor in Darlington's life and career: his method—both of discovery and of expression. Our story has shown that paradoxically, more than *what* Darlington said, *how* Darlington said it, determined the reaction to his views. I say "paradoxically" because words are intended to connote meaning, but the meaning of Darlington's words was often scarcely attended to. "People rarely hear or even read correctly a statement which they think they are going to dislike," Darlington once wrote in his diary, and he was right.[8] To be sure, words connote more than just meaning.

They are never just words alone for they reflect tone, intention, timbre, and character—so the paradox must be tempered. Still, unifying Darlington's life's work in all fields was a deductive method that extolled hypothesis at the expense of fact, principle at the expense of exceptions, and generalities at the expense of particulars. At its core lay both a remarkable scientific intuition and a passion to break down the disciplinary divides that impede human understanding. Desperately seeking to destroy these divides, Darlington came against a cold wall of resistance, even contempt. In the process, he sometimes allowed his enthusiasm and contrariness to lead him to ideas unsupported by well-established fact. This had a deep effect on how he was viewed by contemporaries, who despite his obvious brilliance, never afforded him the stature and respect equal to that of Haldane, Huxley, and Fisher. Darlington felt this and, somewhere beneath his arrogance, was deeply wounded. "You can't do anything," Darlington wrote in his diary three years before his death, "unless you set out to do more than you think possible, more than you think you can."[9]

Perhaps Darlington tried too hard. Perhaps he overreached. Surely, he made many mistakes, some less forgivable than others. But maybe it is the failure of imagination on the part of those around him that deserves our more careful attention. Maybe it's the unwillingness to dare intellectually, to connect, to look beyond the encumbering compartments of knowledge imposed by authority and lack of ability alike, into deeper relations in this world. Then again, considering where such excursions ultimately led Darlington, maybe it is not. Or maybe it is a matter of degree. Whatever position one takes, that a major scientific figure of the twentieth century has been virtually forgotten until now forces us to ask such questions with renewed urgency. After all, they are about nothing less than how we think, and about how we understand the world in which we live.

Ultimately the question of method, which has surfaced as central in Darlington's career, touches upon the very definition of science itself. Darlington may have been pugnacious, contrary, and deeply enamored by his ability to shock, but seeking to advance more than he could actually prove was, in terms of his understanding of the dynamics of knowledge, *by definition* a scientific activity. The strategy of overstatement characterized Darlington's activities in his scientific work as well as in his work on man. Having achieved scientific success at a young age, it stands to reason that he should have adopted the tactic as a general method. To him, since science advances through "pulling of drawers," and through destruction of current knowledge, a bold methodology throwing caution to the wind was in fact very much *scientific*. Few agreed with this judgment. To most, Darlington made daring incursions into the realm of the unreal without renouncing his residence in the partly surveyed and charted region of what we are pleased to call certitude; most did not view such daring kindly. This is understandable. For many scientists, philosophers, and commentators, science separates itself from other kinds of intellectual activity—such as fantasy, or the moral sphere—precisely because it is the art of answering the answerable. Exponents of this outlook point to Newton, who could not discover the nature of gravity, but who could, instead, devise a mathematics that unified the motion of a carriage with the revolution of the moon. They point to Darwin, who did not try to crack the problem of the meaning of life, but rather settled for developing a theory to explain the reason for its change over time. The best science, exponents claim, proceeds by putting aside the overarching generality and focusing instead on a smaller question that can be reliably answered. In these terms, Darlington's method, and his entire career, must suffer under close scrutiny. But to those, like him, who believe that one of the prices to be paid for revolutions in knowledge is a simplification of complexities that necessarily leads to a loss of a modicum of truth, Darlington's life will stand as an illuminating tale. Whether or not one believes that the mysterious dance of the chromosomes is profoundly linked both to man's physical and spiritual future, no one can remain indifferent to Darlington's story and the lessons it offers about the interplay of ideas and the way we express and act on them. The passionate expression and vigorous challenge of new ideas, and their application to society, is where the future of mankind lies.

Darlington is unsettling both in what he got right and what he got wrong. Perhaps he would have considered this his greatest victory, that he unsettles us, his legacy to a world that to him seemed unwilling to listen. On the windowsill of Clare Passingham's house on a quiet Oxford lane, a row of glorious fancy-leaved Pelargoniums nods in the breeze, a reminder of the difficult father she and the rest of Darlington's children are still stuggling to understand. Blessed with a genuine social conscience and a curious intellect, and cursed with the certainty of knowing better than others what was right and true, Darlington lived a controversial life, for which he paid his own personal toll. "All my life," he wrote,

> I have been trying to tie things together, using or finding ideas that will help to make small bits into larger bundles. . . . It makes friends critical, critics angry and enemies furious. And it brings

understanding that is little and late. Why, they ask, doesn't he stick to the facts instead of offering us spurious fiction? But my school master understood when he reprimanded me at the age of ten. Darlington you are not attending; you are dreaming! I was. It had been a lifelong obsession; the one which (together with the boundless appeal of the senses) led me to the chromosomes.[10]

Notes

Abbreviations

AIBS   American Institute of Biological Sciences
AScW   Association of Scientific Workers
CDD   Cyril Dean Darlington
DP   Darlington Papers, Bodleian Library, Oxford
DSIRS   Division for the Social and International Relations of Science
EFD   Ellen Frankland Darlington
HE   hybridity equilibrum
IAAEE   International Association for the Advancement of Ethology and Eugenics
IOC   International Organizing Committee
LUM   Little Universe of Man
RPA   Rational Press Association
SFS   Society for the Freedom of Science
UNESCO   United Nations Educational, Scientific and Cultural Organization
WHRD   William Henry Robertson Darlington

Introduction

1. The Honorable Dame Miriam Rothschild, F.R.S., interview by author, October 24, 1999.

2. "A Geneticist's Interpretation of the Ways of Man," profile in *The New Scientist*, April 7, 1960.

3. See Hermann J. Muller's review of *The Evolution of Genetic Systems*, "How Genetic Systems Come About," *Nature*, 144 (1939), p. 648; obituary in the *Times* (London), March 27, 1981; "Cytologist Supreme," *The New Scientist*, April 16, 1981.

4. Dan Lewis, "Cyril Dean Darlington, 1903–1981," *Biographical Memoirs of Fellows of the Royal Society*, 29 (1983), pp. 113–157. There is also a short piece by Robert Olby in the *Dictionary of Scientific Biography, Supplement II*, ed. F. L. Holmes, vol. 17 (New York: Charles Scribner's Sons, 1990), pp. 203–209.

5. Diary (May 13, 1979), DP:g.44:A.96.
6. Cyril Darlington, *Darwin's Place in History* (Oxford: Blackwell, 1959).

1. An Improbable Birth

1. The following account is based on Darlington, "Family History," DP:e.10:A.72, written in 1946; "A Short History of the Darlington Family," by Oliver Darlington, son of CDD (unpublished); and family trees given to the author by Oliver and his sister, Clare Passingham. Darlington himself was to stress in later life the significance of the fact that he was a result of outbreeding, of a genetic recombination of characters from Yorkshire, Scotland, and Wales, and from the upper and lower tiers of society.

2. William Darlington (WHRD) letter to Ellen Frankland Darlington, December 1893. A large collection of these letters exists in DP:C.5:A.158–163.

3. Autobiographical notes (November 14, 1977), DP:C.1:A.3.

4. DP:C.1:A.3.

5. DP:e.10:A.72.

6. Ibid.

7. DP:e.19:D.31.

8. WHRD letter to A. E. Hillard (December 2, 1916), DP:C.2:A.105.

9. DP:e.10:A.72.

10. Dan Lewis, "Cyril Dean Darlington, 1903–1981," *Biographical Memoirs of Fellows of the Royal Society*, 29 (1983), p. 115.

11. DP:e.19:D.31.

12. Diary, 1924, DP:f.9:A.68.

13. DP:C.1:A.1.

14. Darlington letter to Alfred (March 30, 1918), DP:C.5:A.166.

15. DP:e.10:A.72.

16. During the war Rutherford had famously sent a message to the antisubmarine warfare committee to say that he could not attend the meeting because he had split the atomic nucleus—which, if true, was of greater importance than the war. Darlington's proclamation was therefore slightly belated.

17. Darlington letter to WHRD (1922), DP:C.5:A.169.

18. Darlington letter to Nellie (1922), ibid.

19. Darlington letter to WHRD, ibid.

20. Philadelphia: J.B. Lippincott, 1920.

21. Profile of Darlington in *The Literary Guide*, January 1948, p. 3.

22. DP:g.41:A.96; DP:C.1:A.4; DP:C.9:B.1.

23. "John Innes Horticultural Institution: A Brief History," written and circulated by Darlington as a confidential report to the John Innes Council, (July 16, 1953), DP:C.11B.50. A number of memoirs of scientists who worked at the John Innes exist in the institution's archive, but a proper history of the John Innes Horticultural Institution has yet to be written, and would be an important contribution to the history of biology in Britain. Robert Olby's "Scientists and Bureaucrats in the Establishment of the John Innes Horticultural Institution under Bateson," *Annals of Science*, 46 (1989), pp. 497–510, traces the competing vested interests that ultimately decided the institution's direction, not least through the

choice of its first director. For further reading on plant-breeding institutions in Britain, and the important role of genetic research in their development, see Paolo Palladino, "The Political Economy of Applied Research: Plant Breeding in Great Britain, 1910–1940," *Minerva*, 28 (1990), pp. 446–468.

24. See his widow Beatrice's *William Bateson: His Essays and Addresses Together with a Short Account of His Life* (Cambridge: Cambridge University Press, 1928). While many articles have been written on different aspects of Bateson's science and thought, a comprehensive biography is lacking, other than the above and J. G. Crowther's politically tainted article in *British Scientists of the Twentieth Century* (London: Routledge and Kegan Paul, 1952), pp. 248–310.

25. See Donald A. Mackenzie's chapter "Biometrician Versus Mendelian," in his *Statistics in Britain 1865–1930: The Social Construction of Knowledge* (Edinburgh: Edinburgh University Press, 1981), pp. 120–152, for a social constructionist approach to this debate focusing on the participants' social and political commitments; William Provine, *The Origins of Theoretical Population Genetics* (Chicago: University of Chicago Press, 1971) for an interpretation centering around personal relationships of the scientists involved in the debate; Bernard Norton, "Metaphysics and Population Genetics: Karl Pearson and the Background to Fisher's Multifactorial Theory of Inheritance," *Annals of Science*, 32 (1975), pp. 269–301, for an approach focusing on philosophical and scientific traditions underscoring the debate; Lyndsay Farrall, "Controversy and Conflict in Science: A Case Study—The English Biometric School and Mendel's Laws," *Social Studies of Science*, 5 (1975), pp. 269–301, for a focus on different methodologies and research techniques; and Robert Olby, "The Dimensions of Scientific Controversy: The Biometric–Mendelian Debate," *British Journal for the History of Science*, 22 (1988), pp. 299–320, for an emphasis on different training and the diverging interests of the principal actors in different types of variation. More recently, the debate has been revisited in Kyung-Man Kim's *Explaining Scientific Consensus: The Case of Mendelian Genetics* (New York: Guilford Press, 1994). Here it is argued that the natural world plays a central role in the resolution of scientific controversies. Unsurprisingly, its foreward is by Olby. The biometrician–Mendelian debate thus remains a fighting ground for "internalist" and "externalist" accounts of science even when the extreme positions of Mertonian "scientific" ethos, on the one hand, and Strong Programme constructivism, on the other, are nuanced.

26. The story of rediscovery has been treated extensively in the literature. Two interesting studies are Conway Zirkle, "The Role of Liberty Hyde Bailey and Hugo de Vries in the Rediscovery of Mendelism," *Journal of the History of Biology*, 1, no. 2 (1968), pp. 205–218, in which he shows how the three rediscoverers did not form a "mutual admiration society"; and Curt Stern's account in *The Origin of Genetics: A Mendel Source Book*, eds. C. Stern and R. Sherwood (San Francisco: W. H. Freeman, 1966), in which Tschermak's contribution is minimized. Onno Meijer, "Hugo de Vries No Mendelian?" *Annals of Science*, 42 (1985), pp. 189–232, argues for Corren's primacy. An interesting interpretation of the reasons for the neglect of Mendel's experiments and the creation of the many "Mendel myths" is given in Jann Sapp's "The Nine Lives of Gregor Mendel," in *Experimental Inquiries* (Dordrecht: Kluwer Academic Publishers, 1990), pp. 137–166.

27. Cambridge: Cambridge University Press, 1902.

28. J. B. S. Haldane, *Possible Worlds and Other Essays* (London: Harper and Brothers, 1928), p. 135.

29. See the colorful personal reminiscences of R. C. Punnett in "Early Days of Genetics," *Heredity*, 4 (1950), pp. 1–10.

30. Taped interview of CDD by Brian Harrison (July 25, 1979), John Innes Historical Collection, the John Innes Institute, Norwich; DP:C.88:F.82.

31. Darlington–Bateson correspondence (November 1923), DP:C.9:B.1.

32. Brian Harrison interview.

33. Unpublished draft, "How I First Saw the Chromosomes," DP:C.42:E.150.

2. A Rising Tide

1. For standard introductory accounts see L. C. Dunn, *A Short History of Genetics* (New York: McGraw-Hill, 1965); A. H. Sturtevant, *A History of Genetics* (New York: Harper and Row, 1965); and E. A. Carlson, *The Gene: A Critical History* (Philadelphia: Saunders, 1966).

2. William Waldeyer, "Ueber Karyokinese und ihre Beziehung zu den Befruchtungsvorgängen," *Archiv für mikroskopische Anatomie*, 32 (1888), pp. 1–122. (English translation entitled "Karyokinesis and its relation to the process of fertilization," *Quarterly Journal of Microscopic Science*, 30 (1889), pp. 159–281. Quote on p. 181.)

3. Giving the world the now forgotten *gemmules* (Darwin), *ideoplasm* (Naegeli), *biosphores* and *ids* (Weismann), and *pangenes* (de Vries). See Paul Julian Weindling, *Darwinism and Social Darwinism in Imperial Germany: The Contribution of the Cell Biologist Oscar Hertwig (1849–1922)* (New York: G. Fisher, 1991).

4. See Arthur Hughes, *A History of Cytology* (London: Abelard-Schuman, 1959) and William Coleman, "Cell Nucleus and Inheritance: An Historical Study," *Proceedings of the American Philosophical Society*, 109 (1965), pp. 124–158, for historical accounts of early cytology and its relation to genetics. On Mendel see Robert Olby, "Mendel no Mendelian?" *History of Science*, 17 (1979), pp. 53–72, and V. Orel, *Mendel* (Oxford: Oxford University Press, 1984).

5. See Hermann J. Muller, "Edmund B. Wilson—An Appreciation," *American Naturalist*, 77 (1943), pp. 5–37, 142–172.

6. This instance can be added to those discussed in Robert K. Merton's "Singletons and Multiples in Scientific Discovery: A Chapter in the Sociology of Science," *Proceedings of the American Philosophical Society*, 105 (1961), pp. 470–486.

7. This school gained adherents in the United States as well, and was led by Edward Murray East on the plant side and William Ernest Castle on the animal side.

8. Bateson, Punnett, and Saunders, "Experimental Studies in the Physiology of Heredity," *Reports to the Evolution Committee of the Royal Society*, 3 (1905), pp. 1–131.

9. Years later Punnett explained that not linking "linkage" to the chromosomes was due to his and Bateson's impression of Boveri's 1888 paper on the individuality of chromosomes, for they interpreted it to mean that breakage and recombination were forbidden (see Punnett, "Early Days of Genetics," *Heredity*, 4 (1950), pp. 1–10). Curt Stern pointed out, in "Boveri and the Early Days of Genetics," *Nature*, 166 (1950), p. 446, that this was due to a misreading of Boveri, who had in 1904 discussed the pairing and exchange of parts of homologous chromosomes and

presented the breeding evidence required to prove it.

10. See Garland Allen, "Hugo de Vries and the Reception of the Mutation Theory," *Journal of the History of Biology*, 2 (1969), pp. 55–87.

11. See Darlington, "Morgan's Crisis," *Nature*, 278 (1979), pp. 786–778. Garland Allen deals with Morgan's initial skepticism in his biography *Thomas Hunt Morgan: The Man and His Science* (Princeton, NJ: Princeton University Press, 1979), and again in "The Evolutionary Synthesis: Morgan and Natural Selection Revisited," in *The Evolutionary Synthesis: Perspectives on the Unification of Biology*, eds. Ernst Mayr and William B. Provine (Cambridge, MA: Harvard University Press, 1998), pp. 356–380.

12. Allen fails to mention this episode. The letter from Woods Hole exists in the John Innes Archive, and in copied form in DP:C.23:D.32.

13. For a comprehensive account of the Fly Room enterprise, see Robert Kohler, *Lords of the Fly: Drosophila Genetics and the Experimental Life* (Chicago: University of Chicago Press, 1994).

14. Morgan was aware that the only true confirmation that existed was strictly genetical, coming exclusively from breeding experiments. Nevertheless, he expressed the hope that although Janssens had not shown it to be the case, the physical mechanism of the recombination phenomenon would in the end prove accessible to decisive cytological demonstration.

15. Morgan did, however, remain cautious in his assertions, speaking in terms of associations and correlations, and of an instrumental gene rather than an outright physical one. See Raphael Falk's discussion of the difference between realism and instrumentalism in "What Is a Gene?" *Studies in the History and Philosophy of Science*, 17 (1986), pp. 133–173.

16. For a close reading of Johanssen's 1909 paper and explanation of its significance, see Jan Sapp, "The Struggle for Authority in the Field of Heredity, 1900–1932: New Perspectives in the Rise of Genetics," *Journal of the History of Biology*, 16, no. 3 (1983), pp. 311–342.

17. Morgan, T. H., Sturtevant, A. H., Muller, H. J., Bridges, C., *The Mechanism of Mendelian Heredity* (New York: Holt) 1915. On the reception of the chromosomal theory of heredity, see Stephen G. Brush, "How Theories become Knowledge: Morgan's Chromosome Theory of Heredity in America and Britain," *Journal of the History of Biology* 35, 2002, 471–535.

18. See Daniel Kevles, "Genetics in the United States and Great Britain, 1890–1930: A Review with Speculations," *Isis*, 71, no. 258 (1980), pp. 441–455. Also, Darlington's (draft) "The Survival of Genetics in Britain 1900–1950," DP:C.43:E.178.

19. Quoted in Michael Ruse's *Monad to Man: The Concept of Progress in Evolutionary Biology* (Cambridge, MA: Harvard University Press, 1996), p. 340.

20. Punnett, the first and only genetics professor in Britain at that time, reigned in his chair at Cambridge until the 1940s and by then had not even introduced *Drosophila* to his research program!

21. See Charles E. Rosenberg, "The Social Environment of Scientific Innovation: Factors in the Development of Genetics in the U.S.," in his *No Other Gods: On Science and American Social Thought* (Baltimore: Johns Hopkins University Press, 1978), pp. 196–209.

22. See Darlington's "Genetics and Plant Breeding, 1910–80," *Philosophical Transactions of the Royal Society of London, Series B*, 292 (1981), pp. 401–402.

23. See Robert Olby, "William Bateson's Introduction of Mendelism to England: A Reassessment," *British Journal for the History of Science*, 20 (1987), pp. 399–420. An interesting comparison to Bateson (Darlington's teacher) is Edward Charles Jeffrey (G. Ledyard Stebbins's teacher), who also rejected the chromosomal theory of heredity but for completely different reasons. See Vassiliki Betty Smocovitis, "Botany and the Evolutionary Synthesis: The Life and Work of G. Ledyard Stebbins, Jr." (Ph.D. dissertation, Cornell University, 1988), pp. 60–61. A more often cited comparison is that between Bateson and William Johannsen, who also rejected the chromosomes. See F. B. Churchill, "William Johannsen and the Genotype Concept," *Journal of the History of Biology*, 7 (1974), pp. 5–30. For a general discussion of the rejection of the chromosome theory, focusing on one of its premier critics, see Garland Allen's "Opposition to the Mendelian–Chromosome Theory: The Physiological and Developmental Genetics of Richard Goldschmidt," ibid., pp. 49–92.

24. "The Mechanism of Mendelian Heredity: A Review," *Science*, 44 (1916), pp. 536–543.

25. The writings of Weismann, Haeckel, and Hertwig, three of the towering figures of late-nineteenth-century theoretical biology (who nonetheless saw themselves as experimentalists as well), although very different from one another, are all examples of the integrated biological approach.

26. In his own history of the gene, written years later, Darlington claimed that such a naïve notion could never have arisen in philosophically sophisticated Europe. The naked, material gene was far too crass and indecent a candidate for the bearing of heredity. The world therefore remains eternally beholden to American youthfulness. *The Facts of Life* (London: Allen and Unwin, 1953), pp. 131–132. It would be interesting to interpret the Morgan school's understanding and use of the gene in the context of American pragmatism, the philosophy of which was being formulated by William James, Oliver Wendell Holmes Jr., and John Dewey at about the same time. See Louis Menand's *The Metaphysical Club* (New York: Farrar, Straus and Giroux, 2001).

27. See his "The Method and Scope of Genetics: An Inaugural Lecture Delivered 23 October, 1908" (London), p. 22.

28. W. E. Agar wrote in 1947 in retrospect to E. B. Babcock: "I think it was Bateson's and Punnett's stubbornness in this matter that caused the decline of importance of English genetics for so many years after such a promising start." Quoted in Allen, *Thomas Hunt Morgan*, p. 282.

29. See Gregory's "On Gametic Coupling and Repulsion in *Primula sinensis*," *Proceedings of the Royal Society of London, Series B*, 84 (1911), pp. 12–15.

30. The Annual Reports of the John Innes 1910–1935, published for the "council only" on the occasion of the twenty-fifth anniversary of the institution, testify to the fecund research program before the war, and to the direction of Gregory's investigations under Bateson.

31. Robert H. Lock, *Recent Progress in the Study of Variation, Heredity and Evolution* (London: John Murray, 1907); and Charles C. Hurst, *Experiments in Genetics* (Cambridge: Cambridge University Press, 1925); see Marsha Richmond's

"British Cell Theory on the Eve of Genetics," *Endeavour*, 25 (2001), pp. 55–60, for a description of their work.

32. John Innes Reports, p. 3.

33. See Dan Lewis, "The Genetical Society—the First Fifty Years," in *Fifty Years of Genetics: Proceedings of a Symposium of the Genetic Society on the Fiftieth Anniversary of Its Foundation*, ed. J. Jinks (Edinburgh: Edinburgh University Press, 1969). In 1924, forty-two of the 108 members of the society were private individuals and plant or animal breeders.

34. For a discussion of the debates surrounding the entrance examinations and the related culture battles, see D.S.L. Cardwell, *The Organisation of Science in England* (London: Heinemann Educational, 1972).

35. Alfred North Whitehead, *Science and the Modern World* (New York: Macmillan, 1926), p. 103.

36. See William Coleman, "Bateson and the Chromosomes: Conservative Thought in Science," *Centaurus*, 15 (1970), pp. 228–314. Despite A. G. Cock's attempt to discount his thesis in "William Bateson's Rejection and Eventual Acceptance of Chromosome Theory," *Annals of Science*, 40 (1983), pp. 19–59, Coleman's interpretation still remains the best attempt at explaining the seeming paradox of Bateson's strict empiricism versus his anti-materialism, his discovery of linkage versus rejection of crossing over, and the connections between his social and political sensibilities and scientific thought.

37. Bateson letter to Beatrice (December 24, 1921), in DP:C.23:D.34. Copies of this correspondence were kept by Darlington and betray Bateson's complicated feelings for Morgan, whom he saw as philosophically naïve, narrow-minded, uncultured, and a bore, but who had nonetheless had the insight Bateson had failed to grasp. Bateson was the landed aristocrat scientist-philosopher; Morgan the country bumpkin from Kentucky. "I wish I liked Morgan better," he wrote. "I think he has made a great discovery, but I can't see in him any quality of greatness."

38. Ibid.

39. Bateson letter to Beatrice (January 1, 1922), DP:C.23:D.34.

40. Ford recounted this story to Darlington, DP:g.41:A.93.

3. Auspicious Beginnings

1. Taped interview of Darlington by Brian Harrison (July 25, 1979), John Innes Historical Collection, the John Innes Institute, Norwich; DP:C.88.F.82.

2. Darlington's memoir of Horace Newton Barber, *Biographical Memoirs of Fellows of the Royal Society*, 18 (1972), p. 23. A count of the personnel listed in the John Innes Reports for Bateson's tenure shows an extremely high 30.1 percent of women workers.

3. Diary, 1923–1924, DP:d.1:A.64.

4. William J. C. Lawrence's autobiographical "Catch the Tide: Adventures in Horticultural Research," 1968. Unpublished manuscript deposited in the John Innes Archive, SS1 B2, quote on p. 27.

5. DP:g.41:A.93.

6. See his "Somatic Segregation in Plants," *Report of the International Horticultural Congress* (Amsterdam, 1923), pp. 155–156; DP:C.107:J.62.

7. DP:C.23:D.40.

8. Darlington's replies to an American psychologist's questionnaire on creativity, 1964, DP:C.1:A.12.

9. "Personal Record," DP:C.1:A.8.

10. DP:C.23:D.40.

11. Diary, 1924, ibid.

12. New York: Macmillan, 1896.

13. See Hermann J. Muller, "Edmund B. Wilson—An Appreciation," *American Naturalist*, 77 (1943), pp. 5–37, 142–172; L. C. Dunn, *A Short History of Genetics* (New York: Harper and Row, 1965), pp. 50–58.

14. Walther Flemming, *Zellsubstanz, Kern, und Zelltheilung* (Leipzig: Vogel, 1882).

15. Weismann, "On the Significance of the Polar Globules," *Nature*, 36 (1887), pp. 607–609.

16. This term was first used in J. B. Farmer and J. E. S. Moore, "On the Maiotic *[sic]* Phase (Reduction Divisions) in Animals and Plants," *Quarterly Journal of Microscopic Science*, 48 (1905), pp. 489–557.

17. The first to rigorously elucidate the stages of meiosis was Hans von Winiwarter, "Recherches sur l'ovogenèse et l'orgenogenèse de l'ovaire des mammifères (lapin et homme), *Archives de Biologie*, 17 (1900), pp. 33–200.

18. L. Hogben, "Studies on Synapsis, I, II, III," *Proceedings of the Royal Society of London, Series B*, 91 (1921), pp. 268–293, 305–329, and 92 (1921), pp. 60–80. On Hogben see *Lancelot Hogben, Scientific Humanist*, eds. Adrian and Anne Hogben (Woodbridge, Suffolk: Merlin Press, 1998).

19. Lancelot Thomas Hogben, *Biographical Memoirs of Fellows of the Royal Society*, 24 (1978), pp. 183–221.

20. *P. kewensis* was a hybrid that had then undergone chromosome doubling. Farmer and Digby interpreted the doubling to be caused by crosswise splitting of the "continuous spireme," a feature of telesynapsis. See J. B. Farmer and L. Digby, "On the Dimensions of Chromosomes Considered in Relation to Phylogeny," *Philosophical Transactions of the Royal Society of London, Series B*, 205 (1912), pp. 1–25.

21. E. B. Wilson, *The Cell in Development and Heredity*, third edition (New York: Macmillan, 1925), p. 550.

22. E. Reuter, "Beiträge zu einer einheitlichen Auffassung gewiser Chromosomenfragen," *Acta Zoologica Fennica*, 9 (1930), pp. 1–487. At stake was the universality of chromosome behavior. Some researchers claimed that while parasynapsis may be true of flies, it was most definitely untrue of plants and could not therefore be a universal principle. Gates, however, argued for universal telosynapsis. See his "Some Points on the Relation of Cytology and Genetics," *Journal of Heredity*, 13 (1922), pp. 75–76.

23. F. A. Janssens, "La Théorie de la Chiasmatypie," *Cellule*, 25 (1909), pp. 387–411.

24. Darlington, "Recent Advances in Cytology: A Retrospective, 1952–1932," a recorded seminar at the John Innes, November 7, 1952, DP:C.85:f.21.

25. Paul Kammerer, *The Inheritance of Acquired Characters* (New York: Boni and Liveright, 1924). Unknowingly, Kammerer was actually performing a Darwinian experiment: inducing a genetic potentiality by creating a strong selection pressure.

26. Bateson, *Nature*, 103 (1919), p. 344; 111 (1923), p. 327; 112 (1923), pp. 391, 899; Kammerer, *Nature*, 111 (1923), p. 637; 112 (1923), p. 237.

27. MacBride, *Nature*, 112 (1923), p. 737.

28. Noble, "Przibram," *Nature*, 118 (1926), p. 209.

29. A copy of the suicide letter was published in *Science*, 64 (1926), p. 359. Years later Arthur Koestler revived the story in *The Case of the Midwife Toad* (London: Hutchinson, 1971). In this apologia Koestler portrayed Kammerer's Viennese charm and typical middle European womanizing in the tradition of Arthur Schnitzler's novellas, arguing from psychological grounds that his guilt was impossible. The Austrian's behavior, anti-positivist tinge, and physical resemblance to Pearson were anathema to the ascetic and aristocratic Bateson, portrayed here as the villain. Koestler exposed his own teleological beliefs by exhorting contemporary scientists to replicate Kammerer's experiments, and by writing a book poor in history. Darlington read the book when it came out and marked it "rubbish."

30. DP:e.10:A.72.

31. New York: G. P. Putnam's Sons, 1916.

32. Edward Murray East and Donald F. Jones, "Inbreeding and Outbreeding: Their Genetic and Sociological Significance," in the series *Monographs on Experimental Biology*, eds. Jacques Loeb, T. H. Morgan, and W. J. V. Osterhout (Philadelphia: J. B. Lippincott, 1919).

33. Diary, June 5, 1925, ibid.

34. Beatrice Bateson, *William Bateson: His Essays and Addresses Together with a Short Account of His Life* (Cambridge: Cambridge University Press, 1928), p. 159.

35. Darlington's "Chromosome Studies in the Scilleae," *Journal of Genetics*, 16 (1926), pp. 237–251; Bateson's "Segregation," ibid., pp. 201–235, quote on p. 212.

36. Diary, 1926, DP:d.1:A.64.

37. Karl Sabbagh has recently uncovered most unpleasant exploits of Heslop-Harrison in *A Rum Affair: A True Story of Botanical Fraud* (London: Allen Lane, 1999).

38. DP:C.23:D.40.

4 In Search of Tulips and Truth

1. DP:e.10:C.72. See E. J. Russel, "Alfred Daniel Hall, 1864–1942," *Biographical Memoirs of the Fellows of the Royal Society*, 4 (1942), pp. 229–250.

2. Dan Lewis, "Cyril Dean Darlington, 1903–1981," *Biographical Memoirs of Fellows of the Royal Society*, 29 (1983), p. 118.

3. See Ronald W. Clark, *J. B. S.: The Life and Work of J. B. S. Haldane* (New York: Oxford University Press, 1968), pp. 112–116.

4. Bateson and Gairdner, "Male Sterility in Flax, Subject to Two Types of Segregation," *Journal of Genetics*, 11 (1921), pp. 269–276.

5. Chittenden, "Cytoplasmic Inheritance in Flax," *Journal of Heredity*, 18 (1927), pp. 337–343, quotes on p. 342.

6. Darlington's account of the story in a taped interview by Brian Harrison (25.7.1979), John Innes Historical Collection, the John Innes Institute, Norwich; DP:C.88:F.82.

7. Published as "Biology and Human Life," *Nature*, 118 (1926), pp. 844–846.

8. DP:C.21:D.6.

9. William Bateson, "Common Sense in Racial Problems," The Galton Lecture, 1919, published in Beatrice Bateson, *William Bateson: His Essays and Addresses Together with a Short Account of His Life* (Cambridge: Cambridge University Press, 1928), pp. 371–387, quote on pp. 371–372.

10. Quoted in J. B. S. Haldane, "Forty Years of Genetics," in *Background to Modern Science*, eds. Joseph Needham and Walter Pagel (Cambridge: Cambridge University Press, 1938), p. 238.

11. "Personal Record," DP:C.1:A.8.

12. Darlington, "Genetics and Plant Breeding, 1910–80," *Philosophical Transactions of the Royal Society of London, Series B*, 292 (1981), p. 402.

13. Darlington letter to Alfred (February 20, 1927), DP:C.5:A.166.

14. Darlington letter to Chittenden (November 12, 1927), DP:C.9:B.3.

15. W. C. F. Newton and C. D. Darlington, "Meiosis in a Triploid Tulip," *Nature*, 120 (1927), p. 13.

16. DP:e.10:A.72.

17. Diary, 1927, DP:g.32.A.65.

18. Darlington had gained his Ph.D., "Genetical Studies in *Prunus:* The Cytology of Domestic Cherries," from London University, with which the John Innes was affiliated, in 1927. Chromosome morphology of the many polyploid varieties was shown to be a most fecund and useful tool with which to solve outstanding classification problems.

19. "Personal Record."

20. Darlington letter to Newton (June 17, 1927), DP:C.112:J.169.

21. Darlington letter to Chittenden, ibid.

22. Clark, *J. B. S.*, p. 78.

23. Boris Ephrussi on Haldane, quoted ibid., p. 109.

24. Sadly, Sprunt was killed in the war. Darlington found this letter among Bateson's papers and included it in "Recollections of J. B. S. Haldane" (draft), DP:C.108:J.86.

25. Ibid. Haldane eventually won the appeal and was reinstated.

26. Clark, *J. B. S.*, p. 20.

27. Darlington's replies to an American psychologist's questionnaire on creativity, 1964, DP:C.1:A.12.

28. Diary, 1927, DP:g.32:A.65.

29. Similar observations were made by others for the triploids *Zea, Tulipa,* and *Lilium.* Further, it had been shown in *Secale, Mathiola,* and *Tradescantia* that chromosome fragments remain unpaired, although they have identical mates with which to pair.

30. J. Belling and A. F. Blakeslee, "The configurations and sizes of the chromosomes in the trivalents of 25-chromosome *Daturas,* " *Proceedings of the National Academy of Sciences of the United States of America*, 10 (1924), pp. 116–120.

31. Spencer Brown and Daniel Zohari proved in 1955 that the exact ratio between chiasmata and crossing over is two to one. See their "The Relationship of Chiasmata and Crossing-Over in *Lilium formosanum,*" *Genetics*, 40 (1955), pp. 850–873.

32. Both papers appeared in the *Journal of Genetics,* 21 (1929), pp. 1–16 and 17–56, respectively.

33. Hall actually decided to send Darlington because he was not getting along satisfactorily with other members of staff. Darlington replied that he would not go if his going was to be regarded as a punishment, and that if complaints were made, he would be happy to share some of his. Sir Daniel capitulated. DP:e.10:A.72.

34. Hall's application to the Marketing Board, DP:C.96:H.3.

35. Diary, 1929, DP:g.32:A.65.

36. Darlingon letters to parents (March 4, 1929), DP:C.6:A.180.

37. Letters home, different dates, ibid.; description of journey, DP:C.96:H.9.

38. Mark Popovsky, *The Vavilov Affair* (Hamden, CT: Archon Books, 1984).

39. Darlington letter to family (June 11, 1929), DP:C.6:A.182.

40. W.H.R.D. letter to Darlington (June 20, 1929), DP:C.6:A.190.

5. From Cytology to Evolution

1. Darlington, "Meiosis (precocity theory)," *Biological Review,* 6 (1931), pp. 221–264.

2. Darlington, "Recent Advances in Cytology: A Retrospective, 1952–1932," a recorded seminar at the John Innes, November 7, 1952, DP:C.85:f.21.

3. See Elof Axel Carlson, *Genes, Radiation, and Society: The Life and Work of H. J. Muller* (Ithaca, NY: Cornell University Press, 1981).

4. The most comprehensive account of the story of *Oenothera* was written by one of the men most responsible for its exposition, and published posthumously: see Ralph E. Cleland, *Oenothera: Cytogenetics and Evolution* (London: Academic Press, 1972).

5. De Vries's *Die Mutationstheorie* was translated by J. B. Farmer and A. D. Darbishire, *The Mutation Theory: Experiments and Observations on the Origin of Species in the Vegetable Kingdom* (London: Kegan Paul, 1910). De Vries had been greatly influenced by Bateson's *Materials,* in which he concluded that since species in nature were discontinuous, the variations giving rise to them must also be discontinuous.

6. De Vries, *University of Chicago Record,* October 1904, pp. 201–207. Quoted in fuller form in Vassiliki Betty Smocovitis, "Botany and the Evolutionary Synthesis: The Life and Work of G. Ledyard Stebbins, Jr." (Ph.D. dissertation, Cornell University, 1988).

7. See Garland Allen, "Hugo de Vries and the Reception of the Mutation Theory," *Journal of the History of Biology,* 2 (1969), pp. 55–87.

8. See Darlington, "Otto Renner: 1885–1960," *Biographical Memoirs of Fellows of the Royal Society,* 7 (1961), pp. 207–220.

9. Darlington, "The Evolution of Genetic Systems: Contributions of Cytology to Evolutionary Theory," in *The Evolutionary Synthesis: Perspectives on the Unification of Biology,* eds. Ernst Mayr and William B. Provine (Cambridge, MA: Harvard University Press, 1998), p. 77. Cleland had discovered the phenomenon of ring formation in *Oenothera* in 1922.

10. Cleland, *Oenothera,* pp. 57–58; John Belling, "A Working Hypothesis for

Segmental Interchange Between Homologous Chromosomes in Flowering Plants," *University of California Publications in Botany*, 14, no. 8 (1928), pp. 283–291.

11. E. B. Wilson, in the third edition of *The Cell* (1925), p. 565, had written: "In the case of Oenothera . . . a 'telosynaptic' association of the chromosomes in early diakinesis seems indubitable."

12. Darlington, "Ring Formation in *Oenothera* and Other Genera," *Journal of Genetics*, 20 (1929), pp. 345–363. A. Håkansson had done similar work to Darlington's but from a telosynaptic perspective. Only after Darlington's paper did he become convinced of parasynapsis. See "Die Reduktionsteilung in de Samenanlagen einiger Oenotheren," *Hereditas*, 11 (1928), pp. 225–304, versus "Zur Zytologie trisomatischer Mutanten aus *Oenothera* lamarckiana," *Hereditas*, 14 (1930), pp. 1–32. Both men's work was theoretical and ultimately proven by the experiments of Emerson and Sturtevant (1931) and Cleland and Blakeslee (1930–1931).

13. Darlington, "Chiasma Formation and Chromosome Pairing in *Fritillaria*," *Proceedings of the Fifth International Botanical Congress* (Cambridge, 1930), pp. 189–191; "Meiosis in Diploid and Tetraploid *Primula sinensis*," *Journal of Genetics*, 24 (1931), pp. 65–96.

14. Darlington, "Personal Record," DP:C.1:A.8.

15. Ibid.

16. Darlington, *Recent Advances in Cytology* (London: Churchill, 1932), p. viii.

17. Ibid., pp. 448–485.

18. Ibid., p. 457.

19. Ibid., p. 456.

20. Ibid., p. 478.

21. Ibid., p. 482.

22. Darlington remembered: "Bateson was not confused about one very great issue, namely that Lamarckism was always advanced on fraudulent evidence. He fought against the Lamarckian thing and it was a tremendous strength to us that we had him behind us. We knew that he was impregnable, so long as he was there we were perfectly safe, we could go on working on our lines." Taped interview of Darlington by Brian Harrison (July 25, 1979), John Innes Historical Collection, the John Innes Institute, Norwich; DP:C.88:F.82.

23. Darlington, *Recent Advances*, p. 484.

24. "[F]or nothing is born in the body in order that we may use it, but rather, having been born, it begets a use." *De Rerum Natura*, IV.

25. Darlington, *Recent Advances*, pp. 484–485.

26. Rockefeller Foundation Archives, Record Group 10, Fellowship Recorder Cards—Natural Sciences.

27. DP:C.98:H.45. Paul Dirac shared the 1933 Nobel Prize with Erwin Schrödinger for his work on wave mechanics.

28. Darlington letter to his brother Alfred (March 1932), DP:C.5:A.166.

29. Hampton L. Carson, in *Evolutionary Synthesis*, p. 91.

30. "Personal Record."

31. Brian Harrison interview.

32. Darlington letter to mother (August 11, 1932), DP:C.6:A.186.

33. Diary, DP:A.70:f.11.

34. Letters between Darlington and his mother (August 1932 to June 1933), DP:C.6:A.184–190.

35. Darlington letter to parents (June 18, 1933), DP:C.6:A.90.

6. Roots of a Scientific Controversy

1. "Personal Record" (November 21, 1941), DP:C.1:A.8.

2. This point is made as well by Evelyn Fox Keller in *A Feeling for the Organism: The Life and Work of Barbara McClintock* (New York: W. H. Freeman, 1998).

3. See Michael Polanyi, *The Tacit Dimension* (Garden City, NY: Doubleday, 1967).

4. Mathew Meselson interview by author (November 29, 2000).

5. Hampton L. Carson, "Cytogenetics and the Neo-Darwinian Synthesis," in *The Evolutionary Synthesis: Perspectives on the Unification of Biology*, eds. Ernst Mayr and William B. Provine (Cambridge, MA: Harvard University Press, 1998).

6. Schultz letter to Darlington (August 12, 1935), DP:C.106:J.37.

7. Karl Sax, "Meiosis and Chiasma Formation in *Paeonia suffruticosa*," *Journal of the Arnold Arboretum*, 13 (1932), pp. 375–385.

8. Curt Stern, "Zytologische-genetische Untersuchungen als Beweise für die Morgansche Theorie des Fuktorenaustausch," *Biologisches Zentralblatt*, 51 (1931), pp. 547–587.

9. Harriet Creighton and Barbara McClintock, "A Correlation of Cytological and Genetical Crossing Over in Zea Mays," *Proceedings of the National Academy of Sciences of the United States of America*, 17 (1931), pp. 492–497.

10. *Proceedings of the Sixth International Congress of Genetics, 1932* (Menasha, WI: Brooklyn Botanic Garden, 1932), pp. 256–273.

11. Darlington himself wrote of this tellingly ten years later: "It would of course have been incorrect of me (on Occam) to have made, a priori, a prediction of such complexity as the events ultimately proved to require. I had to make an a priori prediction because there were no preparations available to me for testing the facts directly, and the technical difficulties were insuperable until I went to America." ("Personal Record.")

12. Franz Schrader, *Mitosis* (New York: Columbia University Press, 1944), p. 158.

13. A. H. Sturtevant and G. W. Beadle, *An Introduction to Genetics* (Philadelphia: W. B. Saunders, 1939), p. 364.

14. A. H. Sturtevant, *A History of Genetics* (New York: Harper and Row, 1965), p. 78.

15. See Elof Axel Carlson, *Genes, Radiation, and Society: The Life and Work of H. J. Muller* (Ithaca, NY: Cornell University Press, 1981). See also Debra Kamrat Lang, "Genetics Humanized: H. J. Muller Between Science and Politics 1915–1932" (unpublished Ph.D. dissertation, Hebrew University of Jerusalem, 1993); chapter 4 discusses Muller's attempt to distance himself from Morgan on this issue.

16. See Darlington's "Natural Populations and the Breakdown of Classical Genetics," *Proceedings of the Royal Society of London, Series B*, 145 (1956), pp. 350–364.

17. See Mayr's preface in Mayr and Provine, *Evolutionary Synthesis*. See also Vassiliki Betty Smocovitis, *Unifying Biology: The Evolutionary Synthesis and Evolutionary Biology* (Princeton, NJ: Princeton University Press, 1996).

18. See Joan Fisher Box, *R. A. Fisher: The Life of a Scientist* (New York: Wiley, 1978); Krisha R. Dronamraju, *Haldane and Modern Biology* (Baltimore: Johns Hopkins Press, 1968); William B. Provine, *Sewall Wright and Evolutionary Biology* (Chicago: University of Chicago Press, 1986). See also Provine's *The Origins of Theoretical Population Genetics* (Chicago: University of Chicago Press, 1971).

19. John Maynard Smith, F.R.S., interview by author (March 20, 2000).

20. Wright's idea of genetic drift and his heuristic "adaptive landscape" to explain the theory of shifting balance were more dynamic than the static models of Fisher and Haldane. Nevertheless, these had no influence on Darlington, who by his own admission didn't read a single Wright paper until 1960 (Darlington letter to William Provine, January 23, 1975, DP:C.104:H.161). Haldane, however, was well aware that his equations were necessary oversimplifications of complex systems of genetic change resulting from shuffling, recombination, loss of whole chromosomes (and not only point mutations), and he could only hope that a future mathematical theory of natural selection could take them all into account. See Vassiliki Betty Smocovitis, "Botany and the Evolutionary Synthesis: The Life and Work of G. Ledyard Stebbins, Jr." (Ph.D. dissertation, Cornell University, 1988), pp. 264–265. Of the three, Fisher was the most reductionist in his thinking, equating genes in a gene pool to gas molecules in a flask. Applying analogues to gas laws from classical physics, Fisher emphasized randomness and chance events that could yield statistically predictable results. This type of theoretical population thinking ignored the immense amount of natural variation and was later famously referred to as "beanbag genetics" by the ornithologist and evolutionist Ernst Mayr. Darlington himself would write in 1975: "I think that Fisher, Haldane, and Wright, in talking about isolated genes and isolated organisms, were not talking about evolution as I understand it: they were just doing very interesting exercises" (DP:C.104:H.164).

21. Indeed, many years later Ernst Mayr would write to Darlington: "Your thinking simply did not fit into the atmosphere of what I have dubbed 'beanbag genetics.'" Mayr letter to Darlington (May 13, 1974), DP:C.104:H.160. Thus Darlington's concept can be understood to be premature in that it was not capable of being extended experimentally for technical reasons. See H. V. Wyatt, "Knowledge and Prematurity: The Journey from Transformation to DNA," *Perspectives in Biology and Medicine*, 18 (1975), pp. 149–156, for a discussion of this concept applied to the discovery of the transforming principle by Avery, MacLeod, and McCarty. See also Maclyn McCarty's discounting reply in *The Transforming Principle: Discovering That Genes Are Made of DNA* (New York: W. W. Norton, 1985), pp. 213–235.

22. Apparently, Fisher's aphoristic style was an improvement on what appears in print. This quote is from Darlington, in Mayr and Provine, *Evolutionary Synthesis*, p. 74.

23. See *Proceedings of the Fifth International Botanical Congress* (Cambridge, 1930), pp. 165–172, 185–189.

24. Darlington, in Mayr and Provine, *Evolutionary Synthesis*, p. 79.

25. Darlington, *Recent Advances in Cytology* (London: Churchill, 1932), p. vi.

26. John Belling, "Critical Notes on Darlington's *Recent Advances in Cytology*" (posthumous paper), *University of California Publications in Botany*, 17, no. 5 (1933), pp. 75–110, quote on p. 76.

27. Ibid.

28. Belling was referring to quotes from Darlington, *Recent Advances*, p. viii.

29. Ibid.

30. Darlington, "Recent Advances in Cytology: A Retrospective, 1952–1932," a recorded seminar at the John Innes (November 7, 1952), DP:C.85:f.21. It was common to say after his death that Belling's theory had been applied to *Oenothera*, but in fact Belling rejected the application himself. The difficulty of application was the difficulty of seeing the selective advantage of the whole series of changes necessary to produce the rings of fourteen chromosomes.

31. For another interesting case of such a conflict see Jan Sapp, "What Counts as Evidence or Who Was Franz Moewus and Why Was Everybody Saying Such Terrible Things About Him?" *History and Philosophy of Life Science*, 9 (1987), pp. 277–308.

32. See Appendix I, *Recent Advances*, pp. 486–490.

33. "Reply by C. D. Darlington," in Belling, "Critical Notes," p. 110. Despite their differences, Darlington showed his gratitude by dedicating *The Handling of Chromosomes* (London: Allen and Unwin, 1942), coauthored with Len La Cour, to Belling, "whose ingenious invention brought the chromosomes within reach of every inquiry."

34. Darlington, "Meiosis," *Biological Reviews*, 6, no. 3 (1931), pp. 221–252, quote on p. 221.

35. See Ernst Mayr's discussion in "Weismann and Evolution," *Journal of the History of Biology*, 18, no. 3 (1985), pp. 295–329. Other recent work on Weismann includes Fred Churchill, "Weismann's Continuity of the Germ-Plasm in Historical Perspective," *Freiburger Universitätsblätter*, 24 (1985), pp. 107–124, and "From Heredity Theory to Vererbung, the Transmission Problem, 1850–1915," *Isis*, 78 (1987), pp. 337–364; John Maynard Smith, "Weismann and Modern Biology," *Oxford Surveys in Evolutionary Biology*, 6 (1989), pp. 1–12; J. R. Griesemer and W. Wimsatt, "Picturing Weismannism: A Case Study of Conceptual Evolution," in *What the Philosophy of Biology Is: Essays for David Hull*, ed. Michael Ruse (Dordrecht: Kluwer Academic Press, 1989), pp. 75–137.

36. August Weismann, *Die Bedeutung der sexuellen Fortpflanzung für die Selektionstheorie* (Jena, 1886), p. 295. Quoted in Mayr, "Weissmann and Evolution," pp. 323–324.

37. Weismann, *Studies in the Theory of Descent*, R. Mendola trans. and ed. (New York: AMS Press, 1975), p. xv. Quoted in Mayr, "Weissmann and Evolution," p. 324.

38. Weismann, *Die Continuität des Keimplasmas als Grundlage einer Theorie der Vererbung*, (Jena: Verlag von Gustav Fisher, 1885), p. 17. Quoted in Mayr, "Weissmann and Evolution," p. 324.

39. "Personal Record," DP:C.1:A.8.

40. Just like Weismann before him, Darlington cautioned his students: "To the research worker I would say that here he has a great many hypotheses presented before him which, even if incorrect, may serve to suggest some new directions of inquiry." *Recent Advances*, p. vii.

41. I shall be using the terms "induction" and "deduction" as they were used by contemporary biologists, namely, the one being a scientific method in which conclusions are *described* strictly and only from observation, the other a method by which conclusions are inferred *logically* from observation *and* theoretical premises.

42. Darlington, *Recent Advances*, p. v. The use of the terms "inductive" and "deductive" corresponds in some respects to the classical distinction between empiricism and rationalism. Descartes was a deductivist because he thought of all sciences as deductive systems, while the English empiricists, from Bacon on, all conceived the sciences as collecting observations from which generalizations are obtained by induction.

43. C. L. Huskins, "Contrasts in Cytology," *Journal of Heredity*, 36 (1945), pp. 43–46, quote on p. 43.

44. Schrader, *Mitosis*. Schrader included in his book a discussion of "mitotic meiosis."

45. Huskins, "Contrasts," p. 44.

46. Ibid., p. 46.

47. Koller letter to Darlington (October 30, 1937), DP:C.110:J.126. In particular, Koller was referring to Barbara McClintock, whom he said was "strongly against you." One is here immediately reminded of Bateson's letters to his wife describing his disappointment and disdain for Morgan's "philosophical naiveté" when he visited his laboratory in 1921. Darlington stood firm in a European tradition theoretically and philosophically more sophisticated than its American counterpart. The methodological appendix at the end of *Recent Advances* baffled his American readers, who found it not only problematic, but in general superfluous.

7. Method, Discipline, and Character

1. Darlington, "The Chromosomes and the Theory of Heredity," *Nature*, 187 (1960), p. 893.

2. See Karl Belar, *Die cytologischen Grundlagen der Vererburg* (Berlin: Borntraeger, 1928).

3. See Erlanson letters to Darlington, DP:C.107:J.44.

4. "Personal Record," DP:C.1:A.8.

5. A. H. Sturtevant and G. W. Beadle, *An Introduction to Genetics* (Philadelphia: W. B. Saunders, 1939), p. 364.

6. "Three Quarter-Centuries of Cytology," Franz Schrader, address by the vice president and chairman of the Section of Zoological Sciences (F), AAAS, delivered at the Zoologists' Dinner, December 30, 1947, Chicago, in *Science*, 107 (1948), pp. 155–159, quote on p. 156.

7. Darlington made clear in his preface that he intended *Recent Advances in Cytology* (London: Churchill, 1932) to be read by the "general biologist" as well as the student and specialist. See p. vii.

8. Quoted in Jonathan Harwood, *Styles of Scientific Thought: The German Genetics Community, 1900–1933* (Chicago: University of Chicago Press, 1993), p. 99.

9. See Garland Allen, *Thomas Hunt Morgan: The Man and His Science* (Princeton, NJ: Princeton University Press, 1979), chapter V. Transmission genetics,

genes, and their physical relation to chromosomes could be approached experimentally and quantitatively, allowing for hypothesis development and testing. On the other hand, the study of gene function in development and embryogenesis, the more overtly evolutionary questions, could not yet be approached experimentally at that time. Tim J. Horder points out correctly in "The Organizer Concept and Modern Embryology: Anglo-American Prespectives," *International Journal of Developmental Biology*, 45 (2001), pp. 97–132, that Morgan never lost his interest in development, but for all intents and *practical* purposes, he had all but dropped it during this period. On Morgan's theory of development, see Raphael Falk and Sara Schwartz, "Morgan's Hypothesis of the Genetic Control of Development," *Genetics*, 134 (1993), pp. 671–674.

10. Sturtevant and Beadle, *An Introduction to Genetics*, p. 363.

11. See William Bateson, "Evolutionary Faith and Modern Doubts," *Science*, 55 (1922), pp. 55–61.

12. See Dunn's and Dobzhansky's "Reminiscences," Columbia University Oral History Project typescripts (1987), pp. 345, 865–867.

13. See Garland Allen's *Life Sciences in the Twentieth Century* (New York: Wiley, 1975).

14. Bateson, "Evolutionary Faith," p. 60.

15. For a delineation of the points of contention contributing to the gap between experimentalists and naturalists, see Mayr's prologue in *The Evolutionary Synthesis: Perspectives on the Unification of Biology*, eds. Ernst Mayr and William B. Provine (Cambridge, MA: Harvard University Press, 1998).

16. Mayr points out the paradox that in the first two decades of the century many held Lamarckian views in concert with a belief in natural selection, so that often the staunchest defenders of natural selection were in fact Lamarckian naturalists. See Peter Bowler's *The Eclipse of Darwinism: Anti-Darwinian Evolutionary Theories in the Decades around 1900* (Baltimore: Johns Hopkins University Press, 1983) for a full description of the assault on natural selection.

17. For the most general contemporary account of the synthesis, see Julian Huxley's *Evolution: The Modern Synthesis* (London: Allen and Unwin, 1942). More recent, analytical accounts are Mayr and Provine, *Evolutionary Synthesis;* Betty Smocovitis, *Unifying Biology: The Evolutionary Synthesis and Evolutionary Biology* (Princeton, NJ: Princeton University Press, 1996); and William B. Provine, "The Role of Mathematical Population Geneticists in the Evolutionary Synthesis of the 1930s and 1940s," *Studies in the History of Biology*, 2 (1978), pp. 167–192.

18. Theodosius Dobzhansky, *Genetics and the Origin of Species* (New York: Columbia University Press, 1937). For a description of the cooperation between Dobzhansky and Wright, see William B. Provine, *Sewall Wright and Evolutionary Biology* (Chicago: University of Chicago Press, 1971), pp. 327–365. For a general intellectual biography of Dobzhansky, see Provine, "Origins of the Genetics of Natural Populations Series," in *Dobzhansky's Genetics of Natural Populations I–XLIII*, ed. R. C. Lewontin et al. (New York: Columbia University Press, 1981), pp. 5–83. For a description of his Russian intellectual origins, see Mark Adams, "The Founding of Population Genetics: Contributions of the Chetverikov School, 1924–1934," *Journal of the History of Biology*, 1 (1968), pp. 22–39.

19. For a full list of the participants in the synthesis in the United States and abroad, see Mayr and Provine, *Evolutionary Synthesis*. In the American context,

Morgan's *Evolution and Genetics* (Princeton, NJ: Princeton University Press, 1925) and *The Scientific Basis of Evolution* (New York: W. W. Norton, 1932) signaled a return of the former embryologist to problems evolutionary, *before* what is now considered the time of synthesis. Nevertheless, it is widely held among historians that these books represent a description of genetics and a description of evolution, rather than a true synthesis, which would have been beyond Morgan's ability. Morgan attempted to relate development to evolution and genetics some years later in *Embryology and Genetics* (New York: Columbia University Press, 1934), but here again his case fell short of a true synthesis.

20. See Harwood, *Styles.*

21. Adams, "Founding of Population Genetics."

22. We know less of the interaction between the British pair, who tended to meet in person and left no written record.

23. E. B. Ford, *Mendelism and Evolution* (London: Methuen, 1931); J. B. S. Haldane, *The Causes of Evolution* (London: Harper and Brothers, 1932).

24. See M. A. Sleigh and J. F. Sutcliffe, "The Origin and History of the Society for Experimental Biology (Comprising the Origins of the Society by Lancelot Hogben, F.R.S. Aspects of the History of the Society [1923–1966])," catalogued in the Huxley Papers.

25. *The New Systematics* (Oxford: Oxford University Press, 1940), edited by Huxley, was the principal fruit of this engagement. (It was Darlington who suggested this name to Huxley.) For a discussion of the aims of the ASSRGB, see *Nature*, 141 (1938), p. 163.

26. See R. A. Fisher, *The Genetical Theory of Natural Selection* (Oxford: Clarendon Press, 1930); Julian Huxley, "Natural Selection and Evolutionary Progress," presidential address at annual meeting, *Report of the British Association for the Advancement of Science*, 1936.

27. Like his grandfather before him, Huxley put much energy into popularizing his science. His collaboration with H. G. Wells and son G. P. Wells on *The Science of Life* (London: Amalgamated Press, 1930) was an impressive, influential, and early statement of neo-Darwinism.

28. The First Annual Report of the Association for the Study of Systematics in Relation to General Biology, summarized in *Nature*, 142 (1938), pp. 1069–1070, reported an unusually large membership of 132. By the time of the Second Annual Report, *Proceedings of the Linnean Society*, Session 152 (1939–1940), pp. 399–403, war had already broken out in Europe, and membership declined to seventy-nine. In contrast, it was precisely at this time that the organization of evolutionary biology along disciplinary lines was taking place in the United States. See V. B. Smocovitis, "Organizing Evolution: Founding the Society for the Study of Evolution (1939–1950)," *Journal of the History of Biology*, 27 (1994), pp. 241–309.

29. Darlington's conjectural style, to which many cytologists objected, was to a large extent a result of the fact that his real interest was ultimately evolutionary: "Observations from all these points of view I have had to consider separately, but in order to show their relationship with one another and with the properties of the organism-as-a-whole they have been considered finally in their *bearing on the central problem of evolution. This involves hypotheses, some of them of a speculative kind.*" (Emphasis added.) From *Recent Advances*, p. ix.

30. Belling's refusal to extrapolate to *Oenothera* has already been discussed; Darlington, "Otto Renner: 1885–1960," *Biographical Memoirs of Fellows of the Royal Society*, 7 (1962), pp. 207–220.

31. Dutch botanist Jan Paulus Lotsy's hybridization theory, detailed in *Evolution by Means of Hybridization* (The Hague, 1916), was an exception and had an impact on Darlington early on (see his reply to "Questionnaire Concerning the Evolutionary Synthesis," sent by Mayr and Provine, 1974, DP:C.104:H.164). Nevertheless, Lotsy's writings were prolix, were considered vague and confusing, and made little impression. See V. B. Smocovitis, "Botany and the Evolutionary Synthesis: The Life and Work of G. Ledyard Stebbins, Jr. (Ph.D. dissertation, Cornell University, 1988), chapter 3.

32. Ernst Mayr, in the prologue to *The Evolutionary Synthesis: Perspectives on the Unification of Biology*, Ernst Mayr and William B. Provine eds. (Cambridge, MA: Harvard University Press, 1998), p. 23.

33. Sturtevant and Beadle, *Introduction to Genetics*, p. 363.

34. Darlington was cited fourteen times in Dobzhansky's 1937 book. Moreover, chapter 4, "Chromosomal Changes," was fifty pages long and the centerpiece of the work. In the 1941 second edition, the Darlington citations grew to eighteen, and chapter 4 was augmented. Nevertheless, Darlington is almost never mentioned as an important player in the evolutionary synthesis literature. The exception is Mayr and Provine, *Evolutionary Synthesis*, and even here Provine states in the epilogue: "The role of cytologists is especially difficult to assess. They seem to have played no major role in the synthesis" (p. 408). This is significant, for elsewhere he calls Darlington's *Evolution of Genetic Systems* "enormously influential" (p. 70), meaning that his contribution is not understood as coming from cytology.

35. An immediate effect was the birth of a major branch of *Drosophila* work dealing with cytogenetics, evolution, and natural populations, pioneered at Texas. This work ultimately gave rise to M. J. D. White's *Animal Cytology and Evolution* (Cambridge: Cambridge University Press, 1945).

36. Hampton L. Carson, in Mayr and Provine, *Evolutionary Synthesis*, p. 89.

37. This is a point Bently Glass makes in "The Establishment of Modern Genetical Theory as an Example of the Interaction of Different Models, Techniques, and Inferences" in *Scientific Change: Historical Studies in the Intellectual, Social, and Technical Conditions for Scientific Discovery and Technical Invention, from Antiquity to the Present*, ed. A. C. Crombie (London: Heinemann, 1963), pp. 521–541.

38. The reason for this, as Darlington explained, is that if one magnifies the small chromosome, it becomes apparent that, just like in the sex chromosomes, there are no genes in the region where the chiasma occurs.

39. Koller letter to Darlington, DP:C.110:J.128.

40. Darlington, "Chromosome Study and the Genetic Analysis of Species," *Annals of Botany*, 47 (1933), pp. 811–814.

41. See Smocovitis, "Botany and the Evolutionary Synthesis."

42. The Russian school of cytology founded by S. G. Navashin had pioneered the use of karyology—chromosome number and appearance—as a diagnostic technique to determine systematic relationships already at the turn of the century. Babcock was friendly with Navashin's son Michael, one of the leaders of this school.

43. Joel B. Hagen has told the story of this taxonomic reform in "Experimentalists and Naturalists in Twentieth-Century Botany: Experimental Taxonomy 1920–1950," *Journal of the History of Biology*, 17, no. 2 (1984), pp. 249–270. Here the work of F. E. Clements, H. M. Hall, Göte Turesson, W. B. Turrill, and the team of Jens Clausen, David Keck, William Heisey and the Bay Area systematists is discussed. Hagen argues that the experimentalist/naturalist dichotomy of the synthesis narrative is a simplification in the case of experimental taxonomy, in which cytology, genetics, systematics, and old-school morphology and herbarium work were all combined.

44. E. B. Babcock, "Investigations in the Genus *Crepis*," *Carnegie Institution Year Book*, no. 29 (1928), p. 352.

45. Rockefeller Foundation Archives, Record Group 10, Fellowship Recorder Cards—Natural Sciences. In 1930 Michael Navashin had written to the young Darlington: "I must say, you are developing wonderfully" (January 18, 1930, DP:C10:B.24).

46. This is most evident early on in Darlington's work on *Rubus*, *Prunus*, and *Campanula*.

47. See Kim Kleinman, "His Own Synthesis: Corn, Edgar Anderson, and Evolutionary Theory in the 1940's," *Journal of the History of Biology*, 32 (1999), pp. 293–320, for a discussion of Anderson's career.

48. Anderson letter to Darlington (April 1, 1930), DP:C.106:J.2. Kleinman fails to make the Darlington connection.

49. Edgar Anderson and Robert E. Woodson, Jr., "The Species of *Tradescantia* Indigenous to the United States," *Contributions of the Arnold Arboretum*, 9 (1935), pp. 1–132.

50. It is true that the distinction between those experimentalists using plants as convenient tools and those who studied plants as plants was beginning to blur. Nevertheless, Darlington can be placed firmly within the first group.

51. Incredibly, the first nineteen chapters of *The Evolution of Genetic Systems* (Cambridge: Cambridge University Press, 1939) were written without notes on board ship, December 1–17, 1937, when Darlington was on his way to a lecture tour of India ("Personal Record").

52. G. L. Stebbins, review in *Chronica Botanica*, 6, no. 17/18 (1941), pp. 429–430.

53. See Muller's review of *EGS*, "How Genetic Systems Come About," *Nature*, 144 (1939), p. 648.

54. Stebbins letter to Darlington (August 7, 1946), DP:C.113:J.213.

55. Darlington letter to Stebbins (August 14, 1946), ibid.

56. E. W. MacBride, "Embryology and Evolution," *Nature*, 126 (1930), pp. 918–919.

57. At the height of the controversy over *Oenothera* in 1930, Gates was one of Darlington's examiners for the D.Sc., and used the occasion to try to convince him to recant. The meeting broke up in anger ("Personal Record").

58. "Personal Record." Indeed, Darlington had no instructors in an ordinary sense, and considered Darwin, Weismann, and the chromosomes themselves to be his true teachers. See his "Reply to Questionnaire Concerning the Evolutionary Synthesis," sent by Mayr and Provine, 1974, DP:C.104:H.164.

59. Taped interview of CDD by Brian Harrison (July 25, 1979), John Innes

Historical Collection, the John Innes Institute, Norwich; DP:C.88:F.82. MacBride wrote to Miss Pellew immediately to complain of the young researcher's rudeness.

60. Reviews, *The New Phytologist*, 33 (1934), pp. 320–322.

61. C. L. Huskins, "Contrasts in Cytology," *Journal of Heredity*, 36 (1945), p. 45.

62. Darlington to Alfred (October 26, 1934), DP:C.5:A.166.

63. See Ronald W. Clark, *J. B. S.: The Life and Work of J. B. S. Haldane* (New York: Oxford University Press, 1968), for myriad examples.

64. Diary entries, 1930s, not all dated, DP:g.33:A:66–70, g.34:71; DP:e.11:A.82 (1962).

65. Interviews by author of Dan Lewis (March 28, 2000), Stan Woodell (March 31, 2000), and Lionel Clowes (February 24, 2000).

66. Letter from Harlan Lewis to author (June 24, 2000).

67. Darlington letter to parents (January 17, 1933), DP:C.6:A.187.

68. Schrader, "Three Quarter-Centuries of Cytology," quote on p. 156.

69. An important advancement was made in 1951 by Gunnar Östergen, who showed, contra Darlington, that chiasmata were not sufficient to explain the difference between chromosome movements in mitosis and meiosis, but rather that the key difference lay in the arrangement of the kinetochore regions. See his "The Mechanism of Co-orientation in Bivalents and Multivalents," *Hereditas*, 37 (1951), pp. 85–156.

70. Brian Harrison interview.

71. For an appreciation of the current understanding of chromosome behavior, see D. Zickler and N. Kleckner, "Meiotic Chromosomes: Integrating Structure and Function," *Annual Review of Genetics*, 33 (1999), pp. 603–754.

72. Dan Lewis, "Cyril Dean Darlington, 1903–1981," *Biographical Memoirs of Fellows of the Royal Society*, 29 (1983), p. 123.

73. *Novum Organum*, vol. VIII of *The Works of Francis Bacon*, eds. J. Spedding, R. L. Ellis, and D. D. Heath (New York: Garrett Press, 1968), p. 210; Quoted in Stephan J. Gould, *Hen's Teeth and Horse's Toes* (New York: Norton, 1980), p. 83.

Interlude

1. "Personal Record," DP:C.1:A.8. See Darlington, "The Internal Mechanics of the Chromosomes," *Cytologia*, 7 (1936), pp. 248–255; "The External Forces Acting on Chromosomes," *Nature*, 138 (1936), p. 366.

2. Taped interview of Darlington by Brian Harrison (July 25, 1979), John Innes Historical Collection, the John Innes Institute, Norwich; DP:C.88:F.82; biographical fragment (May 28, 1978), DP:C.24:D.60; "Haldane's memorandum on problems of the organization of the John Innes," (draft), DP:C.108:J.76.

3. The council included two army officers, three active and one retired civil servants, one retired Indian civil servant, two retired university professors (Farmer and MacBride), and one country gentleman. Tansley was the only active scientist, and Middleton the only sensible civil servant.

4. "Haldane's memorandum."

5. DP:C.10:B.19 contains drafts and copies of the signed proposal, sent to the secretary of the council on December 5, 1936. See also the Annual Reports of the

John Innes for 1936–1939. At stake were the relationship between the director and his staff, participation of the scientific staff in the government of the institution, conditions of employment and appointment, recruitment, and the question of the institution's connection to the University of London. A policy of association with UL had been adopted in 1931, and it was now being debated whether the institution should move to be admitted as a school of the university. Haldane opposed; Darlington encouraged.

6. President Robert Robinson on the occasion of the awarding of the Darwin Medal to Darlington by the Royal Society, November 1946, DP:C.8.

7. DP:C:7:A:196.

8. Interviews by author with Clare Passingham (March 3, 2000) and with Oliver Darlington (April 1, 2000); memoir of Debby, written in 1986; letter from Rachel Ritchie (June 9, 2002).

8. The Lysenko Affair

1. There is a large literature on the Lysenko controversy in Russia. The first generation of historiography, represented by P. S. Hudson and R. S. Richens, *The New Genetics in the Soviet Union* (Cambridge: Imperial Bureau of Plant Breeding and Genetics, 1946), Conway Zirkle, *Death of a Science in Russia* (Philadelphia: University of Pennsylvania Press, 1949), and Julian Huxley, *Soviet Genetics and World Science* (London: Chatto and Windus, 1949), explained Lysenko's rise to power as inherently linked to Marxist philosophy and traditions. The next generation of professional historians, such as David Joravsky, *The Lysenko Affair* (Cambridge, MA: Harvard University Press, 1970), Zhores Medvedev, *The Rise and Fall of T. D. Lysenko*, trans. Michael Lerner (New York: Columbia University Press, 1969), and Loren Graham, *Science and Philosophy in the Soviet Union* (New York: Knopf, 1972), revised this interpretation, stressing political and social conditions, political authoritarianism, and abuse of state power over ideological explanations. Diane Paul later pointed out the continuity of two key concepts of Lysenkoism—the doctrine of two sciences and Lamarckian inheritance—with the Marxist tradition, in "Marxism, Darwinism, and the Theory of Two Sciences," *Marxist Perspectives*, 2 (1979), pp. 116–143. Marxist-oriented commentators, such as Dominique Lecourt, *Proletarian Science? The Case of Lysenko* (London: NLB, 1977), and Richard Lewontin and Richard Levins, "The Problem of Lysenkoism," in *The Radicalisation of Science: Ideology of/in the Natural Sciences*, eds. H. Rose and S. Rose (London: Macmillan, 1976), attempted to use Lysenkoism as a lesson about how Marxist principles can be distorted and abused. More recently, Valery N. Soyfer, *Lysenko and the Tragedy of Soviet Science*, trans. Leo and Rebecca Gruliow (New Brunswick, NJ: Rutgers University Press, 1994), and Nikolai Krementsov, *Stalinist Science* (Princeton, NJ: Princeton University Press, 1998), have shown how scientific debates within Russian plant physiology and genetics, the crisis in Soviet agriculture, Stalinist terror, and Marxist philosophical traditions all played a role in Lysenko's rise to power.

2. Quoted in Hudson and Richens, *The New Genetics.*

3. See Robert Conquest, *The Harvest of Sorrow: Soviet Collectivisation and the Terror-Famine* (Oxford: Oxford University Press, 1987).

4. Mark Popovsky, in his *The Vavilov Affair* (Hamden, CT: Archon Books, 1984), has shown that ironically, and tragically, Vavilov can in fact be viewed as one of Lysenko's earliest and weightiest proponents, and in that capacity, responsible to a degree for his ascent.

5. Hudson and Richens, *The New Genetics*, p. 18.

6. Quoted in Medvedev, *Rise and Fall*, pp. 81, 125.

7. R. O. Whyte, "History of Research in Vernalization," in *Vernalization and Photoperio-dism*, eds. A. E. Murneek and R. O. Whyte (Waltham, MA: Chronica Botanica, 1948), pp. 1–38.

8. Hudson and Richens, *The New Genetics*, p. 48.

9. Quoted in *Nature*, 148 (1941), p. 497.

10. In that year Sergei Chetverikov's group at the Kol'tsov Institute was arrested and broken up. Nevertheless, Kol'tsov himself devised a strategy to keep the work going, and a greatly expanded genetics section was set up under Nikolai Dubinin. With an almost complete change of personnel, the enterprise continued to flourish. See Adams in *The Evolutionary Synthesis: Perspectives on the Unification of Biology*, eds. Ernst Mayr and William B. Provine (Cambridge, MA: Harvard University Press, 1998), pp. 242–278.

11. Nikolai Krementsov, "A Second Front in Soviet Genetics: The International Dimension of the Lysenko Controversy, 1944–1947," *Journal of the History of Biology*, 29, no. 2 (1996), p. 232.

12. Krementsov, *Stalinist Science*, pp. 93–128.

13. Krementsov, "Second Front," pp. 233–240.

14. Joravsky, *Lysenko Affair*, pp. 110–111.

15. Lerner letter to Muller, cited in Krementsov, "Second Front," p. 239.

16. Zhebrak letter to Party Secretary Malenkov, in Krementsov, "Second Front," p. 246.

17. Krementsov, "Second Front," p. 237.

18. Julian Huxley, "Evolutionary Biology and Related Subjects," *Nature*, 156 (1945), p. 256.

19. Karl Sax, "Soviet Biology," *Science*, 99 (1944), p. 298–299; Anton Zhebrak, "Soviet Biology," *Science*, 102 (1945), p. 357.

20. Ibid., p. 358.

21. Huxley letter to Darlington (December 31, 1945), marked as confidential, DP:C.109:J.108.

22. Hudson and Richens, *The New Genetics*.

23. T. D. Lysenko, *Heredity and Its Variability*, trans. T. Dobzhansky (New York: King's Crown Press, 1946).

24. Dunn to Lerner (1945), in Krementsov, "Second Front," p. 241.

25. Theodosius Dobzhansky, "Lysenko's Genetics," *Journal of Heredity*, 37, no. 1 (1946), pp. 5–9.

26. Dunn's review of *Heredity and Its Variability*, *Science*, 103 (1946), pp. 180–181.

27. Eric Ashby, *Scientist in Russia* (Harmondsworth: Penguin Books, 1947), p. 114.

28. Ibid., p. 205.

29. See Peter Bowler, "E. W. MacBride's Lamarckian Eugenics," *Annals of Science*, 41 (1984), pp. 245–260, in which he shows how there are no necessary intrinsic links between scientific theories and social views. Theory may acquire different ideological links in different social environments.

30. J. B. S. Haldane, *Science and Everyday Life* (London: Lawrence and Wishart, 1939), p. 243.

31. J. B. S. Haldane, *The Inequality of Man and Other Essays* (Harmondsworth: Penguin Books, 1932), p. 136.

32. Haldane, *Heredity and Politics* (New York: W.W. Norton, 1938), p. 14.

33. See Gary Werskey, *The Visible College* (London: Allen Lane, 1988), pp. 208–210.

34. Haldane, *Science in Peace and War* (London: Lawrence and Wishart, 1940), p. 83.

35. Joseph Needham, "Biological Science in the U.S.S.R.," *Nature*, 148 (1941), pp. 362–363.

36. J. L. Fyfe, "The Soviet Genetics Controversy," *Modern Quarterly*, 3 (1948), p. 348.

37. Ashby, *Scientist in Russia*, p. 202.

38. See Haldane's "A Note on Genetics in the U.S.S.R." and Needham's "Genetics in the U.S.S.R." in *Modern Quarterly*, 1 (1938), pp. 393–394 and 370–371. The enormous excitement over this issue will be discussed in context in the following sections.

39. T. H. Morgan, "Genetics and the Physiology of Development," *American Naturalist*, 60 (1926), pp. 489–515, quote on p. 491; W. Bateson, "Segregation: Being the Joseph Leidy Memorial Lecture of the University of Pennsylvania, 1922," *Journal of Genetics*, 16 (1926), pp. 201–235, quote on p. 234.

40. Darlington, *The Evolution of Genetic Systems* (Cambridge: Cambridge University Press, 1939), p. 114. Here Darlington first spoke of what he called "plasmagenes"—gene-like substances in the cytoplasm whose interaction with the nuclear genes and the environment played an important role in development.

41. DP:C.113:J.190.

42. Not that this was anything new. German researchers had traditionally studied the role of the cytoplasm in heredity and had between the wars led the way in both experimental and theoretical work on cytoplasmic inheritance. Leading workers included Carl Correns, Fritz von Wettstein, Otto Renner, Peter Michaelis, Victor Jollos, and Friedrich Oehlkers. See Jonathan Harwood, *Styles of Scientific Thought: The German Genetics Community, 1900–1933* (Chicago: University of Chicago Press, 1993) for a description of their enterprise. Schools in France and the United States, in particular in the labs of Boris Ephrussi and Tracy Sonneborn, joined the challenge to the "nuclear monopoly" of the Morgan school of transmission genetics. For a "sociology of power" approach to the battles between theorists of cytoplasmic and nuclear inheritance, which includes a description of Darlington's work (pp. 99–100), see Jan Sapp, *Beyond the Gene: Cytoplasmic Inheritance and the Struggle for Authority in Genetics* (Oxford: Oxford University Press, 1987).

43. Letter from Darlington to Popovsky (10.5.1965), DP:C.114: J.226.

44. C. D. Darlington and S. C. Harland "Nikolai Ivanovich Vavilov, 1885–1942," *Nature*, 156 (1945), p. 621.

45. Sax letter to Darlington (December 28, 1945), DP:C.114: J.225.

46. Muller letter to Darlington (September 9, 1946), DP:C.92:G.27.

47. Darlington letter to Michael Polanyi (14.3.1947), DP:C.25:D.87. Polanyi will be discussed in more detail later in the chapter.

48. Darlington letter to Armitage (December 23, 1946), DP:C.25:D.77.

49. Armitage letter to Darlington (January 6, 1947).

50. Lionel Brimble-Darlington correspondence (January 16, 1947 and February 7, 1947), ibid.

51. C. D. Darlington, "A Revolution in Russian Science," *Discovery*, February 1947, p. 43.

52. N. P. Dubinin, "Work of Soviet Biologists: Theoretical Genetics," *Science*, 165 (1947), pp. 109–112.

53. Darlington letter to Orwell (March 27, 1947), DP:C.35:D.78. Orwell himself had had great difficulty getting *Homage to Catalonia* published in the climate of a London intelligentsia secretly or openly in sympathy with Stalinism. When *Animal Farm* was completed some years later, Orwell again had some difficulty publishing, as firms cited the duty to support the Russian ally so long as the war was on. See Jeffrey Meyer, *Orwell: Wintry Conscience of a Generation* (New York: W. W. Norton, 2000).

54. Muller letter to Darlington (March 12, 1947), DP:C.111:J.163.

55. C. D. Darlington, "Retreat From Science in Soviet Russia," *The Nineteenth Century and After*, 142 (October 1947), p. 159.

56. Ibid., p. 168.

57. In fact, Darlington had sent Dobzhansky a draft prior to publication, to which the latter responded that it would be dangerous to include in the eulogy the names of men whose death was not absolutely certain. Darlington agreed and shortened the list. DP:C.93:G.57.

58. Dobzhansky letter to Darlington (November 11, 1947), DP:C.25:D.83.

59. Theodosius Dobzhansky "N.I. Vavilov, a Martyr of Genetics," *Journal of Heredity*, 38, no. 8 (1947), pp. 227–232.

60. Huxley letter to Darlington (March 27, 1947), DP:C.109:J.108.

61. "Genetics and Science in the USSR: Correspondence," *British Medical Journal*, (November 15, 1947), p. 792.

62. Letters to the Editor, *Picture Post*, October 9, 1948.

63. *The New Statesman and Nation*, September 11, 1948. In fact, reliable news of Vavilov's arrest and fate had long since reached the West, as early as 1944.

64. Ashby letter to Darlington (November 10, 1947), DP:C.25:D.83.

65. Karl Sax, "Soviet Science and Political Philosophy," *The Scientific Monthly*, 65, no. 1 (1947), pp. 43–47.

9. Marxism and the Slaying of a Mentor

1. Quoted in M. B. Crane, "The Moscow Conference on Genetics," *Heredity*, 3 (1949), pp. 252–261.

2. Stephen Jay Gould, *Hen's Teeth and Horse's Toes* (New York: W. W. Norton, 1983), p. 135.

3. Darlington letter to Victor Frank (August 18, 1948), DP:C.25:D.91.

4. Gary Werskey, *The Visible College* (London: Allen Lane, 1988); John Langdon Davis, *Russia Puts the Clock Back: A Study of Soviet Science and Some British Scientists*

(London: Gollancz, 1949); Diane Paul, "A War on Two Fronts: J. B. S. Haldane and the Response to Lysenkoism in Britain," *Journal of the History of Biology*, 16, no. 1 (1983), pp. 1–37.

5. J. L. Fyfe, *Lysenko Is Right* (London: Lawrence and Wishart, 1950).

6. J. D. Bernal, "The Biological Controversy in the Soviet Union and Its Implications," *Modern Quarterly*, 4 (1949), p. 204.

7. Haldane finally stopped writing his conciliatory, and often outright supportive, weekly articles on Lysenko in the *Daily Worker* at the end of 1949, and stopped attending Communist Party meetings less than a year later. His final break with Lysenkoism, "In Defence of Genetics," *Modern Quarterly*, 4 (1949), appeared in the same issue as Bernal's endorsement of the August conference. Here Haldane proclaimed that "wholly unjustifiable attacks have been made against my profession, and one of the most important lessons I have learnt as a Marxist is the duty of supporting my fellow workers" (p. 202). For a defense of Haldane's actions, see N. W. Pirie, "The Wrong Penumbra," *The Listener* (January 9, 1969), pp. 53–54.

8. It has been suggested that Lysenko acted as a catalyst while wider disagreements between Haldane and the Party were emerging. See John Maynard Smith, "J. B. S. Haldane," in *The Founders of Evolutionary Genetics*, ed. S. Sarkar (Dordrecht: Kluwer Academic Press, 1992), pp. 37–52.

9. "Resignation of Professor Muller from Academy of Sciences of the USSR," *Science*, 108 (1948), p. 436. The presidium of the academy retorted: "Without a feeling of regret the Academy parts with its former member, who has betrayed the interests of true science and openly gone over into the camp of the enemies of progress and science, of peace and of democracy," *Pravda*, December 14, 1948.

10. Julian Huxley, "Soviet Genetics: The Real Issue," *Nature*, 163 (1949), p. 937.

11. Ibid., p. 976.

12. "Science Without Freedom," BBC Third Programme (November 1, 1948). Appeared in *The Listener*, November 4, 1948, pp. 677–678.

13. Theodosius Dobzhansky, "The Suppression of a Science," *Bulletin of Atomic Scientists*, May 1949.

14. See Conway Zirkle, ed., *Death of a Science in Russia* (Philadelphia: University of Pennsylvania Press, 1949), p. 277.

15. "A Statement of the Governing Board of the American Institute of Biological Sciences," *Science*, 110 (1949), p. 124.

16. Jan Sapp, *Beyond the Gene: Cytoplasmic Inheritance and the Struggle for Authority in Genetics* (Oxford: Oxford University Press, 1987), discusses this debate, but he misconstrues the evolutionary views of some of his German examples.

17. Sapp has argued that the projected non-negotiability of the claims made in the official American statement mirrored Lysenkoist dogmatism (see ibid., chapter 6); David Sonneborn interview by author (June 1, 2001).

18. See his "Heredity, Development and Infection," *Nature*, 154 (1944), pp. 164–169; "The Plasmagene Theory of the Origin of Cancer," *British Journal of Cancer*, 2 (1948), pp. 118–126; "Genetic Particles," *Endeavor*, 13 (1949), pp. 51–61; "Les Plasmagénes," *Unités Biologiques Douées de Continuité Génétique* (Paris: Edition du Centre National de la Recherche Scientifique, 1949), pp. 123–130; *The Elements of Genetics*, with Ken Mather (London: Allen and Unwin, 1949); "Mendel and the Determinants," in *Genetics in the 20th Century*, ed. L. C. Dunn (New York: Macmillan, 1951), pp. 315–332.

19. Astbury letter to Darlington (July 19, 1949), DP:C.106:J.3.

20. Dobzhansky letter to Darlington (February 1, 1949), DP:C.26:D.101.

21. Tracy Sonneborn, "Heredity, Environment, and Politics," *Science*, 111 (1950), pp. 529–539.

22. Darlington letter to Huxley (April 13, 1949), DP:C.26:D.99.

23. BBC Third Programme (November 30, 1948), reported in *The Listener*, December 9, 1948, p. 873–876, quote on p. 874.

24. DP:C.1:A.1 (1917–1921).

25. From Ronald W. Clark, *J. B. S.: The Life and Work of J. B. S. Haldane* (New York: Oxford University Press, 1968), p. 115.

26. Darlington, "Family History," DP:e.10:A.72.

27. "After marching for 12 hours at a thousand feet an hour (standard Baedeker rate, he said) he would be singing—a terrible dirge—as we arrived home long after dark. It nearly killed me." Darlington, "Recollections of J. B. S. Haldane" (draft), DP:C.108:J.86.

28. Sahotra Sarkar, "Science, Philosophy, and Politics in the Work of J. B. S. Haldane, 1922–1937," *Biology and Philosophy*, 7 (1992), pp. 385–409, argues that it was Haldane's increasingly mechanistic and reductionist scientific work that influenced his philosophy and metaphysics, not the other way around: his adoption of Marxism was a consequence of his gradual commitment to dialectical materialism. For a study of J. S. Haldane's philosophy, see S. Sturdy, "Biology as Social Theory: John Scott Haldane and Physiological Regulation," *British Journal of the History of Science*, 21 (1988), pp. 315–340. For J. S. Haldane's own writings, see *The Philosophical Basis of Biology* (Garden City, NY: Doubleday, 1931), and *The Philosophy of a Biologist*, second edition (Oxford: Clarendon Press, 1935).

29. J. B. S. Haldane, *The Inequality of Man* (London: Chatto and Windus, 1932), p. 121.

30. Diary, 1933, DP:f.8:A.67.

31. DP:g.32:A.66.

32. Karpechenko wrote Miss Schafer at the John Innes to point out some trivial bibliographical errors: "Of course I do not consider these of any serious moment for such a book as D's. It is just like looking for mistakes in grammar in Pushkin's poetry" (June 2, 1935), DP:C.38:E.40.

33. Darlington, "Preface to the Russian Edition of Darlington's 'Cytology'" (draft), DP:C.38:E.40.

34. It is not a straightforward problem to explain why Darlington never made such strong claims in his English-language publications. Perhaps he felt that what would be appreciated by a philosophically disposed Russian audience would fall on deaf ears in the West, and thus couched his methodological arguments in more mundane terms. It is also possible that he did not want to be labeled a Marxist. On the other hand, perhaps the Russian foreword is what needs explaining.

35. See Haldane, "A Dialectical Account of Evolution," *Science and Society*, 1 (1937), pp. 473–486. Economist A. P. Lerner retorted with a scathing attack in "Is Professor Haldane's Account of Evolution Dialectical?" *Science and Society*, 2 (1938), pp. 232–239. For other British expositions of the dialectical approach, see J. D. Bernal, *Marx and Science* (London: Lawrence and Wishart, 1952), Hyman Levy, *Philosophy for a Modern Man* (New York: A. A. Knopf, 1938), and Joseph Needham, *Order and Life* (New Haven, CT: Yale University Press, 1936).

36. Although he became a Marxist before that, too. See his *The Marxist Philosophy and the Sciences* (London: Allen and Unwin, 1939).

37. They thereupon instructed Haldane that such a position was logically and practically impossible. See Charlotte Haldane, *The Listener,* January 30, 1969.

38. Cambridge: Cambridge University Press, 1940.

39. For two of the best appreciations of Haldane's mind, scientific legacy, and psychology, see Darlington, "Determined but Lonely," *Nature,* 220 (1968), pp. 933–934, and "Haldane's Influence," *Nature,* 222 (1969), pp. 56–57. These pieces, and the replies to them (DP:C.108:J.83), make clear the need for a further look at Haldane beyond the standard Clark, *J. B. S.;* K. R. Dronamjaru, *Haldane and Modern Biology* (Oxford: Oxford University Press, 1969); and *Haldane's Daedalus Revisited* (Oxford: Oxford University Press, 1995). Mark B. Adams has begun to redress this neglect in "Last Judgment: The Visionary Biology of J. B. S. Haldane," *Journal of the History of Biology,* 33 (2001), pp. 457–491.

40. Haldane's foreword to *Recent Advances* was in this respect remarkably candid.

41. Notes, DP:C.108:J.87.

42. See Elof Axel Carlson, *Genes, Radiation, and Society: The Life and Work of H. J. Muller* (Ithaca, NY: Cornell University Press, 1981).

43. See Diane Paul, "Eugenics and the Left," in her *The Politics of Heredity* (Albany: State University of New York Press, 1998), pp. 11–35.

44. See Mark Adams, "Eugenics as Social Medicine in Revolutionary Russia," in *Health and Society in Revolutionary Russia,* eds. Susan Gross Solomon and John F. Hutchinson (Bloomington: Indiana University Press, 1990), pp. 200–223.

45. DP:C.24:D.63.

46. See the fascinating "Lenin's Doctrines in Relation to Genetics" in *Lenin Decennary Memorial Volume* (Moscow: Press of the Academy of Sciences of the U.S.S.R., 1934), pp. 565–579, in which Muller turns the Marxist Lamarckian thesis on its head. Muller's reductionism and materialism were held equally strongly by Darlington.

47. In a letter from Crew to Darlington (December 17, 1937), DP:C.111: J.162.

48. Muller had in fact stayed with Darlington in May 1937, on his way from Spain to the United States. During this visit, he made clear to his host the plight of genetics and geneticists in Russia.

49. Darlington letter to J. Todd, DP:C.25:D.72.

50. Draft letter to the *Times* (London) (October 1, 1936), DP:C.108:J.75.

51. S. C. Harland letter to Darlington, DP:C.25:D.74.

52. See Mark Popovsky, *The Vavilov Affair* (Hamden, CT: Archon Books, 1984), for details of the events.

53. Clarendon Press, somewhat amazed by Darlington's unorthodox and excited request at such a time, demanded a ready document that did not exceed the promised page limit. "Wartime is no time for long books," Kenneth Sisam of Oxford University Press wrote to Darlington, "and there is little labour to go through alterations." Notwithstanding, the book was ultimately published. See DP:C.114:J.224.

54. Vavilov to Darlington, DP:C.39:E.72. Krementsov, in his *Stalinist Science* (Princeton, NJ: Princeton University Press, 1998), p. 82, shows that Vavilov's correspondence with Darlington was cited, among other Western contacts, in the case against him.

55. See "International Genetics Conference," *Nature*, 157 (1946), pp. 35–38.

56. Darlington letter to *Nature* (May 31, 1946), DP:C.99:H.70. Darlington did not yet know that his friend Karpechenko had been arrested in 1941, and died in prison in 1942.

57. DP:C.25:D.84. Ashby jokingly rejoined: "I think you will soon have to be asking the trustees of the John Innes to provide a private detective for you!" Letter to Darlington (January 29, 1948), DP:C.25:D.90.

58. J. S. Alexandrowicz letter to Darlington (July 14, 1947), DP:C.25:D.88.

59. Darlington letter to Miss I. M. James (March 12, 1949), DP:C.94:C.69.

60. See Arthur Koestler's *Darkness at Noon* (New York: Macmillan, 1941), and *Arrow in the Blue* (New York: Macmillan, 1969) for particularly sensitive and classic descriptions of the psychology of disillusion.

10. Science in a Changing World

1. Clare Passingham interview by author (March 3, 2000).

2. Blacker letter to Darlington (December 17, 1948), DP:C.26:D.94.

3. Darlington letter to Blacker (December 30, 1948), ibid.

4. J. R. Baker, "Counterblast to Bernalism," *New Statesman and Nation*, 18 (1939), pp. 174–175.

5. See, for a sample, John Strachey's *The Coming Struggle for Power* (New York: Covici, 1933); Geoffrey Gorer's *Nobody Talks Politics* (London: M. Joseph, 1936); Solly Zuckerman, "Science and Society," *New Statesman and Nation* (February 25, 1939), p. 298; Harold Lasky's "The Danger of Being a Gentleman: Reflections on the Ruling Class in England" (1932), in his *The Danger of Being a Gentleman, and Other Essays* (London: Allen and Unwin, 1939); and Eric Hutchinson, "Scientists as an Inferior Class," *Minerva*, 8 (1970), pp. 396–411.

6. Neal Wood was the first to give the movement its name, in *Communism and British Intellectuals* (New York: Columbia University Press, 1959), p. 121. See also Michael D. King, "Science and the Professional Dilemma," in *Penguin Social Sciences Survey 1968*, ed. Julius Gould (Harmondsworth: Penguin Books, 1968); Norman J. Vig, *Science and Technology in British Politics* (Oxford: Pergamon Press, 1968); Hilary and Steven Rose, *Science and Society* (London: Allen Lane, 1969); Gary Werskey, *The Visible College* (London: Allen Lane, 1988), and "British Scientists and 'Outsider' Politics, 1931–1945," *Science Studies*, 1 (1971), pp. 67–83; and William McGucken, *Scientists, Society, and State: The Social Relations of Science Movement in Great Britain 1931–1947* (Columbus: Ohio State University Press, 1984).

7. Huxley, *A Scientist Among the Soviets* (London: Chatto and Windus, 1932), p. 60.

8. J. B. S. Haldane, *Possible Worlds and Other Essays* (London: Harper and Brothers, 1928), pp. 220–221.

9. Sidney and Beatrice Webb, *Soviet Communism: A New Civilization*, second edition (New York: Charles Scribner's Sons, 1937), pp. 1132–1133.

10. Werskey, *Visible College*, p. 178.

11. See Diane Paul, *The Politics of Heredity* (Albany: State University of New York Press, 1998).

12. J. G. Crowther, *Soviet Science* (New York: E. P. Dutton, 1936), p. 14.

13. See, for example, R. Brightman, "The Protection of Scientific Freedom," *Nature*, 137 (1936), pp. 963–964, and "Social Responsibilities of Science," *Nature*, 139 (1937), p. 689; B. P. Uvarov, "Genetics and Plant Breeding in the USSR," *Nature*, 140 (1937), p. 297.

14. See Joseph Needham, *The Nazi Attack on International Science* (London: Watts and Co., 1941).

15. See Joseph Needham and Jane Sykes Davies, eds., *Science in Soviet Russia* (London: Watts and Co., 1942).

16. J. D. Bernal, *The Social Function of Science* (London: G. Rutledge and Sons, 1939).

17. Brightman, "Protection of Scientific Freedom."

18. Bernal, *Social Function of Science*, pp. 63–64. By war's end Britain was spending 2.3 percent of its GNP on scientific research and development. Arthur Marwick, *A History of the Modern British Isles, 1914–1999* (Oxford: Oxford University Press, 2000), p. 205.

19. McGucken, *Scientists, Society, and State*, pp. 155–216.

20. Solly Zuckerman, *From Apes to Warlords: The Autobiography (1904–1946) of Solly Zuckerman* (London: Hamilton, 1978), p. 109.

21. On the club, a gathering of eminent scientists that would often host leading journalists, politicians, and intellectuals, see Zuckerman, ibid., pp. 394–404, and J. G. Crowther, *Fifty Years with Science* (London: Barrie and Jenkins, 1970), pp. 210–222.

22. "Men of Science and the War," *Nature*, 146 (1940), p. 107.

23. *Science in War* (Harmonsworth: Penguin Books, 1940); P. M. S. Blacket, "Operational Research," *Brassey's Annual* (1953), pp. 88–106.

24. See *Nature*, 150 (1942), p. 302.

25. *The Scientific Worker*, September 14, 1943, p. 3.

26. "Draft of Sir William Bragg's Broadcast on 5th January, 1942," DP:C.88:F.70.

27. "John Randal Baker (1900–1984)," *Biographical Memoirs of the Royal Society*, 31 (1985), pp. 33–63.

28. J. R. Baker, *Science and the Planned State* (London: Allen and Unwin, 1945). See also his *The Scientific Life* (London: Allen and Unwin, 1942).

29. "Michael Polanyi (1891–1976)," *Biographical Memoirs of Fellows of the Royal Society*, 23 (1977), pp. 413–448; Michael Polanyi, *The Contempt of Freedom* (London: Watts and Co., 1940); *The Logic of Liberty* (Chicago: University of Chicago Press, 1951).

30. Michael Polanyi, *The Tacit Dimension* (Garden City, NY: Doubleday, 1966), p. 3.

31. Baker, *Science and the Planned State*, p. 64.

32. *Nature*, 148 (1941), p. 497.

33. J. G. Crowther, *The Social Relations of Science* (New York: Macmillan, 1941); *The New Statesman and Nation*, August 15, 1943, p. 114.

34. Werskey, *Visible College*, p. 272.

35. Historian of science Everett Mendelsohn recalls that, as a young graduate student in the 1950s, Polanyi told him: "We have won all the battles but lost the war." (Communications with the author.)

36. *Bulletin of the Society for the Freedom of Science*, no. 3, July 1947, p. 1.

37. "Science and the Real Freedoms," *Nature*, 159 (1947), pp. 511–512; Werskey, *Visible College*, p. 285.

38. Notes, DP:C.99:H.66. Darlington's first broadcast, followed by many, was on April 20, 1939, on the topic "Heredity and Evolution," and managed to spark angry mail from the British Bible Association. *"You must have done much towards remoulding the attitude of hundreds of thousands, or millions, of listeners," an excited Muller wrote to him (DP:C.88:F.67).*

39. See his "The Cleavage in Biology," *Nature*, 147 (1941), p. 544.

40. See Darlington's "The Revolution in Education," *World Review*, October 1942, pp. 33–39.

41. Darlington, "The Future Task of Science," *The Listener*, August 13, 1942, pp. 213–214.

42. Darlington, *The Conflict of Science and Society,Delivered at Conway Hall, Red Lion Square, on April 20, 1948 by C. D. Darlington, F.R.S.* (London: Watts and Co., 1948), p. 41. Compare R. M. Pike, "The Growth of Scientific Institutions and the Employment of Natural Science Graduates in Britain, 1900–1960" (M.Sc. thesis, University of London, 1961), a description of the crisis in British science at the time, to Alexander G. Korol, *Soviet Research and Development: Its Organization, Personnel and Funds* (Cambridge, MA: MIT Press, 1965), an account of its rapid growth.

43. "Planning of Science" conference organized by the Association of Scientific Workers, of which Darlington was a member, January 31, 1943. Notes, DP:C99:H.65.

44. Session on "Science and Radio" in the conference "Science and the Citizen: The Public Understanding of Science," March 20, 1943. The transcript was published by the British Association for the Advancement of Science in *Advancement of Science*, 2, no. 8 (1943), pp. 299–301.

45. See Huxley's *UNESCO: Its Purpose and Its Philosophy* (Washington: Public Affairs Press, 1947), and Hogben's *Dangerous Thoughts* (London: Allen and Unwin, 1939) for representations of the scientific humanist outlook. It is also interesting to note the continuity with Bateson, who wrote in 1917: "The fate of nations hangs literally on the issue of contemporary experiments in the laboratory, but those who govern the Empire are quite content to know nothing of all this." See "The Place of Science in Education," reprinted in *Naturalist, op. cit.*, pp. 425–440, quote on p. 426.

46. J. B. S. Haldane, "Darwin and Slavery," *The Daily Worker*, November 14, 1949.

47. Darlington, *Conflict of Science and Society*, p. 28.

48. Ibid., p. 32–33. It is interesting to note the similarity between Darlington's views on Marxism and those expressed years later by the Harvard sociobiologist E. O. Wilson. Marxism was "mortally threatened by the discovery of human sociobiology," he would write, for it was nothing but "sociobiology without the biology." *On Human Nature* (Cambridge, MA: Harvard University Press, 1978), p. 191.

49. Darlington, *Conflict of Science and Society*, p. 28.

50. Sir William Beveridge letter to Darlington (October 3, 1944), DP:C.3:A.130.

51. Darlington, *Conflict of Science and Society*, p. 30.

52. This was doubly true for the reason that never before had the power of

science been so high and the public confidence in it so low. See Greta Jones, *Science, Politics, and the Cold War* (London: Routledge, 1988).

53. "Dead Hand on Discovery" (draft), delivered December 2, 1948, DP:C.84:F.5.

54. In a BBC broadcast in the spring of 1946, three scientists, Haldane, Polanyi, and Waddington, were pitted against two historians, E. L. Woodward and A. D. Ritchie, and two novelists, Arthur Koestler and E. M. Forster, to speak about "The Challenge of Our Time." The "challenge" was the gulf between the scientific and humanistic approach to life. All the humanists lined with the Right against the Soviet Union and in favor of individualistic ideals. All the scientists (except Polanyi, who argued that social problems were not technical but rather political, and could therefore not be solved by science) leaned to the Left, in support of the notion of the people's control of the material foundations of society. The humanists all feared the scientists were endangering personal freedoms, if not the "human spirit" itself. A particularly vicious attack from Koestler on Bernal ensued. Some days later Orwell wrote to Koestler to thank him, claiming that scientists were "much more subject to totalitarian habits of thought than are writers." S. Orwell and I. Angus, eds., *The Collected Essays, Journalism, and Letters of George Orwell*, vol. 4 (Harmondsworth: Penguin Books, 1970), p. 69.

55. See Gary McCulloch, "A Technocratic Vision: The Ideology of School Science Reform in Britain in the 1950s," *Social Studies of Science*, 18 (1988), pp. 703–724, for an assessment of the ideological background to, and a history of the political roots of, technocratic school reform.

56. Darlington, *Conflict of Science and Society*, p. 45.

57. Henry Dale-Darlington correspondence (September 28 and 30, 1948), DP:C.107:J.32.

58. Darlington, *Conflict of Science and Society*, p. 50.

59. "The notion of equality is one of the three chief illusions promoted by the great Semitic religions," he wrote, "and it is perhaps the most comforting." Ibid., p. 28.

60. Darlington letter to B. R. Atkins (June 7, 1950), DP:C.91:G.9. The peak postwar membership numbers, exceeding 15,000, had by 1954 dwindled to 11,382. Jones, *Science, Politics, and the Cold War*, p. 50.

61. Darlington, *The Evolution of Genetic Systems*, second edition (Edinburgh: Oliver and Boyd, 1958).

62. Lancelot Hogben, *The New Authoritarianism*, Conway Memorial Lecture (London: Watts and Co., 1949), p. 41.

11. The Conflict of Science and Society

1. Darlington, *The Conflict of Science and Society, Delivered at Conway Hall, Red Lion Square, on April 20, 1948 by C. D. Darlington, F.R.S.* (London: Watts and Co., 1948), p. viii.

2. Ibid., p. 1.

3. Ibid., p. 3.

4. BBC Third Programme, "The Conflict of Science and Society," October 20, 1948.

5. Darlington, *Conflict of Science and Society*, p. 5.

6. Ibid., p. 28.

7. Diary, 1924, DP:d.1:A.64.

8. Eugenics Society letter to Darlington (February 5, 1927), DP:C.91:G.18.

9. Darlington letter to Alfred (February 20, 1927), DP:C.5:A.166.

10. Diary, 1927, DP:g.32:A.65.

11. Ibid.

12. Diary, DP:g.32:A.67; DP:g.32:A.68; DP:f.10:A.69.

13. J. B. S. Haldane, "Determinism," *Nature*, 129 (1932), pp. 315–316.

14. Darlington, "The Relations Between Science and Ethics," *Nature*, 148 (1941), pp. 344–345.

15. Diary, 1942, DP:g.34:A.71.

16. See Darlington, *The Evolution of Genetic Systems* (Cambridge: Cambridge University Press, 1939).

17. D. P. Riley letter to Darlington (April 6, 1943), DP:C.41:E.120.

18. Darlington, "Race, Class and Mating in the Evolution of Man," *Nature*, 152 (1943), pp. 315–319.

19. Kenneth Mather, "Polygenic Inheritance and Natural Selection," *Biological Reviews of the Cambridge Philosophical Society*, 18 (1943), pp. 32–64.

20. Darlington, "Race, Class, and Mating," p. 316.

21. Ibid.

22. Ibid., p. 317. For an interesting comparison, see the last third of R. A. Fisher's *Genetical Theory of Natural Selection* (Oxford: Clarendon Press, 1930).

23. Ibid.

24. Ibid., p. 318.

25. G. Dahlberg, *Race, Reason and Rubbish* (London: Allen and Unwin, 1929).

26. Darlington, "Race, Class and Mating," p. 318.

27. Ibid., pp. 318–319.

12. On the Determination of Uncertainty

1. His and La Cour's book *The Handling of Chromosomes* (London: Allen and Unwin, 1942) was intended to prepare the way for the extension of chromosome studies to those physicists who were now getting in touch with them.

2. See Conrad Hal Waddington, "Some European Contributions to the Prehistory of Molecular Biology," *Nature*, 221 (1969), pp. 318–321. Waddington, one of the nine participants at Klampenborg, argued that Darlington's explanation of the chromosomes to the physicists (Bernal, Astbury, Auger) must be considered an important origin of the biological-physical cooperation on the transition road from classical genetics to molecular biology. Darlington's role in this endeavor has been described by some, but completely neglected by others: Robert Olby's account in *The Path to the Double Helix: The Discovery of DNA* (New York: Dover Publications, 1994) is the most complete description of Darlington's role. (Olby knew Darlington well, having worked as librarian of the Oxford Botany School during his tenure). In contrast, Darlington is not mentioned either in James D. Watson, *The Double Helix* (New York: Atheneum, 1968), or in Horace Judson, *The Eighth Day of Creation:*

Makers of the Revolution in Biology (Plainview, NY: Cold Spring Harbor Laboratory Press, 1996).

3. For a history of the protein model and eventual acceptance of the DNA model, see Lily Kay, *The Molecular Vision of Life: Caltech, the Rockefeller Foundation, and the Rise of the New Biology* (Oxford: Oxford University Press, 1993).

4. "Nucleic Acid and the Chromosomes" (draft), DP:C.41:E.137.

5. See Dan Lewis's description of the collaborative work on nucleic acid starvation in "Leonard Francis La Cour, 1907–1984," *Biographical Memoirs of Fellows of the Royal Society*, 33 (1986), pp. 357–375. Also Darlington, "Nucleic Acid Starvation of Chromosomes in *Trillium*," *Journal of Genetics*, 40 (1940), pp. 185–213. For his work on the physical basis of heredity, see Darlington, "The Chemical Basis of Heredity and Development," *Discovery*, 6 (1945), pp. 79–86; "The Chromosome as a Physico-Chemical Entity," *Nature*, 176 (1955), p. 1143; "Nucleic Acid: The Midwife Molecules," *Advancement in Science*, 12 (1955), p. 355.

6. Darlington, "On the Origins and Development of *Heredity*," a note for the archive of *Heredity* (April 10, 1970), DP:C.92:G.24–65.

7. Introduction to first issue, *Heredity*, 1, 1947 (draft), DP:C.92:G.32.

8. Significantly, Darlington did not publish his pieces on Lysenko in *Heredity*, which was a professional scientific journal; Lysenko was a political matter.

9. Darlington, "Origins and Development of *Heredity*." Oliver Darlington remembers his father saying of the editorial collaboration with Fisher: "He does 50% of the work, and I do 150%!" Oliver Darlington interview by author (April 1, 2000).

10. "Heredity, Development, and Infection," *Nature*, 154 (1944), pp. 164–169.

11. C. D. Darlington and K. Mather, *The Elements of Genetics* (London: Allen and Unwin, 1949), pp. 359–360.

12. "The Genetic Component of Language," *Heredity*, 1 (1947), pp. 269–286.

13. Ever the Bible quoter, Darlington reminded his readers that this was known already in biblical times, when the Jews were divided according to those who could, and those, numbering 42,000, who "could not frame to pronounce" the word *shibboleth* right (Judges 12:4–6).

14. "The great problems of history will be solved by geography and genetics," he wrote in his diary in 1951. DP:g.35:A.73.

15. Darlington had helped move the John Innes from Merton to a more spacious location at Bayfordbury in Hertfordshire in the late 1940s. While providing better facilities, this led to the weakening of links with London University, and to a growing feeling of isolation.

16. Darlington notes, "Purpose at Oxford," DP:C.14:C.11.

17. Eric Ashby, F. A. E. Crew, C. D. Darlington, E. B. Ford, J. B. S. Haldane, E. J. Salisbury, W. B. Turrill, C. H. Waddington, "Genetics in the Universities," *Nature*, 138 (1936), pp. 972–973; Darlington, "The Unity and Power of Biology," *Quarterly Journal of the British Association for the Advancement of Science*, 3 (1945), pp. 124–126.

18. Darlington, *The Facts of Life* (London: Allen and Unwin, 1953), pp. 17–174.

19. A similar account of biology was later presented by Jacques Monod in *Chance and Necessity: An Essay on the Natural Philosophy of Modern Biology* (New York: Knopf, 1971).

20. Darlington, *Facts of Life*, pp. 177–216.

21. Ibid., p. 242.

22. See Herbert Spencer, *The Principles of Biology* (London: Williams and Norgate, 1863); *The Principles of Sociology* (London: Williams and Norgate, 1877). Greta Jones, *Social Darwinism in English Thought* (Brighton, Sussex: Harvester Press, 1980) shows how Darwinism was actually used to support a wide range of political ideas and social theories. See also Philip Abrams, *The Origins of British Sociology, 1834–1914* (Chicago: University of Chicago Press, 1968).

23. Arnold Toynbee, *A Study of History* (London: Oxford University Press, 1934).

24. Ibid., p. 250.

25. Ibid., p. 252.

26. On Galton, see Nicholas Wright Gilham, *A Life of Sir Francis Galton: From African Exploration to the Birth of Eugenics* (Oxford: Oxford University Press, 2001); Derek Williams, *Francis Galton: The Life and Work of a Victorian Genius* (London: Elek, 1974); Karl Pearson, *Francis Galton, 1822–1922: A Centenary Appreciation* (London: Cambridge University Press, 1922); and R. S. Cowan, "Nature and Nurture: The Interplay of Biology and Politics in the Thought of Francis Galton," *Studies in the History of Biology*, 1 (1977), pp. 133–208.

27. O. von Verschuer, "Twin Research from the Time of Francis Galton to the Present Day," *Proceedings of the Royal Society of London, Series B*, 128 (1939), pp. 62–81; Johann Lange, *Crime as Destiny*, trans. Charlotte Haldane (London: Allen and Unwin, 1931).

28. Darlington, *Facts of Life*, pp. 292–293.

29. Ibid., pp. 251–357.

30. Ibid., p. 269.

31. Ibid., p. 356–357.

32. Ibid.

33. Arthur Eddington, *The Nature of the Physical World* (Cambridge: Cambridge University Press, 1928); R. A. Fisher, "Indeterminism and Natural Selection," *Philosophy of Science*, 1 (1934), pp. 99–117.

34. Ibid., p. 426.

35. Ibid., p. 416.

36. Ibid., p. 427.

13. The Breakdown of Classical Genetics

1. Stan Woodell interview by author (March 31, 2000).

2. Ford was excited to gain a desperately needed ally in an institution that had yet to create a chair for his subject. "It will be grand to have you and to push things together," he wrote to Darlington after hearing of his appointment. E. B. Ford letter to Darlington (February 3, 1953), DP:C.14:C.2.

3. Lionel Clowes interview by author (February 2, 2000). See also J. B. Morell, "The Non-Medical Sciences, 1914–1939," and John Roche, "The Non-Medical Sciences, 1937–1970," in *The History of the University of Oxford, Vol. VIII: The Twentieth Century*, ed. Brian Harrison (Oxford: Oxford University Press, 1994).

4. Sinclair letter to Darlington (February 3, 1953), DP:C.14:C.3.

5. Paul and Yvonne Harvey interview by author (April 6, 2000).

6. Letter to the author from Brian Cox (November 16, 2000); Stan Woodell

interview by author (March 31, 2000). Interviewing Woodell in 1959, Darlington said: "One of your referees said you aren't a pure ecologist," adding with a slight smile, "I don't like pure ecologists."

7. Darlington jotting on Clark letter of congratulation to Darlington (February 3, 1953), DP:C.14:C.3. Ernst Mayr recalls the story that E. J. Salisbury, upon congratulating Darlington, asked: "By the way, since when have you been interested in botany?" Mayr interview by author (January 7, 2001)

8. Joy Boyce interview by author (April 4, 2000).

9. Darlington letter to Pontecorvo (February 3, 1953), DP:C.14:C.3.

10. An interesting comparison to Darlington's arrival at Oxford is the arrival of James Watson at Harvard in the mid-1950s. Here, too, the winds of change met with strong departmental resistance by old-school scientists. See E. O. Wilson, "The Molecular Wars" in *Naturalist* (Washington, DC: Island Press/Shearwater Books, 1994), pp. 218–237.

11. Darlington, "The Control of Evolution in Man," The Woodhull Lecture, March 21, 1958, *Proceedings of the Royal Institute of Great Britain*, 37 (1958), pp. 77–92.

12. Diary, 1962, DP:F.18:A.81.

13. See Darlington's critique of classical genetics in "Natural Populations and the Breakdown of Classical Genetics," *Proceedings of the Royal Society of London, Series B*, 145 (1956), pp. 350–364.

14. See Arthur Mourant, *The Distribution of the Human Blood Groups* (Oxford: Oxford University Press, 1954); R. R. Race and R. Sanger, *Blood Groups in Man*, third edition (Oxford: Oxford University Press, 1958); E. B. Ford, "Polymorphism in Plants, Animals, and Man,"*Nature*, 180 (1957), pp. 1315–1319; J. Fraser Roberts, "Blood Groups and Susceptibility to Disease: A Review," *British Journal of Preventive and Social Medicine*, 11 (1957), pp. 107–125; A. C. Allison, "Protection Afforded by Sickle-Cell Trait Against Subtertian Malarial Infection," *British Medical Journal*, 1 (1954), p. 298.

15. See Darlington, "Psychology, Genetics and the Process of History," *British Journal of Psychology*, 54 (1963), pp. 299–307. Contemporary studies cited by Darlington as evidence included F. J. Kallmann, *The Genetics of Mental Illness* (New York, [Company??] 1959); Cyril Burt, "Intelligence and Social Mobility," *British Journal of Statistical Psychology*, 14 (1961), pp. 3–25, and *The Gifted Child* (London: Hodder and Stoughton, 1962); J. Fraser Roberts, "The Genetics of Mental Deficiency," *Eugenics Review*, 44 (1952), pp. 71–83; and John L. Fuller and W. Robert Thompson, *Behavior Genetics* (New York: Wiley, 1960). Less recent, but nonetheless endorsed, studies included Richard Dugdale, *The Jukes* (New York: Putnam, 1877); Francis Galton, *Hereditary Genius* (London: Macmillan, 1869); and "The History of Twins as a Criterion of the Relative Power of Nature and Nurture," *Journal of the Anthropological Institute of Great Britain and Ireland*, 5 (1876), pp. 391–406.

16. C. D. Darlington and L. F. La Cour, "Hybridity Selection in *Campanula*," *Heredity*, 5 (1950), pp. 217–218.

17. Darlington defined fertility as the property of leaving descendants. This was an integral property of the sequence: desire for sexual intercourse, ability to beget and bear offspring, viability of embryos and offspring.

18. Darlington, "Cousin Marriage and the Evolution of Breeding in Man," *Heredity*, 14 (1960), pp. 297–232. Also, "Marriage Makes History," *New Scientist*, 7 (1960), pp. 942–944.

19. Darlington, "Control of Evolution in Man," *Nature*, 182 (1958), p. 16.

20. Ibid., p. 17.

21. Indeed, when Eugenics Society president C. P. Blacker asked Darlington to become a member of its council, he declined, using the (true) excuse that "all committees are anathema to me." Darlington letter to Blacker (January 27, 1948), DP:C.91:G.18.

22. Darlington, "The Chromosomes and the Theory of Heredity," *Nature*, 187 (1960), p. 895.

23. Darlington, "Freedom and Responsibility in Academic Life," *Bulletin of Atomic Science*, 13 (1957), pp. 131–134, quote on p. 134.

24. Oxford University Archive (OUA), Department of Botany, 1942–1963, UC/FF/549/1–3; Registry file relating to Botany and Forestry Building, 1949–1960, UR6/BG/FI, files 2–3.

25. Darlington, "The Cleavage in Biology," *Nature*, 147 (1941), and "The Unification of Biology," *New Scientist*, 13 (1962), pp. 72–74.

26. See J. W. S. Pringle, "Biology at Oxford," *The School Science Review*, 154 (1963), pp. 1–3, and C. D. Darlington, "Teaching Biology," *Nature*, 199 (1963), pp. 117–119. That same year Darlington co-edited a book with A. D. Bradshaw, *Teaching Genetics* (Edinburgh: Oliver and Boyd, 1963).

27. Oxford University Archive, Reports of the Biological Sciences Faculty Board, 1951–1972, FA4/3/2/10–17. Darlington's absence from meetings became so acute that he was removed in 1967 from the board.

28. See A. H. Halsey, "The Franks Commission," in *The History of the University of Oxford, Vol. VIII: The Twentieth Century*, ed. Brian Harrison (Oxford: Oxford University Press, 1994), pp. 721–736.

29. See Darlington's "Proposals for Reform in the University of Oxford, 1964," DP:C.18:C.92. Also, "Oxford Reformed," submitted to *Encounter*, May 4, 1965 (unpublished) in DP:C.18:C.96–102, "Oxford Unreformed," *Encounter*, 28 (1967), pp. 93–94, and "The Evolution of Oxbridge," *Question*, 3 (1970), pp. 37–51. Darlington kept a file marked "CDD vs. Oxford United" of press cuttings from all the major British newspapers and the *New York Times* (October 13, 1964).

30. *The Daily Mail*, October 30, 1964.

31. Darlington, "The Case of Oxford," notes, DP:C.18:C.86. See C. P. Snow, *The Two Cultures*, with introduction by Stefan Collini (Cambridge: Cambridge University Press, 1993) for an appreciation of the humanities versus science battles of the 1950s and 1960s in Britain.

32. Reply to an American psychologist (April 6, 1964), DP:C.1:A.12.

33. Diary (August 7, 1965), DP:f.23:A.86.

34. Darlington correspondence; interviews by author with Professor Paul Harvey (March 15, 2000 and April 4, 2000) and Sir Allan Bullock (March 2, 2000); visits, DP:C.100:H.90; C.102.

35. Diary notes, August 1967, DP:f.24:A.88.

36. Darlington, *The Evolution of Man and Society* (London: Allen and Unwin, 1969).

37. "We have to face the fact that we are committed to a launching of a book which quite inevitably will be compared to Charles Darwin's *Origins of Species*." Memorandum to Sir Stanley Unwin (January 3, 1968) in DP:C.62:E.466.

38. Darlington, *Evolution of Man and Society*, pp. 302, 398, 459, 167, 429, 379–380, 464–466, 499, 512, 596, 129.

39. Ibid., pp. 671–673.

40. Diary, 1953, DP:g.36:A.74.

41. Ibid., p. 679.

42. See also Darlington's "The Impact of Man on Nature," 1970, reprinted in Ivo Mosely, ed., *Dumbing Down: Culture, Politics and the Mass Media* (Thorverton: Imprint Academic, 2000), pp. 301–314.

14. On the Uncertainty of Determination

1. Notes (June 7, 1969), DP:C.61:E.460.

2. "Statement on Nazi Race Theory," draft, signed by Huxley, Haldane, and Darlington (May 10, 1945), DP:C.41:E.130. See letter of thanks from Julian Huxley, ibid.

3. *New York Times*, July 18, 1950. A second, more conservative Statement on Race was issued two years later. By this time physical anthropologist Henri Vallois had replaced Ashley Montagu as chair, and the composition of the committee shifted to a greater number of physical anthropologists and geneticists at the expense of cultural anthropologists. Still, their conclusion that the existence of races is possible, but unproven, and the agnosticism with respect to the effects of race on intelligence, remained for Darlington a cop-out. See UNESCO, *The Race Concept: Results of an Inquiry* (Paris: UNESCO, 1952). For discussions of the drafting of the statement, see Ashley Montagu, *Statement on Race*, third edition (New York: Oxford University Press, 1972); Greta Jones, *Science, Politics, and the Cold War* (London: Routledge, 1988), chapter 4 (quote on p. 67); and Pat Shipman, *The Evolution of Racism* (Cambridge, MA: Harvard University Press, 1994), pp. 158–170.

4. Darlington, "The Genetic Understanding of Race in Man," *International Social Science Bulletin*, 2 (1950), pp. 479–488, and 3, pp. 753–754.

5. Darlington letter to Metraux (December 7, 1951), DP:C.44:E.181.

6. Montagu letter to Metraux (November 9, 1951), DP:C.44:E.181.

7. Metraux letter to Hans Nachtsheim (March 8, 1952), DP:C.11:J.164.

8. Muller–Darlington correspondence (April 8 and 12, 1952), DP:C.111:J.164.

9. See Muller's dissent in *The Race Concept*, pp. 52–54. Exemplars of the orthodox view are L. C. Dunn and Theodosius Dobzhansky, *Heredity, Race, and Society* (New York: New American Library, 1946), and Leo Kuper, ed., *Race, Science and Society* (Paris: UNESCO, 1956).

10. Darlington, "Genetics and Society: A Scientist Replies," *Encounter*, 37 (1971), pp. 84–93, quote on p. 85.

11. See Shipman, *Evolution of Racism*, and Elazar Barkan, *The Retreat of Scientific Racism: Changing Concepts of Race in Britain and the United States Between the World Wars* (Cambridge: Cambridge University Press, 1992), both of whom show how

politics, rather more than science, determined the antiracist reaction of scientists to the Nazis' perversion of genetics and evolutionary theory. Barkan shows how, despite the failure of national and international attempts in the 1930s to reach a consensus on the definition of "race," by 1950 "racism" had been refuted. On the political dimensions of the change of rhetoric among geneticists before and after the war, see also Diane Paul, "Eugenics and the Left," in her *The Politics of Heredity* (Albany: State University of New York Press, 1998), and Nancy Stepan, *The Idea of Race in Science: Great Britain, 1800–1960* (Hamden, CT: Archon Books, 1982), pp. 140–190.

12. See Haldane's *Heredity and Politics* (New York: W. W. Norton, 1938), and Julian Huxley and Alfred C. Haddon, *We Europeans, A Survey of 'Racial' Problems, with a Chapter on Europe Overseas by A. M. Carr-Saunders* (London: J. Cape, 1935).

13. Daniel Kevles, *In the Name of Eugenics* (Cambridge, MA: Harvard University Press, 1995), p. 251. Penrose changed the name of his chair in 1954 to the Galton Chair of Human Genetics.

14. See Shipman's discussion of the conference, pp. 173–191.

15. Darlington letter to William Provine (June 21, 1974), DP:C.104:H.16. On the scientific underpinnings of the reform eugenics of the 1930s, and difficulties encountered in early human genetics, see Kevles, *In the Name*, chapters 11, 13. See also Greta Jones, "Eugenics and Social Policy Between the Wars," *The Historical Journal*, 13 (1982), pp. 717–728, and G. R. Searle, "Eugenics and Politics in Britain in the 1930s," *Annals of Science*, 36 (1979), pp. 159–169.

16. Theodosius Dobzhansky and L. S. Penrose, review in *Annals of Human Genetics*, 19 (1954), pp. 75–77. Haldane, too, felt Darlington's determinism exaggerated and without foundation, "The Facts of Life," *British Journal of Sociology*, 5 (1954), pp. 88–89, as did John Fraser Roberts and S. C. Harland, unidentified reviews, DP:C.45:E.210, C. H. Waddington, "The Facts of Life," *Heredity*, 8 (1954), pp. 279–282, and C. Stern, "The Facts of Life," *Population Studies*, 8 (1954), pp. 188–191.

17. Muller letter to Darlington (April 23, 1953).

18. Ibid. (June 18, 1954).

19. See Kevles, *In the Name*. Even the scions of the British Union of Fascists were singing a new, modified tune. See Richard Thurlow, *Fascism in Britain: A History, 1918–1985* (Oxford: Oxford University Press, 1987), pp. 233–259.

20. It was Joe Hin Tjio and Albert Levin who discovered the correct number of human chromosomes in 1956. In England, Charles E. Ford, a cytogeneticist in the Radiobiological Research Unit of the MRC, became in the 1950s the leader in human genetics from the cytological side, together with Paul E. Polani. Darlington was not involved with their work.

21. Darlington letter to Baker (23.2.1953), DP:C.95:G.104.

22. Notes, 1955, DP:C.85:F.30.

23. Public talk, "Dead Hand on Discovery, or The Conflict of Science and Society," London (November 2, 1948), transcript, DP:C.84:F.8; *The Facts of Life* (London: Allen and Unwin, 1952), p. 347.

24. Dobzhansky and Penrose review, p. 76.

25. Ibid.; Alex Comfort, review of *Genetics and Society*, the *Guardian*, March 25, 1965.

26. Stern, "Facts of Life," p. 191; A. C. Allison, "Scientific Calvinism," *Impact*, 5 (1953), pp. 191–200; Oswald Mosley, *Mosley—Right or Wrong?* (London: Lion Books, 1961), pp. 119–120. Mosley was in fact a Lamarckist; nonetheless, he was happy to garner support for his ideas from genetics. See Robert Skidelsky, *Oswald Mosley* (London: Papermac, 1990), and Nicholas Mosley, *Beyond the Pale* (London: Secker and Warburg, 1983).

27. Darlington, "Was Huxley Right?" Huxley Lecture, Birmingham, March 7, 1957. See T. H. Huxley, *Evidence as to Man's Place in Nature* (New York: D. Appleton, 1863).

28. Huxley letter to Darlington (March 20, 1953), DP:C.45:E.207.

29. See Julian Huxley's *New Bottles for New Wine* (London: Chatto and Windus, 1957), and Darlington's review, "Transhumanism," *New Scientist*, 3 (1957), pp. 28–29; Diary (July 9, 1978), DP:g.44:A.96. On Huxley and his evolutionary humanism, see John C. Green, "From Huxley to Huxley: Transformation in the Darwinian Credo," in his *Science, Ideology, and World View* (Berkeley: University of California Press, 1981), pp. 158–193, and "The Interaction of Science and World View in Sir Julian Huxley's Evolutionary Biology," *Journal of the History of Biology*, 23 (1990), pp. 39–55; William B. Provine, "Progress in Evolution and Meaning in Life," in *Evolutionary Progress*, ed. Mathew H. Nitecki (Chicago: University of Chicago Press, 1998), pp. 49–74; Robert S. Gascoigne, "Julian Huxley and Biological Progress," *Journal of the History of Biology*, 24 (1991), pp. 433–455; V. B. Smocovitis, "Unifying Biology: The Evolutionary Synthesis and Evolutionary Biology," *Journal of the History of Biology*, 25 (1992), pp. 40–43; Collin Duval, "From a Victorian to a Modern: Julian Huxley and the English Intellectual Climate," in *Julian Huxley: Biologist and Statesman of Science*, eds. C. Kenneth Waters and Albert Van Helden (Houston: Rice University Press, 1993), pp. 31–44.

30. See P. B. Medawar, *The Future of Man* (London: Methuen, 1960), and Darlington's review in *Heredity*, 16 (1961), pp. 441–446.

31. Diary, August 1967, DP:F.24:A.88.

32. P. B. Medawar, "The Future of Man: A Reply," *Heredity*, 16 (1961), pp. 228–231.

33. On the "New Right" and race, see Greta Jones, *Science, Politics and the Cold War* (London: Routledge, 1988), chapter 7; Arthur Aughey, Greta Jones, and W. T. M. Riches, *The Conservative Political Tradition in Britain and the United States* (London: Pinter, 1992), chapter 4; and Rodney Barker, *Political Ideas in Modern Britain: In and After the Twentieth Century* (London: Routledge, 1997), chapter 8. On the persistence of subterranean bigotry in the eugenics movements into the 1970s, see Martin Barker, *The New Racism: Conservatives and the Ideology of the Tribe* (Frederick, MD: Aletheia Books, 1981).

34. Gregor letter to Darlington (July 9, 1961), DP:C.54:E.364.

35. Darlington–Gini correspondence (October 24 and 27, 1962); Oliver Darlington interview by author (April 1, 2000); Professor M. A. MacConnail, Department of Anatomy, University College, Cork, letter to Tansil (February 4, 1963), DP:C.86:F.39.

36. See Zig Layton-Henry, *The Politics of Race in Britain* (London: Allen and Unwin, 1984), figures on pp. 16–29; Paul Foot, *The Rise of Enoch Powell: An Examination of Enoch Powell's Attitude to Immigration and Race* (London: Cornmarket Press, 1969).

37. The race-intelligence and sociobiology debates provided the main venues. See Mark Snyderman and Stanley Rothman, *The IQ Controversy: The Media and Public Policy* (New Brunswick, NJ: Transaction Books, 1988); Ned Joel Block and Gerald Dworkin, *The IQ Controversy* (New York: Pantheon Books, 1976); Daniel Seligman, *A Question of Intelligence: The IQ Debate in America* (New York: Carol Publication Group, 1992); Ullica Segerastråle, *Defenders of the Truth: The Battle for Science in the Sociobiology Debate and Beyond* (Oxford: Oxford University Press, 2000); Pat Shipman, *The Evolution of Racism* (Cambridge, Mass.: Harvard University Press).

38. A very partial list includes Robert Ardrey, *The Territorial Imperative* (New York: Atheneum, 1966); Konrad Lorenz, *On Aggression* (New York: Harcourt, 1966); Desmond Morris, *The Naked Ape* (New York: Dell, 1967); Arthur Jensen, "How Much Can We Boost I.Q. and Scholastic Achievement?" *Harvard Educational Review*, 39 (1969), pp. 1–123; Christopher Jenks, *Inequality: A Reassessment of the Effect of Family and Schooling* (New York: Basic Books, 1972); Hans J. Eysenck, *The Inequality of Man* (London: Temple Smith, 1973); Richard Herrnstein, *I.Q. in the Meritocracy* (Boston: Little, Brown, 1973); Edward O. Wilson, *Sociobiology* (Cambridge, MA: Harvard University Press, 1975). See Jones, *Science, Politics, and the Cold War*, chapter 7.

39. Max Beloff, "The Genetic Approach to World History," *Encounter*, 35 (1970), pp. 85–88, quote on p. 85.

40. For a careful study of the professional and political variables relevant to the debate see Jonathan Harwood, "The Race-Intelligence Controversy: A Sociological Approach I—Professional Factors," and "The Race-Intelligence Controversy: A Sociological Approach II—External Factors," *Social Studies of Science*, 6 (1976), pp. 369–394, and 7 (1977), pp. 1–30.

41. Sir Cyril Burt, "Biology Joins Hands with History," *New Scientist*, (September 4, 1969), p. 489; Garrett Hardin, "Genetics and History," *Science*, 168 (1970), pp. 1332–1333; Hamilton letter to Darlington (March 10, 1970), DP:C.63:E.487 (see Andrew Brown's interesting portrait of the recently deceased Hamilton in *Prospect*, January 2002, pp. 54–59); Carleton Coon, "The Evolution of Man and Society," *Boston Globe*, June 27, 1970; John Ziman, "Do Some Theories Stink?" *New Humanist*, 88, 1972, pp. 150–152; Robert Young, "Understanding It All," *New Statesman*, 26 (1969), pp. 117–118; P. B. Medawar, "The Volubility of DNA," *Harper's Magazine*, (May 20, 1971), pp. 17–18. See also F. A. Hayek, "Nature v. Nurture Once Again," *Encounter*, 34 (1970), pp. 81–83; Robert Nisbet, "The Evolution of Man and Society," *New York Times Book Review*, August 2, 1970; A. C. Allison, "An Intelligible Flow of Genes," *Science Journal*, December 1969, pp. 85–88.

42. Popper letter to Darlington (May 25, 1970), DP:C.63:E.483.

43. Beloff, "The Genetic Approach."

44. Darlington, "Genetics and Society: A Scientist Replies," *Encounter*, 37 (1971), p. 88.

45. Darlington, *The Conflict of Science and Society* (London: Watts and Co., 1949), p. 6.

46. Popper letter to Darlington.

47. Darlington letter to Popper (June 11, 1970), DP:C.63:E.483.

48. From *The First Circle*, quoted in Darlington, "A Scientist Replies," p. 91.

15. One Final Hurrah

1. Letters to the author from Rachel Ritchie (June 9, 2002) and Clare Passingham (June 18, 2002).

2. Darlington farewell speech at Magdalen College (June 10, 1971), DP:C.20:C.113.

3. Despite the fact that Darlington is generally given credit for bringing botany at Oxford into the modern era, the department's national standing, its annual publications, and the number of its undergraduates all dropped during his tenure, while those for zoology, biochemistry, and physiology rose. For statistics, see John Roche, "The Non-Medical Sciences, 1937–1970," in *The History of the University of Oxford, Vol. VIII: The Twentieth Century*, ed. Brian Harrison (Oxford: Oxford University Press, 1994).

4. See Horace Freeland Judson, *The Eighth Day of Creation: Makers of the Revolution in Biology* (Plainview, NY: Cold Spring Harbor Laboratory Press, 1979).

5. James D. Watson, *Passion for DNA: Genes, Genomes, and Society* (Plainview, NY: Cold Spring Harbor Laboratory Press, 2000), p. 19.

6. Letter to the author from Brian Cox (one of the reviewers).

7. Darlington, "The Perfectibility of Man," *Nature*, 229 (1971), pp. 575–576; "Heredity and Environment in the Development of Human Intelligence," in *Symposium on Ecology of Child and Human Development*, University of Miami, March 1971; "Genetics of Intelligence: Bearing on Education," letter, *Times* (London), November 23, 1976; "Epilogue: The Evolution and Variation of Human Intelligence," in *Human Variation: The Biopsychology of Age, Race and Sex*, eds. R. Travis Osborne, Nathaniel Weyl, and Clyde E. Noble (New York: Academic Press, 1978).

8. Diary (June 15, 1968), DP:C.30:D.147.

9. See Michael Ruse, *Monad to Man. The Concept of Progress in Evolutionary Biology* (Cambridge, MA: Harvard University Press, 1996), pp. 456–484.

10. For a nice article putting this change in attitudes of scientists and the scientific profession in perspective, see Diane Paul, "'Our Load of Mutations' Revisited," *Journal of the History of Biology*, 20 (1987), pp. 321–335.

11. "Resolution of Scientific Freedom Regarding Human Behaviour and Heredity," *American Psychologist*, July 1972, pp. 660–661. The academic environment was so hostile that the Society for the Psychological Study of Social Issues, a division of the American Psychological Association, set up a commission to investigate the motivations of the signatories (among others, Eysenck, Herrnstein, Jensen, Hardin, John Kendrew, Raymond Cattell, Jacques Monod, and Francis Crick), DP:C.30:D.146; Darlington's "Shockley and Leeds," unpublished article written for *Minerva*, DP:C.65:E.516; Darlington, review of Hans Eysenck, *Race, Intelligence and Education* (London: Temple Smith, 1971), *Sunday Times* (London), June 27, 1971; John Baker, *Race* (New York: Oxford University Press, 1974); and Darlington's review, "Members of One Another," *Sunday Times* (London), February 17, 1974. Compare with Amitai Etzioni's review, "Racist Hodgepodge," *The Washington Post*, March 8, 1974; and Darlington, "Cracks in the Gene: The Genetics of Human Populations, by L. L. Cavalli-Sforza and W. F. Bodmer" (review), *New Society*, January 6, 1972, pp. 25–26.

12. Darlington, *The Little Universe of Man* (London: Allen and Unwin, 1978).

13. Ibid., pp. 81–82.

14. Diary (October 3, 1969), DP:g.37:A.89.

15. Ibid., p. 17.

16. Correspondence between Darlington and publishers, and publishers and Field, Fisher, and Martineau, 1977, DP:C.77:E.637.

17. Burt was accused posthumously of doctoring his data to prove the high correlation between twin-pairs for the heritability of intelligence. See Leon Kamin, *The Science and Politics of I.Q.* (Potomac, Md.: Erlbaum, 1974) and Leslie Hearnshaw, *Cyril Burt: Psychologist* (New York: Random House, 1971). Recently, other assessors have dismissed these accusations as unfounded. See R. Fletcher, *Science, Ideology and the Media* (New Brunswick, NJ: Transaction Press, 1991), and R. B. Joynson, *The Burt Affair* (London: Routledge, 1989).

18. Richard C. Lewontin, "The appointment of human diversity." *Evolutionary Biology* 6, 1971, 381-388.

19. Lorenz letter to John Churchill (July 3, 1976), DP:C.77:E.636; Mayr letter to Darling-ton, (November 28, 1978), DP:C.77:E.644; Stuart Sutherland, "A Tough-Minded Scientist," *New Scientist*, September 28, 1978, p. 954; Anthony Storr, "The Professor Who Likes Gardening, Planting, and Shocking People," *Sunday Times* (London), August 27, 1978; Sue Freeman, "Danger Ahead!" *The Sun*, September 1, 1978; "Detailed critique of *The Little Universe of Man* by an unidentified American anthropologist," DP:C.77:E.637; letters to Darlington, DP:C.77:E.645.

20. *Sunday Times*, ibid.

Conclusion: Paradoxes

1. "Chromosomes as We See Them," opening address in *Chromosomes Today, Vol. 1: Proceedings of the First Oxford Chromosome Conference, July 28–31, 1964*, eds. C. D. Darlington and K. R. Lewis (Edinburgh: Oliver and Boyd, 1966), pp. 3–4.

2. See Evelyn Fox Keller, *The Century of the Gene* (Cambridge, MA: Harvard University Press, 2000) for an elegant presentation of this point of view, replete with contemporary evidence. For a more detailed and technical discussion, see Peter J. Beurton, Raphael Falk, and Hans-Jörg Rheinberger, eds., *The Concept of the Gene in Development and Evolution: Historical and Epistemological Perspectives* (Cambridge. Cambridge University Press, 2000). Among the men and women of Darlington's generation, Barbara McClintock and Richard Goldschmidt stand out as having advocated this perspective.

3. See his "The Place of the Chromosomes in the Genetic System," *Chromosomes Today*, 4 (1973), pp. 1–13, and "A Diagram of Evolution," *Nature*, 278 (1979), pp. 447–452.

4. See an intelligent discussion of the matter by John Maynard Smith, *Did Darwin Get It Right?* (London: Penguin Books, 1988), pp. 39–50.

5. See Stephen Jay Gould, "Kropotkin Was No Crackpot," in his *Bully for Brontosaurus* (New York: W.W. Norton, 1991), pp. 325–339.

6. An obsession attested to by numerous diary entries on the subject and by his choice of favorite book: George Plekhanov's *The Role of the Individual in History*.

7. Lately, Hilary Putnam has produced a valuable discussion of this age-old conundrum in *The Collapse of the Fact/Value Dichotomy and Other Essays* (Cambridge, MA: Harvard University Press, 2002).

8. Diary, February 1959, DP:f.16:A.79.

9. Diary (9.7.1978), DP:g.44:A.96.

10. Quoted in Dan Lewis, "Cyril Dean Darlington, 1903–1981," *Biographical Memoirs of Fellows of the Royal Society*, 29 (1983), p. 148.

Index

Academy of Sciences Bulletin, 143
Acton, Lord, 181, 232
Adams, Arthur, 230
Afghanistan, 64
Agar, Wilfred Eade, 28
Agol, Isaak, 142
Agrobiology, 139, 143
Albertus Magnus, 181
Alexandrowicz, J. S., 166
All-Union Institute for Genetics and Plant Breeding, 140
All-Union Society for Cultural Relations with Foreign Countries (VOKS), 143
Allard, Edgar John, 28
Allison, Anthony, 225, 243
America, 22, 27, 28, 30, 83, 87, 89, 115, 121, 142, 145, 160, 219, 230, 233
American Association for the Advancement of Science: Toronto, 30
American Genetics Society, 109
American Institute of Biological Sciences (AIBS), 156, 157
American Soviet Science Society, 143
Anderson, Edgar, 116-117
Anderson, Irma, 32, 48
Animal Farm (Orwell), 150, 169, 292n53
Anisogeny, 50
Anti-Fascist Committee of Soviet Scientist, 143
Apomixis, 80, 161, 195, 197
Armitage, John, 150

Arnold, Arboretum (Harvard), 90
Arnold, Thomas, 10
Ashby, Eric, 144, 146, 153, 155, 167
Aspergillus, 253
Association for the Study of Systematics in Relation to General Biology, 112
Association of Scientific Workers, 174-175, 176, 177, 185, 198
Astbury, William Thomas, 157
Asteria, 64
Ate, 268
Athens, 202
Austen, Jane, 129
Australia, 230
Austria, 45, 159
Avery, Oswald, 207

BBC, 153, 155, 170, 177, 178, 192
Babcock, Ernest Brown, 84, 84, 115-116, 121
Backhouse, William Ormston, 26, 28
Baghdad, 62, 63, 186
Bailey, Arthur, 28
Baker, John Randal: SFS, 170, 175-177, 184; Darlington letter to, 241, 245; *Race*, 257
Balfour, Alfred, 19, 35
Bateson, Beatrice, 30, 35
Bateson, Jonathan, 28

Bateson, William: elected to Royal Society, 17; Mendel's paper, 18; appointment to John Innes, 19; linkage, 23; rejection of chromosomes, 26–30; review of *Mechanism*, 26; trip to America, 30–31; admission to Darlington, 33; 32, 34; method of starvation, 35; dinner parties, 35–36; chess, 44; Kammerer, 44–45; travels to Leningrad, 46; collapses and dies, 47; theoretical disposition, 49; "anisogeny", 50–51, eugenics, 52; after death of, 54; letter from Haldane 55–56; back from Turkistan, 63, 65, 81, 106, 107, 109, 110, 111, 119; and cytoplasm, 148, 194; "the immaterialist," 213, 217, 224, 267
Baur, Erwin, 35
Beadle, George, 84, 93, 108, 110, 114, 145, 222
Beale, Geoffrey, 127
Belar, Karl, 56, 106, 108
Belling, John: Bateson meets, 30; hypothesis of crossing over, 54; *Datura*, 58, 74, 65; trisomics, 74; segmental interchange, 74–75, 83, 84, 99, 107; "Critical Notes," 98–100, 103; polysomics, 107–108; old school cytologists, 108, 113, 160
Beloff, Max, 248, 250
Benzer, Seymour, 265
Beria, L. P., 149
Berlin Blockade, 186
Bernal, John Desmond, 153, 172, 173, 174, 179
Bevan, Alan, 223
Beveridge, William, 183
Biffen, Roland Henry, 47
Biological Prelim, 228, 255
Biometrical School, 18
Biometrika, 18
Blacker, Carlos Paton, 170, 241
Blackett, P. M. S., 172, 175, 176, 178
Blakeslee, Albert F., 30, 58, 74, 122
Bodmer, Walter, 253, 257
Boston Globe, 249
Botanical Garden, 133
Botany, 15, 16, 32, 112, 115, 127, 193, 223, 228, 255
Boveri, Theodor, 22, 36
Brachet, Jean, 206
Brade-Birks, 16
Bragg, Sir William, 175

Breeding habit, 77, 78, 82, 83, 97, 214, 217, 230
Brenner, Sidney, 265
Bridges, Calvin, 24, 29, 31, 65, 83, 106, 142
Brimble, Lionel, 174, 176
British Association for the Advancement of Science, 112, 177
British Council, 165, 176
British Medical Journal, 152
British Science Guild, 51
Brno, 140
Bruno, Giordano, 192
Bukharin, Nikolai, 175
Burbank, Luther, 140
Burnett, John, 223
Burt, Sir Cyril, 248, 255, 259
Bussey Institute, 30
Butler, Samuel, 29

Cain, Arthur, 210
Cairo, 62
California Institute of Technology, 84, 90, 94, 103, 122
Cambridge, 28, 29, 45, 47, 50, 55, 56, 147, 172, 198, 208, 253, 255
Campanula persicifolia, 20, 226
Car-Saunders, Alexander Morris, 232
Carnegie Institute of Washington's Department of Genetics, 30
Carson, Hampton, 84
Caspersson, Torbjorn Oskar, 206
Causes of Evolution, The (Haldane), 112, 114
Cavalli Sforza, Luigi, 257
Cayley, Dorothy, 32, 48
Cell in Heredity and Development, The (Wilson), 36, 42, 101, 106
Chaplin, Charlie, 216
Chiasma, 24, 43; relationship to crossing over, 53; frequency and position, 58–59; chromosome pairing, 56–61; crossing over, 71–72; reciprocal, 76; Sax, 92, 269. *See also* Chromosomes
Chiasmatype hypothesis/theory, 24–25, 43, 58, 90, 105
Childe, Vere Gordon, 232
Chimeras, 27, 33
China, 64

Chittendon, Reginald John, 32, 48, 50, 53, 148
Chromatin, 26
Chromosoma, 207
Chromosomes: naming of, 21; physical bearers of heredity, 25; architecture, 26–27; Hyacinth, 33–34; Darlington begins work on, 40; pairing, 41, 56–61; parasynapsis and telosynapsis, 41–43, 53, 59, 74; "classical theory," 43; crossing over, 71–77; achiasmatype meiosis, 92; and botany, 115–118, 263–266
Chronica Botanica, 119
Civil Right Movement, 247
Clarck, Bennet, 223
Cleland, Ralph E., 122
Cold Spring Harbor, 238
Cold War, 3, 186
Collins, Ernest Jacob, 32, 48
Columbia Series, 111, 117
Columbia University, 22, 23, 102, 146, 239
Comfort, Alex, 243,
Commissariat of Agriculture, 141
Committee to Aid Geneticists Abroad, 143
Communist Manifesto, 220
Communist Party, 143, 154, 155, 156; of Great Britain, 161, 172, 176
Conflict of Science and Society, The (Darlington), 191–193, 251
Constantinople, 65
Conway Memorial Lecture, 181, 184, 191
Coon, Carleton, 232, 249
Copernicus, Nicolas, 192
Correns, Carl, 18,
Cousin marriages, 226
Cowan, John Macqueen, 62–64
Cox, Brian, 223
Crane, Morley Benjamin, 32, 127
Creighton, Harriet, 90
Crepis, 115, 117
Crew, Francis Albert Eley, 47, 164
Crick, Francis, 207, 224, 255, 264, 265
Crossing over, 24–25, 27; problem of, 41, 43, 71–73; chiasmata, 53, 58, 60–61, 115; Drosophila, 76, 218; cytological proof of, 90, 122; Sax, 92; Morgan, 105, 157, 160, 161, 214, 266. *See also* Chromosomes
Crowther, J. G., 172, 176, 179
Cytogenetics: Sutton-Boveri hypothesis, 22, 40, 105; Darlington summarizes state of, 77; reinvigorated, 108–109; and evolution, 109–118

Cytology, 22, "a real thing," 30, 32; slippery character of, 89–90; deductive, 102; history of 105–109; and evolution, 109–118
Cytoplasm, 50; Darlington returns to, 148, 211, 213, 252, 267. *See also* Cytoplasmic inheritance
Cytoplasmic inheritance, 51, 65, 156–158

Dahlberg, Gunnar, 203
Dale, Sir Henry, 154, 184
Dalton, John, 192
Damascus, 62
Darlington, Alfred, 8, 9, 10, 13, 120, 194
Darlington, Andrew Jeremy, 131, 253
Darlington, Cyril Dean: childhood, 8–10, 13–14; St. Paul's, 11–12; Wye, 14–16; John Innes, 16, 19–20, 33–36, 44–48, 127–131, 207; begins work on chromosomes, 40–44; Chittenden, 50–51; Eugenics, 51–51, 219–220, 226; first discovery, 52–53; struggling, 53–55; Haldane, 56, 158–160, 161–163, 164–165; chromosome pairing, 57–61, 69–70; Persia, 62–66; crossing over, 62–66; evolution, 72, 77–77, 113–118, 124–125; Rockefeller Fellow in America, 83–84; marriages, 85–87, 127, 129, 132; Japan and India, 87–88; Morgan and population genetics, 94–97; conjectures, 98–104, 123–124; character, 119–123; spirals and spindles, 126, 206–207; children, 131, 133 135, 253–254; cytoplasmic inheritance, 148, 156–158; Lysenkoism, 148–152, 153, 157–158, 165–168, 170, 177–187; science and society; Man, 194–196, 210–212, 235–237, 243–245, 256–257; unifying themes of life, 262–273. *See also Conflict of Science and Society, The*; *Evolution of Genetics Systems*; *Evolution of Man and Society, The*; *Facts of Life, The*; *Little Universe of Man*; Oxford; *Race, Class and Mating in the Evolution of Man*; *Recent Advances in Cytology*; Sax; Woodhull Lecture
Darlington, Deborah Jane, 131, 262
Darlington, Oliver Franklin, 131
Darlington, Rachel Drew, 132, 253
Darlington, Susan Clare, 134, 253; Clare Passingham, 272

Darlington, William "Will" Henry
  Robertson, 7, 8, 9, 10, 13, 85
Darlington's Rule, 52–53
Dart, Raymond, 230
Darwin, Charles, 2, 4, 23, 95, 100, 101,
  107, 117, 119, 140, 146, 181, 182, 192,
  198, 215, 231, 258, 272
Darwin, Leonard, 31
Das Kapital (Marx), 216
Datura stramonium, 30, 58, 74, 75, 99
Davenport, Charles Benedict, 30
Davis, Bradley, 73
Davis, Nicholas, 256
Dawkins, Richard, 256
de Beer, Gavin, 112
de Coulanges, Fustel, 232
de Vries, Hugo, 18, 41, 72–73, 109, 111
de Winton, Dorothea, 32, 48, 127
Declaration of Independence, 220
Development, 22, 26, 27, 36; Wilson,
  40; cytoplasmic inheritance, 50, 51;
  chromosomes, 77, 99; Morgan, 109;
  Darlington, 148, 157, 211, 213; Wad-
  dington, 161, 249, 266
Diagram of Evolution (Darlington), 255
Dialectical materialism, 140, 160, 161,
  172, 183
Dickens, Charles, 216
Digby, L., 42
Dirac, P. A. M., 83
Discovery, 150, 152
Division for the Social and International
  Relations of Science (DSIRS), 177, 178
DNA, 153, 206, 207, 265
Dobzhansky, Theodosius, 83, 84, 106; and
  synthesis, 111–112, 114, 115, 121; and
  Lysenkoism, 145, 146, 151–152, 155,
  157, 167, 179; UNESCO, 236, 238;
  Facts of Life, 239, 241, 242, 243, 245, 256
Doncaster, Leonard, 18, 28, 32, 41
Dostoyevsky, Fyodor, 156, 216
Driesch, Hans, 35, 100
Driver, G. R., 230
Dubinin, Nikolai, 150, 155, 166
Dunn, Leslie Clarence, 110, 142, 145,
  146, 236
Durham, Florence M., 18

East, Edward Murray, 25, 30, 32, 46,
  90, 104
Eddington, Sir Arthur, 220
Edgar the Atheling, 7
Edinburgh, 28, 43, 47, 129, 164, 165

Einstein, Albert, 187
Embryology, 25, 105, 109
Emerson, Rollins Adams, 25, 122
Engels, Frederick, 147, 159, 160, 186
Erlanson, Eileen, 85, 106
Ethiopia, 64
Eugenics: Eugenics Society, 29, 194;
  Darlington's first reaction to, 51–52;
  Muller, 163; Darlington keeps implicit,
  219, 226; movement, 241; IAAEE,
  245–248
Evolution of Genetic Systems (Darlington), 2,
  114, 117, 118, 130, 161, 162, 186
Evolution of Man and Society, The (Darling-
  ton), 194, 229–234, 248–250, 253
Evolution: Darwinian and saltationist,
  17–18; 71–72; and Mutation Theory, 73;
  Oenothera, 75–76; mathematical popula-
  tion genetics, 94–97; variation in, 97;
  interest to Darlington, 102; and cytoge-
  netics, 109–118; Man, 198–205, 210–212
Evolution; The Modern Synthesis
  (Huxley), 111, 112
Evolutionary Synthesis, 111–114
Eysenck, Hans, 255

Facts of Life, The (Darlington), 194,
  213–221, 239–240, 242, 243
Farmer, John, 40, 41, 44, 48, 54, 207
Fifth International Congress of Genetics,
  57, 72
First Oxford Chromosome
  Conference, 264
Fisher, Eugen, 245, 247
Fisher, R. A., 2; mathematical population
  genetics, 94–97; and synthesis, 111, 113;
  and Lysenkoism, 149, 158; and Eugenics
  Society, 194, 199; *Heredity*, 208–210; free
  will, 220; blood groups, 225, 226, 256,
  271
Flemming, Walther, 36–37
Ford, E. B. "Henry," 25, 31, 112, 113,
  209, 222
Fortnightly Review, 150
Fourier, Jean Baptiste Joseph, 192
France, 10, 50–51, 55, 171
Frankland, Ellen "Nelly," 7, 9, 10,
  85–86, 253
Franklin, Benjamin, 216
Franks Commission, 228–229
Freud, Sigmund, 187, 250

Fritillaria meleagris, 76, 91, 92
Fundamentos, 157
Fyfe, J. L., 147, 153

Gairdner, Alice, 32, 48, 50, 52
Galileo, 104, 165
Galton, Francis, 2, 18, 95, 216, 217, 224, 225, 226, 230, 241, 261
Galton Laboratory of Eugenics, 238
Gametic coupling, 23
Garrett, Henry, 245
Gates, Reginald Ruggles, 33, 42, 44, 47, 74, 85, 119, 207, 245
Gayre, Robert, 247
Gene: "factors," 18, 23, 45, 105; theory of, 24; Morgan's enterprise, 94; Fisher emphasizes level of, 96; Muller speaks of, 118; MacBride, 119; and Man, 199, 216–217, 224–227; giving way to genome, 266
Genetic systems: defined, 78; incompatible with, 94–97; inbreeding and outbreeding, 113; and Man, 214, 263–266
Genetics (Genetical) Society, 28, 35, 47, 131, 135, 165, 166, 210
Genetics and Eugenics (Castle), 109
Genetics and the Origin of Species (Dobzhansky), 111
Genetics Institute of the Academy of Sciences, Leningrad, 140
Genetics: Bateson coins, 17; rewards from, 18; theory of crossing over, 24; in Britain and America, 35–28; as tool of social reform, 24, 32; modern classical born, 40; Morgan's classical, 94; mathematical population, 94–97; and cytology, 105–109; and evolution, 109–118; in Russia, 143–144, 153; and language, 211–212; great paradox of, 214; and method, 224–226; Nazis pervert, 237; critique of classical, 263–267
George, Lloyd, 35
Georgia, U.S.A., 230
Germany, 8, 10, 14, 21, 35, 85, 130, 139, 142, 150, 155, 163, 172, 174, 235, 237, 241
Ghengis Khan, 65
Gibraltar, 62
Gini, Corrado, 245, 247
Gladston, William, 7
Gobineau, Arthur de, 242

Goethe, Johan Wolfgang von, 196, 225
Goldschmidt, Richard, 142, 146, 306n2
Goodrich, Edwin Steven, 26
Gorbunov, Nikolai, 142
Gregor, James, 245, 246
Gregory, Reginald, 26, 27, 28
Gregory, Richard, 172, 191
Grey, James, 207
Guardian, 243
Gurney, O. R., 230

Haeckel, Ernst, 101
Haldane, J. B. S., 2, 18, 47; contracted by Hall, 50, 53; arrives at John Innes, 55–56, 65, 77, 83, 84; mathematical population genetics, 94–97, 101–102; and synthesis, 111, 113; "Oxford School," 112; supports Darlington, 118, 121; nonsensicalness, 127–128; bid for directorship, 130–131; and Lysenko, 142, 146–147, 152, 154, 158; and Marxism, 158–163, 164–165, 167, and Soviet science, 171–172, 177, 179; and class, 180–181, 194; and human behavior, 196; *Journal of Genetics*, 207; UNESCO, 236, 238, 244, 256, 259, 271
Haldane, John Scoot, 159
Hall, Daniel, 14, 49, 51, 53, 54, 107, 127, 130–131, 172
Hall, Harvey Monroe, 116
Hamilton, William, 248, 256
Hardie, Colin, 230
Hardin, Garrett, 248
Harland, Sidney, 142, 149, 158, 165
Harley, Jack, 223
Harvard University, 30, 90, 110, 116, 144
Harvey, Gwendolen Adshead, 132, 133; Darlington, 213, 253
Harvey, Jack, 132
Harvey, Paul, 230
Harvey, William, 192
Heisenberg, Werner Karl, 196
Hereditary Genius (Galton), 216
Heredity and its Variability (Lysenko), 145
Heredity: An International Journal of Genetics, 149, 207–210, 227, 235
Heredity: nuclear theory of, 21; Mendelian, 23; and development, 26–27; and reduction division, 39; chromosomes role in, 44; subject to evolution, 78; whole of, 94; mechanisms of evolving, 124;

and Lysenko, 139–142, 146. *See also* Genetics; Man; Mendelism
Hertwig, Oscar, 21, 36, 100, 105
Heslop-Harrison, John, 48
Heterostyly, 197, 217, 219
Hill, E. Raymond, 245
History of Western Philosophy (Russell), 231
Hitler, Adolph, 150, 173, 174, 187, 235, 243
Hogben, Lancelot, 28, 41–42, 172, 179, 187, 238, 240
Hopkins, Frederick Gowland, 172
Hudson, P. S., 145
Hunger Marches, 171
Huns, 212
Hurst, Charles C., 18, 28, 47
Huskins, Charles Leonard: Hall recruits, 54; offers to punch Darlington, 89; precocity theory, 93; reviews *Recent Advances*, 102–103, 120; old school cytologists, 108
Huxley, Julian, 2, 47; recommends Haldane, 50; Lockyer Lecture, 51–52, 54, 58, 66, 181; and synthesis, 111–113; and Lysenkoism, 142, 144, 145, 152, 153, 155, 158, 167; Muller letter to, 164; on Soviet science, 171; Social Relations of Science Movement, 172, 179; Eugenics Society, 194, 234; UNESCO, 236, 238; cultural evolution, 244–245, 256, 259, 271
Huxley, Thomas Henry, 243–244, 258, 268
Hyacinth, 57–58, 71
Hybridity: meiosis, 79–80; equilibrium, 197; hybrid vigor, 203

I.Q., 147, 255–256
Imperial Bureau of Plant Genetics, 145
Imperial Cancer Research Fund, 253
Incas, 202
India, 87, 126, 200, 201
Indiana University, 156
Indo-China, 186
International Genetic Congress, 142, 149
Introduction to Cytology (Sharp), 119–120
Introduction to Genetics, An (Sturtevant and Beadle), 93
Isfahan, 62

James, William Owen, 223
Janaki-Ammal, Edaveleth Kakkat, 85, 87

Janssens, Frans Alfons, 23, 43, 58, 90, 105
Japan, 9, 64, 87, 126
Jensen, Arthur, 255
Jerusalem, 62, 63
Jews, 230, 231, 259
Johannesburg, 230
Johansens, Wilhelm Ludwig: distinction between phenotype and genotype, 24, 35
John Innes Horticultural Institution: establishment of, 17–18; run like a family affair, 32; Council, 34, 43, 46, 47, 49, 81, 119, 130, 162; Hall arrives at, 49; Haldane, 50, 56; work divided at, 51, 54, 63; Darlington affairs at, 85, 103, 106; Darlington's work at, 116, 119, 122; Darlington's impact on, 124; Hall's directorship, 127–129; bid for directorship, 130–131, 162; move from Merton to Bayfordbury, 132; Darlington increasingly isolated at, 207
Jones, Donald, 32, 46
Joule, James Prescott, 192
Journal of Experimental Botany, 207
Journal of Genetics, 47, 207, 208
Juniper, Barry, 223

Kafiristan, 64
Kaiser Wilhelm Institute, 175
Kammerer, Paul, 44–45
Karpechenko, Georgy D., 160, 166
Kenya, 230
Kettlewell, Bernard, 209
Kihara, 87
Klampenborg, Denmark, 206
Koller, Pio, 103, 115, 129
Kremlin, 153, 172, 182
Kropotkin, Petr, 268
Kurdistan, 64
Kuwada, 87

La Cour, Leonard, 48
Lamarckism, 44, 45, 48, 65, 82–82, 95, 119, 140, 146, 156–158, 215, 244
Lancashire: Leigh, 7; Chorley, 8, 9, 63
Lange, Johann, 216
Lankester, Edwin Ray, 112
Lawrence, William C., 48
Leakey, Louis, 230
Lecky, William Edward Hartpole, 232

Lederberg, Joshua, 253
Lenin All-Union Academy of Agricultural Sciences, 46, 142
Lenin, Vladimir Ilyich, 164
Leningrad, 46, 141, 160, 163
Lerner, Michael, 145
Lesley, James, 28
Levi, Hyman, 172, 178, 179
Levit, Solomon, 142
Lewis, Dan, 127
Lewis, Keith, 223
Lewontin, Richard, 260
Linkage, 23, 24, 27; sex, 23
Little Universe of Man (Darlington), 194, 257–260
Lock, Robert Heath, 24, 28
Loeb, Jacques, 46
Long Sutton, 49
Lorenz, Conrad, 260
Lucretius, 12, 82, 186
Ludwig Feuerbach (Engels), 161
Lutz, Frank, 23
Lysenko Affair: background, 139–142; western reaction to, 144–148; Darlington reaction to, 148–152; August 1948 conference, 153–154; Society for Freedom in Science, 170–177, 185, 186
Lysenko Is Right (Fyfe), 153
Lysenko, Trofim Denisovitch: background, 139–142; western scientists meet, 144; western strategy against, 145–148; Darlington, 148–152; late reaction to, 156–158, 268

MacBride, E. W., 44–45, 47, 48, 81, 119, 146, 165, 194
MacDonald, Ramsey, 10
MacLeod, Colin, 207
Maisky, Ivan, 142
Malaya, 186
Man: evolution of, 198–205, 210–212, 229–234; genetics of, 216–221
Man's Place in Nature (Huxley), 244
Mankind Quarterly, 245, 247
Markel, Karl, 8, 9, 10
Marsh, George Perkins, 232
Marx, Karl, 159, 181, 198, 216, 220, 231, 235
Marxism: "Two Camps" philosophy, 140–141, 145, 152; western scientists,

146–148, 153–154; Haldane, 158–163, 164–165; and science, 171–172, 181–182
Materials for the Study of Variation (Bateson), 17–18
Mather, Kenneth, 127, 199
Maynard Smith, John, 95
Mayr, Ernst, 111, 113, 117, 260
McCarty, Maclyn, 207
McClintock, Barbara, 90, 119, 122, 306n2
McFarlane, K. B., 230
Mechanism of Mendelian Heredity, The, 25, 26P
Medawar, Peter, 244–245
Medico-Genetical Institute, 142
Mein Kampf (Hitler), 243
Meiosis, 38–40, 51, 61; relationship to mitosis, 69–70; advantage of, 79; 82; precocity theory, 70, 92–93. *See also* Chromosomes
Mendel, Gregor, 17, 18, 21, 22, 60–61, 95, 100, 213, 224
Mendel's Principles of Heredity, A Defence (Bateson), 18
Mendelism: segregation, 23, 48; Haldane interested in, 50; Lysenko, 139–148. *See also* Genetics; Heredity
Messina, 62
Metraux, Alfred, 236, 237
Mexico, 230
Michurin, Ivan Vladimirovitch, 140, 156, 166
Middleton, Thomas, 131
Mitosis, 36–37, 39, 69–70. *See also* Meiosis
Modern Quarterly, 147, 172
Molecular biology, 228, 256, 264
Molotov-Ribbentrop Non-aggression Pact, 142
Molotov, Viacheslav, 144
Mongols, 202, 212
Montagu, Ashley, 237
Moore, John, 40, 41
Morel, E. D., 10
Morgan, Thomas Hunt: conversion to Mendelism, 23; chromosomal theory of heredity, 24–27; Croonian Lecture, 31, 33; unsure of chiamatype theory, 43; theory of recombination, 60–61, 65; two greatest discoveries, 84; classical genetics, 94, 105, 106, 109, 111, 116, 160, 224
Moscow, 45, 144, 151, 153, 160, 161, 165, 167, 172, 182
Mosley, Oswald, 243

Muller, Hermann J., 24; X ray, 71, 83, 214; and physical gene, 94; modifier genes, 110; *Evolution of Genetic Systems*, 117; supports Darlington, 118; and Lysenkoism, 145, 149, 154, 155; and Russia, 142, 163–164, 169; and UNESCO, 237; *Facts of Life*, 240, 241, 259
Muralov, Alexander I., 141
Mussolini, Benito, 247
Mutation Theory (de Vries), 23, 73, 109–110
Mutation, 18; Morgan, 23, 111, 113, 224; meiosis and mitosis, 70, 78, 79; *Oenothera*, 73–74, 75, 109–110; and variation, 80, 95; point, 96, 97, 125; and recombination, 113; and Lysenko, 141; hybridity, 197; Muller, 214, 225, 265, 266, 267. *See also* Genetics
Mutual Aid (Kropotkin), 268

National Service League, 10
Nature, 38, 45, 50, 53, 144, 149, 150, 155, 163, 166, 172, 174, 177, 196, 198, 207
Navashin, Sergey Gavrilovitch, 115
Nazi Party, 139, 163, 171, 173, 175, 235, 237, 241, 242, 269
Needham, Joseph, 147, 172
New Right, 245
New York Times, 150, 228, 236
New Zealand, 230,
Newsweek, 156
Newton, Frank: appointment to John Innes, 31; "men's lab," 32; certain of chromosomes, 33; shows chromosomes to Darlington, 33–34, 35, 43; triploid tulip, 44; chess, 44, 48, undergoes operation, 51; in hospital, 53; dies, 54; widow Lily, 61, 65, 74
Newton, Isaac, 272
Newton, Lily, 60
Nilsson-Ehle, Hermann, 110
Nineteenth Century and After, 151, 152, 166, 179, 207
NKVD (People's Commissariat for International Affairs), 149, 165
Noble, Gladwyn Kingsley, 45
Nondisjunction, 29–30, 31, 106

Oakenshott, Walter, 228
Oenothera lamarckiana, 41, 72–76, 77, 95–96, 99, 107, 194

Olby, Robert, 299n2
Operational Research, 174
Organisms and Genes (Waddington), 161
Origin of Species (Darwin), 2, 4, 231
Ormandy, Reggie, 8
Orwell, George, 150–151, 184
Osborn, Fredrick, 241
Osborne, Thomas, 222, 226
Ostertock, Herbert Charles, 48
Out of Night (Muller), 163
Oxford: 2, 13, 56, 133, 212–213, 221, 226–229, 240; Botanical Garden, 133; Sherardian Chair of Botany, 212, 230; Botany Department, 222–223, 227–228; Magdalen College, 2, 133, 230, 235, 254; Center of Biological Studies, 228, Frank Commission, 228–229; Darlington's legacy at, 255; Department of Genetics, 255

Paeonia suffriticosa, 90
Palmyra, 62
Pangenesis, 146
Paramecium, 156
Pareto, Vilfredo, 125
Parker, Geoffrey, 256
Parliamentary and Scientific Committee, 176
Payne, Fernandus, 23
Pearson, Karl, 18, 29,
Pelargoniums, 255, 272
Pellew, Caroline, 32, 48, 50
Penrose, Lionel, 238, 239, 241, 242, 243, 245
People's Commissariat of Enlightenment, 142
Persia, 62–64, 69, 83, 87, 227
Petrie, Flinders, 232
Philipchenko, Iurii Alexandrovitch, 11
Physical Basis of Heredity, The (Morgan), 16
Pinsdorf, Kate, 85–87, 121
Pirenne, Henri, 232
Plasmagenes, 157, 158
Plastids, 50, 148, 157
Poland, 173
Polanyi, Michael: Darlington writes to, 150; SFS, 170, 175–177, 184
Polemic, 150–151
Polyploidy, 34, 52–53, 80, 96, 113
Pontecorvo, Guido, 223
Popper, Sir Karl, 249–150, 251–252, 257
Port Said, 62

Position effect, 96
Powell, Enoch, 248
Prain, David, 47
Pravda, 140, 156
Presence and Absence, 35
Prezent, Isaak I., 140, 151
Primula kiwensis, 43
Primula sinensis, 26, 32, 57, 218
Pringle, John, 228
Proceedings of the Royal Society, 207
Prunus, 62, 65
Przibram, Hans, 45
Ptolmies, 202
Punnett, Reginald Crundall, 18, 19, 23, 24, 47

Race Relations Act, 248, 259
Race, Class and Mating in the Evolution of Man (Darlington), 198–205, 210, 219
Race, Robert, 225
Raglan, Baron Fitzroy Richard Somerset, 232
Rationalist Press Association (RPA), 166–167
Recent Advances in Cytology (Darlington), 2, 77, 83; reception, 83–84, 87, 93, 169; Haldane's forward to, 96; Belling attacks, 98–100; Haldane judges, 102; immediate effect, 108, 113; shock of, 114; prefaces of, 120; second edition, 130; Russian edition, 160–161
Recombination, 25, 60–61, 71; *Oenothera*, 74; meiosis, 77, 79, 82, 97; and mutation, 113, 125, 195, 199, 203, 214, 229, 231, 265; evolution, 233, 267. *See also* Chromosomes
Reduplication, 23
Renner, Otto, 65, 73, 113, 148
Richens, R. S., 145
Riley, D. P., 198
Ring-formation, 80, 96, 122
Rivers, William Halse, 232
Roberts, John Fraser, 225
Robinson, Robert, 132
Rockefeller Fellowship, 83, 87, 103, 116
Rockefeller Institute, 207
Rothamsted Experimental Station, 49
Rothschild, Charles, 1
Rothschild, Miriam, 1–2
Rousseau, Jean Jacques, 198, 216
Royal Botanic Gardens, 62
Royal Botanical Conference, 194
Royal Halloway College, 127

Royal Society of London, 1, 17, 18, 35, 41, 54, 55, 121, 131, 132, 135, 148, 155, 184, 191, 196
Ruskin, John, 106, 108
Russell, Bertrand, 231
Russian Academy of Science, 120
Rutherford, Ernest, 15

Salman, E. S., 16
Saunders, Edith Rebecca, 18
Sax, Karl: objections to Darlington, 90–92; old school cytologists, 108; and Anderson, 116; and Lysenko, 144, 146, 149, 152, 160
Schafer, Brenhilda, 49, 54, 85
Schneider, Friedrich Anton, 36
Schrader, Franz, 102, 103, 108, 123–124
Science and the Planned State (Baker), 175
Science In War, 174, 179
Science, 23, 110, 144, 149, 150, 152, 155, 163
Scientific Worker, 175
Scientist in Russia (Ashby), 146
Seventh International Congress of Genetics, 165
Shakespeare, 129
Shanghai, 186
Sharp, Lester S., 119
Shaw, George Bernard, 36, 52, 172
Sheppard, Philip, 210
Sherardian Chair of Botany. *See* Oxford
Shull, George Harrison, 30
Siam, 202
Simpson, Gaylord, 111
Sinclair, Hugh MacDonald, 222
Sixth International Congress of Genetics, 84, 90, 96
Smith, Geoffrey Watkins, 28
Snow, C. P., 229
Social Function of Science, The (Bernal), 173
Social Relations of Science Movement, 171–177
Society for Cultural Relations Between the Peoples of the British Commonwealth and the U.S.S.R., 165
Society for Experimental Biology, 112
Society for Freedom in Science (SFS): founded, 170; and Soviet Union, 175–177, 183, 245
Socrates, 97

Solzhenitsyn, Alexander, 252
Sonneborn, Tracy, 156–157
South Eastern Agricultural College at Wye, 14–16, 49, 169
Soviet Academy of Sciences, 144, 155
Soviet Science (Crowther), 172
Soviet Union, 3; Lysenko, 139–142, 148; World War II, 142–145; science planning, 154, 171–172; *Recent Advances in Cytology*, 160–161; Muller, 163; Darlington dissolution with, 170; 175; Darlington visit to, 177–178; imperative to attack, 183–184, 257
Spain, 163
Spencer, Herbert, 146, 215, 231
Sprunt, A. D., 55
St. Paul's, 10–12, 14, 169, 230
Stalin, Joseph, 139, 140, 141, 143, 144, 149, 163, 171
Stebbins, G. Ledyard, 110, 111, 117, 118, 122, 146
Stern, Curt, 90, 122, 146
Sterne, Lawrence, 216
Stocks, Mary, 183
Stopes, Marie, 85
Storr, Anthony, 260
Strasburger, Edward, 21, 36
Stromboli, 62
Study of History (Toynbee), 215, 231
Sturtevant, Alfred, 24, 65, 93, 108, 110, 114, 115, 122, 160
Sukhum, 64
Sunday Times, 260
Sutton–Boveri hypothesis, 22, 40, 105
Sutton, Walter Stanborough, 22, 108
Sverdrup, Aslaug, 32, 48
Swann, Donald, 245

Tabriz, 62, 63
Tansil, Charles, 245
Tansley, Arthur George, 131, 170, 222
Tatum, Edward, 222
Terminalization, 59
Texas, 163
Theoretical Basis of Plant Breeding (Vavilov), 165
Tilbury, 62
Tots and Quots Club, 174, 177, 179
Toynbee, Arnold, 215, 231, 250
Tradescantia, 116–117
Triploids, 57, 58, 60, 70, 95
Trisomics, 58, 60, 74, 95
Tulips, 49, 50, 62
Tyler, Philip, 230

Ukraine, 139, 141, 165
UNESCO Statement on Race, 236–237, 239, 240, 247, 269
University College London, 44, 131, 239
University of California at Berkeley, 84, 115, 116, 145
University of Pennsylvania, 84, 103, 114, 156
Upcott, Margaret Blanche, 126–127, 129, 130, 132–133, 253
Upcott, Sir Gilbert, 127, 132
Ur, 202
Urmia, 63

Variation: Bateson, 17–18, 29, 50, 51, 73, 78, 109; origin of, 80–81, 83; and inversions, 97, 111; and recombination, 113, 114, 123, 124; hybridity, 197; Man, 203, 224, 225, 232; Mendelism, 224
Variations of Plants and Animals Under Domestication (Darwin), 216
Vassar College, 127
Vavilov, Nikolai: John Innes, 46, 47; arranges documents for Darlington, 64; Soviet hero, 65, 66; and collectivization, 141; disappears, 142; fate unknown, 148; death confirmed, 149; invites Darlington to U.S.S.R., 160; Muller, 163; letter to Harland, 165; "What is your philosophy?", 194, 231
Vernalization, 140
von Humboldt, Alexander, 232
von Koelliker, Rudolph Albert, 21, 36
von Naugeli, Carl Wilhelm, 21, 36
von Tschermak, Eric, 18
von Vershuer, Otmar, 216
Vosa, Canio, 223

Waddington, Conrad Hal, 158, 161
Waldeyer, William, 21–22
Watson, James, 207, 224, 255, 264
Weismann, August, 21, 34; and reduction division, 38–39; Darlington drawn to, 40; theory of recombination, 60–61; Darlington's hero, 100–101, 214
Weldon, W. F. R., 18
Wells, H. G., 172, 232
Whewell, William, 171
White, Frank, 223
White, M. J. D., 288n35
Whitehall, 181
Whitehead, Alfred North, 29, 63